# *The Fundamentals of Fractional Calculus*

# The Fundamentals of Fractional Calculus

Dharmendra Kumar Singh, PhD
Mehmet Yavuz, PhD
*Editors*

APPLE ACADEMIC PRESS

First edition published 2026

**Apple Academic Press Inc.**
1265 Goldenrod Circle, NE,
Palm Bay, FL 32905 USA

760 Laurentian Drive, Unit 19,
Burlington, ON L7N 0A4, Canada

**CRC Press**
2385 NW Executive Center Drive,
Suite 320, Boca Raton FL 33431

4 Park Square, Milton Park,
Abingdon, Oxon, OX14 4RN, UK



**Library and Archives Canada Cataloguing in Publication**

Title: The fundamentals of fractional calculus / Dharmendra Kumar Singh, PhD, Mehmet Yavuz, PhD, editors.
Names: Singh, Dharmendra Kumar, editor. | Yavuz, Mehmet (Lecturer in mathematics), editor.
Description: First edition. | Includes bibliographical references and index.
Identifiers: Canadiana (print) 2025014090X | Canadiana (ebook) 20250140918 | ISBN 9781779643353 (hardcover) | ISBN 9781003638339 (ebook)
Subjects: LCSH: Fractional calculus.
Classification: LCC QA314 .F86 2025 | DDC 515/.83—dc23

**Library of Congress Cataloging-in-Publication Data**

CIP data on file with US Library of Congress

ISBN: 978-1-77964-335-3 (hbk)
ISBN: 978-1-00363-833-9 (ebk)

# *About the Editors*

**Dharmendra Kumar Singh, PhD**
*Assistant Professor, Department of Mathematics in the School of Basic Sciences, UIET, Chhatrapati Sahau Ji Maharaja University, Kanpur, India*

Dharmendra Kumar Singh, PhD, is an Assistant Professor in the Department of Mathematics, School of Basic Sciences, University Institute of Engineering and Technology (UIET) at Chhatrapati Shahu Ji Maharaj University, Kanpur. His research interests include special functions, fractional calculus, partial fractional differential equations, and distribution theory. Dr. Singh received the Best Publication Award from the Society for Special Functions and Their Applications in 2019. He has 12 years of research and teaching experience in fractional calculus. Dr. Singh has published and submitted various research papers and book chapters on fractional calculus in reputed journals. He has been a reviewer for many well-regarded international journals and has attended and presented research papers at numerous national and international conferences and symposiums.

**Mehmet Yavuz, PhD**

*Associate Professor, Mathematics and Computer Sciences Department, Necmettin Erbakan University, Türkiye*

Mehmet Yavuz, PhD, is an Associate Professor in the Mathematics and Computer Sciences Department at Necmettin Erbakan University, Türkiye. He has published more than 85 research papers in reputed journals and 60 conference papers, and several book chapters in international books. He edited a book titled *Fractional Calculus: New Applications in Understanding Nonlinear Phenomena*, published by Bentham Science Publishing. He has received an award for being the most published scientist with international cooperation scientists (gold medal) and two awards for being the most published scientist in the field of science/engineering for his outstanding research work. His research interests lie in biomathematics, infectious disease dynamics, population dynamics, stochastic analysis, calcium dynamics, fractional mathematical models, fractional mathematical theory and methods, bifurcation analysis, and dynamical control theory. Apart from being an associate editor for several esteemed journals, he is the Editor-in-Chief of the journal *Mathematical Modelling and Numerical Simulation with Applications* and the *Bulletin of Biomathematics.* Dr. Yavuz received his BSc degree from Zonguldak Bulent Ecevit University, Türkiye, and his MSc and PhD degrees in Applied Mathematics from Balıkesir University, Türkiye. He visited the Department of Mathematics at the University of Exeter, UK, for postdoctoral research.

# Contents

# Contributors

**Yasir Nadeem Anjam**

Department of Applied Sciences, National Textile University Faisalabad, Pakistan

**Seda İğret Araz**

Siirt University, Department of Mathematics, Faculty of Education, Turkey; Institute for Groundwater Studies, Faculty of Natural and Agricultural Sciences, University of the Free State, South Africa

**Abdon Atangana**

Institute for Groundwater Studies, Faculty of Natural and Agricultural Sciences, University of the Free State, South Africa; Department of Medical Research, China Medical University Hospital, China Medical University, Taichung, Taiwan

**Amna Batool**

Department of Applied Sciences, National Textile University Faisalabad, Pakistan

**B. Bayraktar**

Bursa Uludag University, Faculty of Education, Görukle Campus, Turkey

**Binayak Dihudi**

Konark Institute Science and Technology (KIST), Techno Park, Jatni, Bhubaneswar, Odisha, India

**Marcela Fabio**

ITECA, UNSAM-CONICET, ECyT, CEDEMA, Buenos Aires, Argentina

**Bengt Fornberg**

Department of Applied Mathematics, University of Colorado, Boulder, CO, USA

**Austin Higgins**

Department of Mathematical Sciences, Michigan Technological University, Houghton, MI, USA

**Rabha W. Ibrahim**

Department of Mathematics, Mathematics Research Center, Near East Boulevard, PC, Nicosia/Mersin, Turkey; Department of Computer Science and Mathematics, Lebanese American University, Beirut, Lebanon; Information and Communication Technology Research Group, Scientific Research Center, Al-Ayen University, Thi-Qar, Iraq

**Mousa Ilie**

Department of Mathematics, Rasht Branch, Islamic Azad University, Iran

**K. Kanagarajan**

Department of Mathematics, Sri Ramakrishna Mission Vidyalaya College of Arts and Science, Coimbatore, Tamil Nadu, India

**Ajay Kumar**

Department of Basic Engineering and Applied Sciences, Mahamaya College of Agricultural Engineering and Technology, Ambedkar Nagar (A.N.D.U.A.T, Kumarganj), Ayodhya, India

**Dilip Kumar**

Department of Mathematics, University of Kerala, Kariavattom Campus, Thiruvananthapuram, Kerala, India

**Hemant Kumar**

Department of Mathematics, DAV Postgraduate College, Kanpur, Uttar Pradesh, India

**Min-Jie Luo**

Department of Mathematics, School of Mathematics and Statistics, Donghua University, Shanghai, People's Republic of China

**J. E. Nápoles-Valdés**

UNNE, FaCENA, Ave. Libertad, Corrientes, Argentina; UTN-FRRE, French, Resistencia, Chaco, Argentina

**Anasuya Nath**

Utkal University, Bhubaneswar, Odisha, India

**Zulfiqar Ahamad Noor**

Department of Mathematics, University of Management and Technology, Lahore, Pakistan

**Cécile Piret**

Department of Mathematical Sciences, Michigan Technological University, Houghton, MI, USA

**Amit Prakash**

Department of Mathematics, National Institute of Technology, Kurukshetra, Haryana, India

**Mati Ur Rahman**

School of Mathematical Sciences, Jiangsu University, Zhenjiang, Jiangsu, P.R. China

**Rahul**

Department of Mathematics, National Institute of Technology, Kurukshetra, Haryana, India

**Ravinder Krishna Raina**

M.P. University of Agriculture and Technology, Udaipur, Rajasthan, India

**Yerra Shankar Rao**

NIST University, Berhampur, Odisha, India

**Muhammad Bilal Riaz**

IT4Innovations, VSB–Technical University of Ostrava, Ostrava, Poruba, Czech Republic; Department of Mathematics, University of Management and Technology, Lahore, Pakistan

**Silvia Seminara**

Department of Mathematics, Faculty of Engineering, University of Buenos Aires, Argentina

**Dharmendra Kumar Singh**

Department of Mathematics, School of Basic Sciences, CSJM University, Kanpur, Uttar Pradesh, India

**Imran Talib**

Nonlinear Analysis Group (NAG), Mathematics Department, Virtual University of Pakistan, Pakistan

**María Inés Troparevsky**

Department of Mathematics, Faculty of Engineering, University of Buenos Aires, Argentina

**D. Vivek**

Department of Mathematics, PSG College of Arts and Science, Coimbatore, Tamil Nadu, India

**R. Vivek**

Department of Mathematics, Sri Ramakrishna Mission Vidyalaya College of Arts and Science, Coimbatore, Tamil Nadu, India

**Mehmet Yavuz**

Department of Mathematics and Computer Sciences, Faculty of Science, Necmettin Erbakan University, Meram Yeniyol, Meram, Konya, Türkiye; Department of Applied Mathematics and Informatics, Kyrgyz-Turkish Manas University, Bishkek, Kyrgyzstan

# *Acknowledgment*

We would like to express our sincere gratitude to all those who contributed, directly and indirectly, to this work. We extend our heartfelt thanks to the various authors for their valuable contributions and wish to record our appreciation for the time and expertise each has dedicated to this book. We warmly welcome and will gratefully acknowledge all comments and suggestions from faculty members and students regarding the inclusion or exclusion of topics, as well as any identified errors.

—***Editors***

# Preface

Fractional calculus is one of the most essential branches of theoretical and applied mathematics. This book aims to present a wide variety of interesting applications of fractional calculus in a flexible and accessible format. It provides a comprehensive theoretical background on the subject, followed by a carefully graded set of detailed solved examples. Nearly all chapters are based on recent research in fractional calculus, and the contributing authors are esteemed academics from around the world, each an expert in their respective field.

Chapter 1 explains the foundational structure of fractional calculus, incorporating illustrations on the application of fractional integral and differential operators through generalized functions and pathway-type integral transforms. Chapter 2 focuses on the interrelationships between the Laplace transform and fractional integrals and derivatives. Chapter 3 proposes a numerical scheme for estimating, from experimental measurements, the order of a general nonlinear fractional model described by the Caputo derivative. Chapter 4 explores a new class of integro-differential equations involving non-singular differential operators. Chapter 5 presents various methods for determining the general solution of second-order linear ordinary differential equations, extending these approaches to second-order linear fractional differential equations. Initially, methods such as the known solution approach, the characteristic scheme, and the technique of solving Euler's equidimensional differential equation are established to determine the general solution ($u_h$) of second-order homogeneous fractional differential equations. Subsequently, the methods of variation of parameters and undetermined coefficient techniques are expanded to determine a particular solution ($u_p$) of second-order linear fractional differential equations. Finally, combining the general solution of second-order homogeneous fractional differential equations and the particular solution of second-order linear fractional differential equations, the general solution of second-order linear fractional differential equations ($u_h + u_p$) is introduced.

Chapter 6 addresses the numerical computation of fractional derivatives of the Caputo type, describing methods to compute fractional-order derivatives of this type with high accuracy using only function values on grids with node spacing $h$. Chapter 7 presents Caputo-weighted integrals and their applications. Chapter 8 focuses on $k$-symbol fractional differential equations, demonstrating the use of the fractional settle operator in a non-compact set and establishing the existence of solutions for fractional boundary value problems (BVPs) with multi-evolution. Chapter 9 explores an extension of the gingerbread-man map using the concept of a $k$-symbol Caputo derivative within the framework of fractional calculus. Chapter 10 develops a highly efficient computational technique based on new operational matrices involving fractional integral and derivative operators of Vieta-Lucas function vectors (VLFVs).

Chapter 11 examines the existence and uniqueness of solutions for implicit fuzzy differential equations with tempered $\Xi$-Hilfer fractional derivatives. Chapter 12 presents a mathematical model developed to analyze social behavior and reduce alcohol addiction in society, employing a non-integer model based on domain discretization and the short memory principle. Chapter 13 examines a deterministic mathematical model of an epidemic using the Wright-type probability density function. Chapter 14 introduces a nonlinear deterministic model capturing the dynamics of divorce rates within a population. Chapter 15 explores the Jaulent-Miodek equation through the Caputo-Fabrizio operator. Chapter 16 is dedicated to the study of the fractional kinetic equation.

This book rigorously and comprehensively analyzes various aspects of fractional calculus. It is a valuable resource for scientific readers and researchers working in the field, guiding them through essential steps in research methodology, concept development, and related processes.

# Chapter 1

## An Outlook on Fractional Calculus via Generalized Functions and Integral Transforms of Pathway Type

**Dilip Kumar**

*Department of Mathematics, University of Kerala, Kariavattom Campus, Thiruvananthapuram, Kerala, India*

**Abstract**

Beginning with a brief history of fractional calculus, we discuss a unified definition of fractional calculus. The historical note presented in the chapter may lack information about the contributions of many researchers in the area, but it provides a brief account of the development of the subject matter to date. The chapter introduces various definitions of fractional integral and differential operators. Illustrations on the application of fractional integral and differential operators are also incorporated via generalized functions and integral transforms of the pathway type. The fractional integrals and differentials of the generalized Krätzel function and $\mathcal{P}$-transforms are obtained via generalized special functions. This chapter is based on two research publications by the author: Kumar [1] and Kumar and Kilbas [2].

### Introduction

Many great theories in literature were developed through extensions of existing theories. For example, the evolution of the number system begins with counting numbers, extends to integers, then to real numbers, and eventually to complex numbers. Similarly, scalar cases evolve into vector, multivariate, and matrix cases. In the same manner, the theory of fractional calculus originated as an extension of the meaning of classical calculus—namely, differential and integral calculus.

When Gottfried Wilhelm Leibniz (1 July 1646–14 November 1716) introduced the notation $\frac{\mathrm{d}^n y}{\mathrm{d}x^n}$, the combination of symbols in the notation perhaps prompted Guillaume de l'Hôpital (1661–2 February 1704) in 1695 to ask Leibniz: What if $n$ equals a half-integer? Leibniz replied, "One can express $\frac{\mathrm{d}^{1/2} y}{\mathrm{d}x^{1/2}}$ by an infinite series." In a subsequent letter, Leibniz explained that $\mathrm{d}^{1/2}x$ would be the same as $x\sqrt{\mathrm{d}x : x}$ and described it as an apparent paradox that would one day yield useful consequences [27]. Later in 1695, Leibniz corresponded with Johann Bernoulli and discussed derivatives of "general order" [28].

In 1697, Leibniz wrote a letter to John Wallis in which he used the notation $\mathrm{d}^{1/2}y$ to denote a derivative of order $\frac{1}{2}$ [29]. In 1730, Euler [9] contributed to the area, and in 1849, J. L. Lagrange made an indirect contribution to fractional calculus [24]. P. S. Laplace, in 1820, defined a fractional derivative via an integral [25]. It was S. F. Lacroix [23] who first explicitly mentioned derivatives of arbitrary order. Lacroix developed a formula for derivatives of arbitrary order, which is easily derivable from the modern Riemann–Liouville fractional derivative. Lacroix obtained

$$\frac{\mathrm{d}^r y}{\mathrm{d}x^r} = \frac{\Gamma(r+s)}{\Gamma(r+s+1)} x^{s-r}, r+s>0$$

by employing Legendre's symbol $\Gamma$, which represents the generalized factorial. This formula arises as a generalization of the formula for the $n$th derivative given by

$$\frac{\mathrm{d}^r y}{\mathrm{d}x^r} = \frac{s!}{(s-r)!}x^{s-r}.$$

Joseph B. J. Fourier [10] was another mathematician who discussed derivatives of arbitrary order, but his definition was based on the integral representation of a function. Although many of the mathematicians mentioned above explored fractional derivatives of arbitrary order, it was Niels Henrik Abel, in 1823, who used fractional calculus to solve integral equations arising from the tautochrone problem [3]. His "elegant" contribution to fractional calculus demonstrated that fractional derivatives of a constant function need not be zero unless the constant itself is zero.

The development of fractional calculus remained relatively dormant for several years until the significant contributions of Joseph Liouville in 1832. Liouville published three extensive memoirs, along with several other works in succession [31]. He applied fractional calculus to problems in potential theory and derived two formulas for fractional derivatives. For a function $f(x)$ represented in series form as $f(x) = \sum_{n=0}^{\infty} c_n \mathrm{e}^{a_n x}$, Liouville's first formula for fractional derivatives is expressed as:

$$D^{\nu} f(x) = \sum_{n=0}^{\infty} c_n {a_n}^{\nu} \mathrm{e}^{a_n x},$$

provided there is a restriction on $\nu$ to ensure the convergence of the series. This result was obtained by naturally extending the result for derivatives of integer order:

$$D^m \mathrm{e}^{ax} = a^m \mathrm{e}^{ax}.$$

Liouville's second formula for a fractional derivative is obtained as:

$$D^{\lambda}x^{-\varrho} = \frac{(-1)^{\lambda}\Gamma(\varrho+\lambda)}{\Gamma(\varrho)}x^{-\varrho-\lambda}, \quad \varrho+\lambda>0, \tag{1.0.1}$$

where $\Gamma(\varrho)$ is the gamma function, introduced by Euler, whose integral representation is:

$$\Gamma(\varrho) = \int_0^{\infty} v^{\varrho-1}\mathrm{e}^{-v}\mathrm{d}v, \quad \varrho>0,,v>0. \tag{1.0.2}$$

Liouville derived (1.0.1) by considering the integral:

$$I = \int_0^{\infty} v^{\varrho-1}\mathrm{e}^{-xv}\mathrm{d}v, \varrho>0, v>0.$$

By changing the variable $xv = t$, he found $I = \frac{\Gamma(\varrho)}{x^{\varrho}}$ or equivalently $x^{-\varrho} = \frac{1}{\Gamma(\varrho)}I$. Applying the operator $D^{\lambda}$ to both sides, he obtained:

$$D^{\lambda}x^{-\varrho} = \frac{(-1)^{\lambda}}{\Gamma(\varrho)}\int_0^{\infty} v^{\varrho+\lambda-1}\mathrm{e}^{-xv}\mathrm{d}v.$$

By substituting $xv = t$ and recognizing the resulting integral on the right-hand side as $\Gamma(\varrho+\lambda)$, he arrived at the result. Liouville was also the first to solve a differential equation using fractional operators. He investigated the complementary function for an ordinary differential equation of the form $\frac{\mathrm{d}^v y}{\mathrm{d}x^v} = 0$, where $v$ is arbitrary, in the same manner as in the case where $v$ is replaced by $n$ for integers [32].

George Peacock and S. S. Greatheed also worked on complementary functions and published many papers on the topic. See, for example, [43]. Two main definitions of fractional operations existed during that period, each with a different domain of usefulness. One, studied by Lacroix, Abel, and Peacock, applied to functions of the form $x^a$ and was referred to as functions of the Riemann class. The other, studied by P. Kelland [16], applied to a different set of functions and was referred to as functions of the Liouville class. In 1850, William Center [5] discovered that when the fractional derivative is applied to a constant function using the Lacroix-Peacock method, the result is not necessarily zero.

August De Morgan also studied fractional calculus and published a paper on the subject in 1840 [7]. Riemann [44] contributed to fractional calculus by obtaining a general version of a Taylor series and deriving the formula:

$$D^{-\nu} f(y) = \frac{2}{\Gamma(\nu)} \int_{c}^{y} (y-t)^{\nu-1} f(t), \mathrm{d}y + \psi(y), \tag{1.0.3}$$

where $\psi(y)$, according to Riemann, is a complementary function that measures exponentiation error. The current formulation of the fractional integral is (1.0.3) without the complementary function $\psi(y)$. Davis later pointed out errors made by Liouville and Peacock regarding the complementary function [6].

C. J. Hargreave [14] was the first to derive the general form of Leibniz's $p^{\text{th}}$ derivative of the product of two functions. In modern form, it is written as:

$$D^{\lambda} f(y) g(y) = \sum_{p=0}^{\infty} \binom{p}{r} D^{(p)} f(y) D^{(\lambda-p)} g(y),$$

where $D^p$ represents ordinary differentiation, $D^{(\lambda-p)}$ denotes a fractional operation, and $\binom{p}{r} = \frac{\Gamma(\lambda+1)}{p,\Gamma(\lambda-p+1)}$ is the general form of the binomial coefficient. The Leibniz rule is available in many textbooks from early times. See, for example, [45].

In addition to these contributors, H. R. Greer, Mikolás, W. Zachartchenxo, H. Holmgren, Spitzer, and others made significant contributions to fractional calculus. For a detailed account of their work, see [46].

The great contributions of many mathematicians ultimately culminated in the present-day definition of the Riemann–Liouville fractional integral, which first appeared in N. Ya. Sonin's 1869 paper titled *On Differentiation with Arbitrary Index* [49]. A. V. Letnikov, along with A. K. Grunwald, published four articles on this topic between 1868 and 1872, expanding upon Sonin's conclusions [30]. Both Letnikov and Sonin sought to express the fractional integral as a closed contour integral similar to Cauchy's integral formula.

The theory of fractional calculus entered the modern era through the contributions of H. Laurent [26]. Subsequent mathematicians, including P. A. Nekrassov [40] and Krug [20], continued refining the foundational definitions of fractional calculus. Oliver Heaviside [15] advanced the field by publishing many papers demonstrating that fractional operators could solve certain linear differential equations. Heaviside's work primarily focused on linear functional operators, and his methods were instrumental in analyzing the transmission of electric currents in cables—a method now known as Heaviside Operational Calculus [13].

T. J. Bromwich [4] extended Heaviside's work, but D. F. Gregory [12] is regarded as the founder of the calculus of operations. Gregory derived the heat equation and its solution in symbolic operator form: $u = Ae^{y\beta^{\frac{1}{2}}} + Be^{-y\beta^{\frac{1}{2}}}$, where $\beta = a^{-1}\frac{\mathrm{d}}{\mathrm{d}x}$. For a detailed historical account of fractional calculus from 1695 to the early 20th century, see [39, 42, 46].

In the 20th century, many researchers further contributed to fractional calculus, including R. E. Moritz, S. Pincherle, H. T. Davis, M. Al-Bassam, G. H. Hardy, A. Erdélyi, E. Schuylar, L. O'Shaughnessy, M. T. Naraniengar, and others. The first international fractional calculus conference was held in 1974 at the University of New Haven, Connecticut. The proceedings were published by B. Ross [45] under Springer-Verlag. The second international conference was held in 1984 under the auspices of the University of Strathclyde, Glasgow, Scotland [38], and the third in 1989 at Nihon University, Tokyo [41]. Oldham and Spanier [42] provide a year-wise bibliography of fractional calculus along with an exposition by Bertram Ross.

Today, numerous researchers continue to work on fractional calculus, applying it across various fields.

---

## 1.1 Unified Definition of Fractional Integral

One of the major challenges in fractional calculus has been the existence of multiple definitions for fractional integral and differential operators, creating significant ambiguity. In a series of papers by Mathai and in the book by Mathai and Haubold [36], a unified definition of fractional integral is presented with a geometric interpretation that is more convincing than prior derivations.

The interpretation begins by considering fractional integrals as fractional parts of the total integral of two kinds—namely, left-sided (or first-kind) and right-sided (or second-kind) integrals. Consider the two shaded regions in Figure 1.1, where one lies on the left side of the line $y_1 = y_2$ and the other on the right side of the same line.

Note that the areas of the shaded regions are fractions of the total area of the square $(b-a) \times (b-a)$. First, let us consider the left-side area of the triangle in a plane, which is assumed to lie in 2-space. The area is given by:

$$\int_{y_1=a}^{x} \mathrm{d}y_1 \int_{y_2=x_1}^{x} \mathrm{d}y_2 = \int_{y_1=a}^{x} (x - y_1)\mathrm{d}y_1.$$

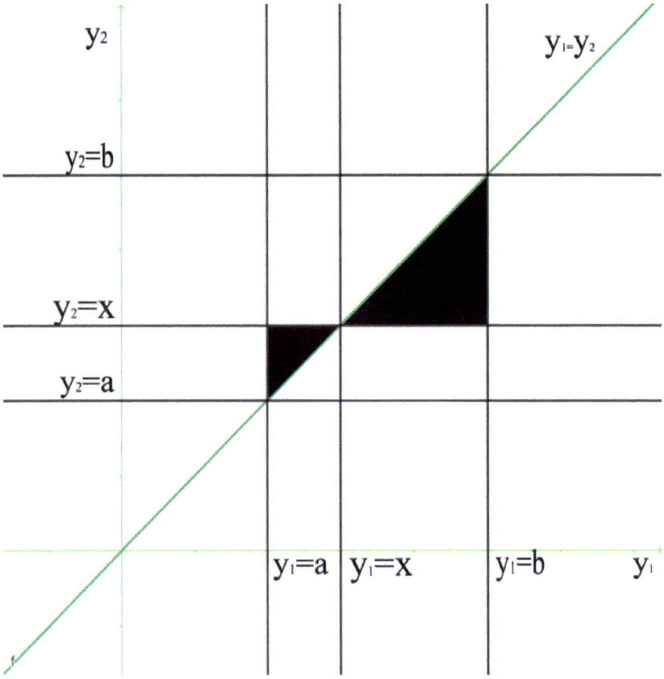


**FIGURE 1.1**. Integration over simplices in 2-space.
*Source:* Modified from Ref. [36].

Correspondingly, in 3-space, the volume of the simplex on the right side may be calculated as:

$$\int_{y_1=a}^{x} \mathrm{d}y_1 \int_{y_2=y_1}^{x} \mathrm{d}y_2 \int_{y_3=y_2}^{x} \mathrm{d}y_3 = \int_{y_1=a}^{x} y_1 \int_{y_2=y_1}^{x} (x-y_2)\mathrm{d}y_2$$
$$= \int_{y_1=a}^{x} \frac{(x-y_1)^2}{2!}\mathrm{d}y_1.$$

Thus, the volume in $m$-space for a simplex on the right side is: $\int_{y_1=a}^{x} \frac{(x-y_1)^{m-1}}{(m-1)!}\mathrm{d}y_1$. If we integrate the function $f(y_1)$ with respect to $y_1$, the result is:

$$\int_{y_1=a}^{x} \frac{(x-y_1)^{m-1}}{(m-1)!} f(y_1)\mathrm{d}y_1 = \int_{t=a}^{x} \frac{(x-t)^{m-1}}{\Gamma(m)} f(t)\mathrm{d}t,$$

where $\Gamma(m)$ is the gamma function defined in (1.0.2). The integral over the simplex on the right side in $m$-space can be denoted as ${}_x\mathcal{D}_b^{-m} f(t)$. Replacing $m$ with a general $\lambda$, the right-sided or second-kind Riemann–Liouville fractional integral of order $\lambda$ is given by:

$$ {}_aI_x^\lambda f(t) = {}_a D_x^{-\lambda} f(t) = \frac{1}{\Gamma(\lambda)} \int_{t=a}^{x} (x-t)^{\lambda-1} f(t)\mathrm{d}t, \qquad (1.1.4)$$

where $\mathrm{Re}\,(\lambda) > 0$. Similarly, the fractional derivative of order $\lambda$ is denoted by ${}_aD_x^\lambda f(t)$. Here, $-\lambda$ indicates integral and $\lambda$ indicates derivative. In the same way, we can consider the area of the right side of the line $y_1 = y_2$ in a plane or 2-space. Then the area is given by:

$$\int_{y_1=x}^{b} \mathrm{d}y_1 \int_{y_2=x}^{y_1} \mathrm{d}y_2 = \int_{y_1=x}^{b} (y_1 - x)\mathrm{d}y_1.$$

In 3-space, the corresponding simplex is the volume given by:

$$\int_{y_1=x}^{b} \mathrm{d}y_1 \int_{y_2=x}^{y_1} \mathrm{d}y_2 \int_{y_3=x}^{y_2} \mathrm{d}y_3 = \int_{y_1=x}^{b} (\frac{(y_1-x)^2}{2!}\mathrm{d}y_1.$$

Thus, the simplex on the right side has a volume in the $m$-space as:

$$\int_{y_1=x}^{b} \frac{(y_1-x)^{m-1}}{(m-1)!}\mathrm{d}y_1 = \int_{y_1=x}^{b} \frac{(y_1-x)^{m-1}}{\Gamma(m)}\mathrm{d}y_1.$$

If we integrate out an arbitrary function of $y_1$ alone, say $f(y_1)$, the result would be:

$$\int_{y_1=x}^{b} \frac{(y_1-x)^{m-1}}{(m-1)!} f(y_1)\mathrm{d}y_1 = \int_{x}^{b} \frac{(t-x)^{m-1}}{\Gamma(m)} f(t)\mathrm{d}t.$$

We denote the right-sided integral as ${}_xD_b^{-m}f(t)$ = the integral over the right simplex in $m$-space. Now, superseding $n$ by a generic $\lambda$, we have the right-sided or second-kind Riemann-Liouville integral of order $\lambda$, which is denoted by:

$$xI_b^{\lambda}f(t) =_x D_b^{-\lambda}f(t) = \frac{1}{\Gamma(\lambda)}\int_{t=x}^{b} (t-x)^{\lambda-1}f(t)\mathrm{d}t, \qquad (1.1.5)$$

where $\mathrm{Re}\,(\lambda) > 0$. This fractional derivative of order $\lambda$ is denoted by ${}_xD_b^{\lambda}f(t)$.For a more detailed overview, see Mathai and Haubold [36].

---

## 1.2 Fractional Integrals

Now let us see some well-known fractional integrals and derivatives that are commonly used in the literature [36].

### *1.2.1 Weyl Fractional Integrals*

A more practically useful fractional integral is the Weyl fractional integral which is defined as follows [36]: The left-sided or the first-kind Weyl fractional integral of order $\lambda$ is identified and defined as:

$$_{-\infty}I_x^{\lambda} f(t) = _{-\infty} W_x^{-\lambda} f(t) = D_{1,x}^{-\lambda} f(t) = \frac{1}{\Gamma(\lambda)} \int_{-\infty}^{x} (x-t)^{\lambda-1} f(t) \mathrm{d}t,$$

where $\mathrm{Re}(\lambda) > 0$.

The Right-sided or Second-kind Weyl fractional integral of order $\lambda$ is identified and defined as:

$$_{x}I_{\infty}^{\lambda} f(t) = _{x} W_{\infty}^{-\lambda} f(t) = W_{2,x}^{-\lambda} f(t) = \frac{1}{\Gamma(\lambda)} \int_{x}^{\infty} (t-x)^{\lambda-1} f(t) \mathrm{d}t,$$

where $\mathrm{Re}(\lambda) > 0$ and the fractional derivative is identified for $x > 0$ by Samko et al. [48] by:

$$(\mathcal{D}_{-}^{\lambda} f)x = \left(-\frac{\mathrm{d}}{\mathrm{d}x}\right)^{n} (\mathcal{I}_{-}^{n-\lambda} f)(x), n = [\mathrm{Re}(\lambda)] + 1, \qquad (1.2.6)$$

where $\mathrm{Re}(\lambda) > 0, -\infty < x < \infty$ and $[\mathrm{Re}(\lambda)]$ identifies as the integer part of $\mathrm{Re}(\lambda)$.

### *1.2.2 Left and Right-Sided Kober Fractional Integral*

The left-sided or first-kind and right-sided or second-kind Kober fractional integrals studied by Kober [36] in 1940 for $\mathrm{Re}(\xi) > 0$ with $\xi, \eta \in \mathbb{C}$ is defined as:

$$(\mathcal{I}_{\eta,\xi}^{+} f)x = \frac{x^{-\xi-\eta}}{\Gamma(\xi)} \int_{0}^{x} (x-t)^{\xi-1} t^{\eta} f(t) \mathrm{d}t, x \in \mathbb{R}_{+} \qquad (1.2.7)$$

and

$$\begin{aligned}(K^{-}_{\eta,\xi}f)(x) &= (I^{\xi}_{-;1,\eta}f)(x)\\ &= \frac{x^{\eta}}{\Gamma(\xi)}\int_x^{\infty}(t-x)^{\xi-1}t^{-\xi-\eta}f(t)\mathrm{d}t, x\in\mathbb{R}_+. \qquad (1.2.8)\end{aligned}$$

### *1.2.3 Left- and Right-Sided Saigo Fractional Operators*

The left-sided or first-kind and right-sided or second-kind fractional integrals given by Saigo [47] for $\mathrm{Re}\,(\xi) > 0$ with $\xi, \delta, \eta \in \mathbb{C}$ is defined as:

$$I^{\xi,\delta,\eta}_{0+}f(x) = \frac{x^{-\xi-\delta}}{\Gamma(\xi)}\int_0^x (x-t)^{\xi-1}{}_2F_1(\xi+\delta,-\eta;\xi;1-\frac{t}{x})f(t)\mathrm{d}t, \qquad (1.2.9)$$

and

$$I^{\xi,\delta,\eta}_{-}f(x) = \frac{1}{\Gamma(\xi)}\int_x^{\infty}(t-x)^{\xi-1}t^{-\xi-\delta}{}_2F_1(\xi+\delta,-\eta;\xi;1-\frac{x}{t})f(t)\mathrm{d}t, \qquad (1.2.10)$$

where ${}_2F_1(\lambda,\beta;\xi;z)$ is the hypergeometric series due to Gauss for $\lambda,\beta,\xi\in\mathbb{C},\xi\neq 0,-1,-2,\cdots$ defined by the series [8]

$${}_2F_1(\lambda,\beta;\xi;z) = \sum_{k=0}^{\infty}\frac{(\lambda)_k(\beta)_k}{(\xi)_k}\frac{z^k}{k!}. \qquad (1.2.11)$$

Moreover,

$${}_2F_1(\lambda,\beta;\xi;1) = \frac{\Gamma(\xi)\Gamma(\xi-\lambda-\beta)}{\Gamma(\xi-\lambda)\Gamma(\xi-\beta)}, \ \mathrm{Re}\,(\xi-\lambda-\beta)>0, \qquad (1.2.12)$$

where $(\lambda)_k$, $(\beta)_k$ and $(\xi)_k$ are the Pochhammer symbols defined by:

$$(\beta)_0 = 1, (\beta)_n = \beta(\beta+1)\cdots(\beta+m-1), m = 1,2,\cdots,\beta \neq 0. \tag{1.2.13}$$

The series in (1.2.11) is converges absolutely for $|z| < 1$, $\mathrm{Re}(\xi - \lambda - \beta) > 0$. When $\delta = -\xi$, the relations in (1.2.9) and (1.2.10) coincides with the classical Riemann-Liouville fractional integrals $I_{0+}^{\xi,-\xi,\eta} f(x) = (\mathcal{I}_{0+}^{\xi} f)x$ and $I_{-}^{\xi,-\xi,\eta} f(x) = (\mathcal{I}_{-}^{\xi} f)x$. When $\delta = 0_-$ in (1.2.9) and (1.2.10) we get the Kober Operators $I_{0+}^{\xi,0,\eta} f(x) = (\mathcal{I}_{\eta,\xi}^{+} f)x$ and $I_{-}^{\xi,0,\eta} f(x) = (K_{\eta,\xi}^{-} f)(x) = (I_{-;1,\eta}^{\xi} f)(x)$.

---

## 1.3 Fractional Derivatives

### *1.3.1 Left-Sided or First-Kind Riemann-Liouville Fractional Derivative*

The left-sided or first-kind Riemann-Liouville fractional derivative is identified and defined as:

$$D_{1,(a,x)}^{\lambda} f(t) = {}_aD_x{}^{\lambda} f(t) = {}_aD_x{}^{\lambda} f(x)$$
$$= \begin{cases} \frac{1}{\Gamma(r-\lambda)} \left(\frac{\mathrm{d}}{\mathrm{d}x}\right)^r \int_{t=a}^{x} (x-t)^{r-\lambda-1} f(t)\,\mathrm{d}t, r-1 < \lambda < r \\ \left(\frac{\mathrm{d}}{\mathrm{d}x}\right)^{r-1} f(x), \text{ if } \lambda = r-1. \end{cases}$$

Note that the structure of the left-sided or first-kind Riemann-Liouville fractional derivative is of the form $D^m$ times a fractional integral of the form ${}_aD_x{}^{-(r-\lambda)}$. That is,

$$D^r\,{}_aD_x{}^{-(r-\lambda)} = {}_aD_x{}^{\lambda}.$$

### *1.3.2 Left-Sided or First-Kind Caputo Fractional Derivative*

If the integer order derivative is operating on $f(t)$ inside the integrand in the above Riemann-Liouville fractional integral, then we obtain the left-sided or first-kind Caputo fractional derivative of order $\lambda$ denoted and defined as:

$$C^{\lambda}_{1,(a,x)}f(t) = {}_a^C D_x{}^{\lambda} f(t)$$
$$= \begin{cases} \frac{1}{\Gamma(r-\lambda)} \int_{t=a}^{x} (x-t)^{r-\lambda-1} f^{(r)}(t)\mathrm{d}t, r-1 < \operatorname{Re}(\lambda) < r \\ \left(\frac{\mathrm{d}}{\mathrm{d}x}\right)^{r-1} f(x), \text{ if } \lambda = r-1, \end{cases}$$

where $f^{(r)}$ is the $r$-th integer order derivative of $f$ with respect to $t$.

### *1.3.3 Left-Sided or First-Kind Weyl Fractional Derivative*

The left-sided or first-kind Weyl fractional derivative of order $\lambda$ is defined as

$$W^{\lambda}_{1,x}f(t) = {}_{-\infty}W_x{}^{\lambda} f(t) = \frac{1}{\Gamma(r-\lambda)} \left(\frac{\mathrm{d}}{\mathrm{d}x}\right)^{r} \int_{-\infty}^{x} (x-t)^{r-\lambda-1} f(t)\, \mathrm{d}t.$$

### *1.3.4 Right-Sided or Second-Kind Riemann-Liouville Fractional Derivative*

The right-sided or second-kind Riemann-Liouville fractional derivative is denoted and defined as:

$${}_xD_b{}^{\lambda} f(t) = \begin{cases} (-1)^r \frac{1}{\Gamma(r-\lambda)} \left(\frac{\mathrm{d}}{\mathrm{d}x}\right)^{r} \int_{t=x}^{b} (t-x)^{r-\lambda-1} f(t)\, \mathrm{d}t, r = [\lambda]+1, \\ \left(\frac{\mathrm{d}}{\mathrm{d}x}\right)^{r-1} f(x), \text{ if } \lambda = r-1, \end{cases}$$

where $[\lambda]$ denotes the integer part of $\operatorname{Re}(\lambda)$.

### 1.3.5 Right-Sided or Second-Kind Caputo Fractional Derivative

The right-sided or second-kind Caputo fractional derivative of order $\lambda$ denoted and defined as:

$$ {}_{x}^{C}D_{b}{}^{\lambda} f(t) = \begin{cases} \frac{1}{\Gamma(r-\lambda)} \int_{x}^{b} (t-x)^{r-\lambda-1} f^{(r)}(t) \mathrm{d}t, r = [\lambda]+1, \\ \left(\frac{\mathrm{d}}{\mathrm{d}x}\right)^{r-1} f(x), \text{ if } \lambda = r-1, \end{cases} \tag{1.3.14} $$

where $f^{(r)}$ is the $r$-th integer order derivative of $f$ with respect to $t$.

### 1.3.6 Right-Sided or Second-Kind Weyl Fractional Derivative

The right-sided or second-kind Weyl fractional derivative of order $\lambda$ is defined as:

$$ {}_{x}W_{\infty}{}^{\lambda} f(t) = \begin{cases} (-1)^{r} \frac{1}{\Gamma(r-\lambda)} \left(\frac{\mathrm{d}}{\mathrm{d}x}\right)^{r} \int_{t=x}^{\infty} (t-x)^{r-\lambda-1} f(t) \ \mathrm{d}t, r = [\lambda]+1, \\ \left(\frac{\mathrm{d}}{\mathrm{d}x}\right)^{r-1} f(x), \text{ if } \lambda = r-1. \end{cases} $$

A detailed account of the various fractional integral and differential operators and their composition with various functions can be seen in numerous books, for example, Mathai and Haubold [36], Miller and Ross [39], etc.

## 1.4 Application of Fractional Integrals and Derivatives to Krätzel Function and Generalizations

As an application to fractional integrals, we consider here a special function with two parameters introduced by E. Krätzel in 1975 [19], known as the Krätzel function in literature, and denoted by $Z_\sigma^\tau(x)$, defined for $x > 0, \sigma \in \mathbb{R} = (-\infty, \infty)$ along with $\tau \in \mathbb{C}$ as:

$$Z_\sigma^\tau(x) = \int_0^\infty t^{\tau-1} \mathrm{e}^{-t^\sigma - xt^{-1}} \mathrm{d}t, x \geq 0, \tag{1.4.15}$$

which is the kernel associated with the transform

$$(\mathcal{K}_\tau^\sigma f)x = \int_0^\infty Z_\sigma^\tau(xt) f(t) \mathrm{d}t, x \geq 0, \tag{1.4.16}$$

defined for every function $f(t)$ with $f(t), f(t)\ln t, t^\tau f(t)$ are locally integrable for $\mathrm{Re}\,(\tau) \geq 0$ and $\tau \neq 0$; $\tau = 0$; $\mathrm{Re}\,(\tau) < 0$, respectively. This transform was introduced as a generalization to the Meijer transform and the well-known Laplace transform.

By utilizing Mathai's pathway concept [34], Kilbas and Kumar generalized the Krätzel function in 2009. Later, two forms of generalized Krätzel functions were obtained. The first form is the type-1 generalized Krätzel function which is a three-parameter generalized special function $Z_\sigma^{\tau,p_-}(x)$ defined for $x \geq 0, \sigma > 0, \tau \in \mathbb{C}$ and $p < 1$ as:

$$Z_\sigma^{\tau,p_-}(x) = \int_0^{\left(\frac{1}{1-p}\right)^{\frac{1}{\sigma}}} t^{\tau-1} \left[1 - (1-p)t^\sigma\right]^{\frac{1}{1-p}} \mathrm{e}^{-xt^{-1}} \, \mathrm{d}t. \tag{1.4.17}$$

The second form which is known as type-2 generalized Krätzel function and denoted by $Z_\sigma^{\tau,p_+}(x)$ is defined as:

$$Z_\sigma^{\tau,p_+}(x) = \int_0^\infty t^{\tau-1}\left[1+(p-1)t^\sigma\right]^{-\frac{1}{p-1}} \mathrm{e}^{-xt^{-1}}\, \mathrm{d}t, \tag{1.4.18}$$

for $x \geq 0, \sigma \in \mathbb{R}, \sigma \neq 0, \tau \in \mathbb{C}$ and $p > 1$. The functions given in (1.4.17) and (1.4.18) are defined for any $\sigma \in \mathbb{R}$ with an extra condition $\mathrm{Re}\,(\tau) < 0$ in the case $\sigma < 0$. As $p \to 1$, by the property that $[1+(p-1)t^\sigma]^{-\frac{1}{p-1}} \to \mathrm{e}^{-t^\sigma}$ and (1.4.17) and (1.4.18) collide with the Krätzel function defined in (1.4.15). That is,

$$\lim_{p\to 1_-} Z_\sigma^{\tau,p_-}(x) = \lim_{p\to 1_+} Z_\sigma^{\tau,p_+}(x) = Z_\sigma^\tau(x),\ \ x > 0. \tag{1.4.19}$$

Especially when $\sigma = 1$ and $x = \frac{t^2}{4}$, then in accordance with Erdélyi et al. [8],

$$\lim_{p\to 1_-} Z_1^{\tau,p_-}(\frac{t^2}{4}) = \lim_{p\to 1_+} Z_1^{\tau,p_+}(\frac{t^2}{4}) = Z_1^\tau(\frac{t^2}{4}) = 2(\frac{t}{2})^\tau K_{-\tau}(t),$$

where $K_{-\tau}(t)$ is usually referred to as the McDonald function, which is defined in Erdélyi et al. [8]. The generalized functions $Z_\sigma^{\tau,p_-}(x)$ and $Z_\sigma^{\tau,p_+}(x)$ defined in (1.4.17) and (1.4.18) are obtained by applying the pathway model [34, 35]. The following theorems give the composition formulae for the two generalized Krätzel functions with Riemann-Liouville fractional operators.

**Theorem 1.4.1.** *Assume* $\kappa,\ \tau \in \mathbb{C},\ \ \mathrm{Re}\,(\kappa) > 0,$ *and* $\sigma > 0$. *Then*

$$\left(\mathcal{I}_-^\kappa Z_\sigma^{\tau,p_-}\right)(x) = Z_\sigma^{\tau+\kappa,p_-}(x),\ \ p < 1,\ \ x > 0 \tag{1.4.20}$$

*and*

$$\left(\mathcal{I}_{-}^{\kappa} Z_{\sigma}^{\tau,p_{+}}\right)(x) = Z_{\sigma}^{\tau+\kappa,p_{+}}(x),\ p > 1,\ x > 0. \tag{1.4.21}$$

*Proof.* Using (1.4.17) for $p < 1$ and (1.1.5) and changing the sequence of integration owing to the uniform convergence, we have

$$\left(\mathcal{I}_{-}^{\kappa} Z_{\sigma}^{\tau,p_{-}}\right)(x) = \int_{0}^{\left(\frac{1}{1-p}\right)^{\frac{1}{\sigma}}} y^{\tau-1}\left[1-(1-p)y^{\sigma}\right]^{\frac{1}{1-p}} (\mathcal{I}_{-}^{\kappa}\mathrm{e}^{-ty^{-1}})(x)\mathrm{d}y.$$

By means of the formula [48], $\left(\mathcal{I}_{-}^{\kappa}\mathrm{e}^{-at}\right)(x) = a^{-\kappa}\mathrm{e}^{-ax}$, $\mathrm{Re}\,(\kappa) > 0$, $\mathrm{Re}\,(a) > 0$, we arrive at:

$$\begin{aligned}\left(\mathcal{I}_{-}^{\kappa} Z_{\sigma}^{\tau,p_{-}}\right)(x) &= \int_{0}^{\left(\frac{1}{1-p}\right)^{\frac{1}{\sigma}}} y^{\tau+\kappa-1}\left[1-(1-p)y^{\sigma}\right]^{\frac{1}{1-p}} \mathrm{e}^{-xy^{-1}}\mathrm{d}y \\ &= Z_{\sigma}^{\tau+\kappa,p_{-}}(x).\end{aligned}$$

Similarly proceeding in the case of $p > 1$, we have

$$\left(\mathcal{I}_{-}^{\kappa} Z_{\sigma}^{\tau,p_{+}}\right)(x) = \int_{0}^{\infty} y^{\tau-1}[1+(p-1)y^{\sigma}]^{-\frac{1}{p-1}} (\mathcal{I}_{-}^{\kappa}\mathrm{e}^{-ty^{-1}})(x)\mathrm{d}y.$$

By employing the power function formulae, we arrive at the result in (1.4.21).

**Corollary 1.4.1.** *If we take the same settings in Theorem 1.4.1, then*

$$\lim_{p\to 1_{-}}\left(\mathcal{I}_{-}^{\kappa} Z_{\sigma}^{\tau,p_{-}}\right)(x) = \lim_{p\to 1_{+}}\left(\mathcal{I}_{-}^{\kappa} Z_{\sigma}^{\tau,p_{+}}\right)(x) = \left(\mathcal{I}_{-}^{\kappa} Z_{\sigma}^{\tau}\right)(x) = Z_{\sigma}^{\tau+\kappa}(x), \tag{1.4.22}$$

*for* $x > 0$.

The next theorem illustrates the composition formula for the Riemann-Liouville fractional derivative with $Z_{\sigma}^{\tau,p_{-}}(x)$ and $Z_{\sigma}^{\tau,p_{+}}(x)$.

**Theorem 1.4.2.** *Assume* $\kappa, \tau \in \mathbb{C}$, $\operatorname{Re}(\kappa) > 0$ *and* $\sigma > 0$. *Then*

$$\mathcal{D}_{-}^{\kappa} Z_{\sigma}^{\tau,p_{-}}(x) = Z_{\sigma}^{\tau-\kappa,p_{-}}(x), p < 1 \tag{1.4.23}$$

*and*

$$\mathcal{D}_{-}^{\kappa} Z_{\sigma}^{\tau,p_{+}}(x) = Z_{\sigma}^{\tau-\kappa,p_{+}}(x), p > 1. \tag{1.4.24}$$

*Proof.* Let us assume that $p < 1$ and $n = [\operatorname{Re}(\kappa)] + 1$. By using (1.4.17) and (1.2.6) and employing the formulae in (1.4.20) with $\kappa$ superseded by $n - \kappa$, we have

$$\begin{aligned}\left(\mathcal{D}_{-}^{\kappa} Z_{\sigma}^{\tau,p_{-}}\right)(x) &= \left(-\frac{\mathrm{d}}{\mathrm{d}x}\right)^{n} (\mathcal{I}_{-}^{n-\kappa} Z_{\sigma}^{\tau,p_{-}})(x) = \left(-\frac{\mathrm{d}}{\mathrm{d}x}\right)^{n} Z_{\sigma}^{\tau+n-\kappa,p_{-}}(x),\\ &= \left(-\frac{\mathrm{d}}{\mathrm{d}x}\right)^{n} \int_{0}^{\left(\frac{1}{1-p}\right)^{\frac{1}{\sigma}}} y^{\tau-\kappa+n-1}[1-(1-p)y^{\sigma}]^{\frac{1}{1-p}} \mathrm{e}^{-\frac{x}{y}} \mathrm{d}y.\end{aligned} \tag{1.4.25}$$

The uniform convergence of the integral in (1.4.25) helps in taking the differentiation inside the integral, and we get

$$\begin{aligned}\left(\mathcal{D}_{-}^{\kappa} Z_{\sigma}^{\tau,p_{-}}\right)(x) &= \int_{0}^{\left(\frac{1}{1-p}\right)^{\frac{1}{\sigma}}} y^{\tau-\kappa+n-1}[1-(1-p)y^{\sigma}]^{\frac{1}{1-p}} \left(-\frac{\mathrm{d}}{\mathrm{d}x}\right)^{n} \mathrm{e}^{-\frac{x}{y}} \mathrm{d}y,\\ &= \int_{0}^{\left(\frac{1}{1-p}\right)^{\frac{1}{\sigma}}} y^{\tau-\kappa-1}[1-(1-p)y^{\sigma}]^{\frac{1}{1-p}} \mathrm{e}^{-\frac{x}{y}} \mathrm{d}y.\end{aligned}$$

Because of (1.4.17), we obtain the result in (1.4.23). Proceeding exactly in a similar fashion, we arrive at the result in (1.4.24) for $p > 1$. Thus the theorem is proved.

**Corollary 1.4.2.** *Following the settings in Theorem 1.4.2, we get*

$$\lim_{p\to 1_-}\left(\mathcal{D}_-^{\kappa} Z_\sigma^{\tau,p_-}\right)(x) = \lim_{p\to 1_+}\left(\mathcal{D}_-^{\kappa} Z_\sigma^{\tau,p_+}\right)(x) = \mathcal{D}_-^{\kappa} Z_\sigma^{\tau}(x) = Z_\sigma^{\tau-\kappa}(x). \tag{1.4.26}$$

**Corollary 1.4.3.** *Assume* $\beta,\kappa,\tau\in\mathbb{C},\ \operatorname{Re}(\beta)>0,\ \operatorname{Re}(\kappa)>0$ *and* $\sigma>0$, *then*

$$(\mathcal{D}_-^{\beta}\mathcal{I}_-^{\kappa} Z_\sigma^{\tau,p_-})(x) = Z_\sigma^{\tau+\kappa-\beta,p_-}(x), p<1, \tag{1.4.27}$$

*and*

$$(\mathcal{D}_-^{\beta}\mathcal{I}_-^{\kappa} Z_\sigma^{\tau,p_+})(x) = Z_\sigma^{\tau+\kappa-\beta,p_+}(x), p>1. \tag{1.4.28}$$

*In particular, when* $p\to 1_-$ *in (1.4.27) and* $p\to 1_+$ *in (1.4.28), we have*

$$(\mathcal{D}_-^{\beta}\mathcal{I}_-^{\kappa} Z_\sigma^{\tau})(x) = Z_\sigma^{\tau+\kappa-\beta}(x). \tag{1.4.29}$$

**Corollary 1.4.4.** *Assume* $\beta,\kappa,\tau\in\mathbb{C},\ \operatorname{Re}(\beta)>0,\ \operatorname{Re}(\kappa)>0$ *and* $\sigma>0$. *Then*

$$(\mathcal{I}_-^{\beta}\mathcal{D}_-^{\kappa} Z_\sigma^{\tau,p_-})(x) = Z_\sigma^{\tau+\beta-\kappa,p_-}(x), p<1. \tag{1.4.30}$$

*and*

$$(\mathcal{I}_-^{\beta}\mathcal{D}_-^{\kappa} Z_\sigma^{\tau,p_+})(x) = Z_\sigma^{\tau+\beta-\kappa,p_+}(x), p>1. \tag{1.4.31}$$

*In particular, when* $p\to 1_-$ *in (1.4.30) and* $p\to 1_+$ *in (1.4.31) we arrive at*

$$(\mathcal{I}_-^{\beta}\mathcal{D}_-^{\kappa} Z_\sigma^{\tau})(x) = Z_\sigma^{\tau+\beta-\kappa}(x). \tag{1.4.32}$$

**Remark 1.4.1.** *Theorem 1.4.1, Theorem 1.4.2 and their corollaries were obtained in Kilbas & Kumar [17].*

## 1.5 Application of Fractional Integrals and Derivatives to $\mathcal{P}$-Transforms

In 2011, Kumar [21] introduced a new integral transform that generalizes the standard Krätzel transform in two different ways. The type-1 and type-2 $\mathcal{P}$-transforms are defined as follows by Kumar [21]:

**Definition 1.5.1.** *A type-1 $\mathcal{P}$-transform of a real variable function $f(x)$ for $x > 0$ is defined as:*

$$(\mathcal{P}_{\tau}^{\sigma,\beta,p_-} f)(x) = \int_0^{\infty} Z_{\sigma,\beta}^{\tau,p_-}(xt) f(t)\mathrm{d}t, x > 0, \tag{1.5.33}$$

*where $Z_{\sigma,\beta}^{\tau,p_-}(x)$ denotes the function*

$$Z_{\sigma,\beta}^{\tau,p_-}(x) = \int_0^{\left(\frac{1}{1-p}\right)^{\frac{1}{\sigma}}} y^{\tau-1}[1-(1-p)y^{\sigma}]^{\frac{1}{1-p}} \mathrm{e}^{-xy^{-\beta}} \mathrm{d}y,\ x > 0, \tag{1.5.34}$$

*with $\tau \in \mathbb{C}, \beta > 0, \sigma > 0, p < 1$.*

**Definition 1.5.2.** *A type-2 $\mathcal{P}$-transform of a real variable function $f(x)$ for $x > 0$ is defined as:*

$$(\mathcal{P}_{\tau}^{\sigma,\beta,p_+} f)(x) = \int_0^{\infty} Z_{\sigma,\beta}^{\tau,p_+}(xt) f(t)\mathrm{d}t, x > 0, \tag{1.5.35}$$

*where $Z_{\sigma,\beta}^{\tau,p_+}(x)$ is defined as:*

$$Z_{\sigma,\beta}^{\tau,p_+}(x) = \int_0^{\infty} y^{\tau-1}[1+(p-1)y^{\sigma}]^{-\frac{1}{p-1}} \mathrm{e}^{-xy^{-\beta}} \mathrm{d}y,\ x > 0, \tag{1.5.36}$$

*for $\tau \in C, \beta > 0, \sigma \in R, \sigma \neq 0, p > 1$.*

The type-1 and type-2 $\mathcal{P}$-transforms are defined in the space $\mathbb{L}_{\tau,r}(0,\infty)$ containing functions $f$ which are Lebesgue measurable and complex-valued for which

$$\|f\|_{\tau,r} = \left\{\int_0^\infty |t^\tau f(t)|^r \frac{\mathrm{d}t}{t}\right\}^{\frac{1}{r}} < \infty, 1 \leq r < \infty, \tau \in \mathbb{R}. \quad (1.5.37)$$

Note that $Z_{\sigma,\beta}^{\tau,p_-}(x)$ and $Z_{\sigma,\beta}^{\tau,p_+}(x)$ are the generalized forms of the functions in (1.4.17) and (1.4.18), respectively. That is $Z_{\sigma,1}^{\tau,p_-}(x) = Z_\sigma^{\tau,p_-}(x)$ and $Z_{\sigma,1}^{\tau,p_+}(x) = Z_\sigma^{\tau,p_+}(x)$. Moreover, when $\beta = 1$,

$$\lim_{p\to 1_-} Z_{\sigma,1}^{\tau,p_-}(x) = \lim_{p\to 1_+} Z_{\sigma,1}^{\tau,p_+}(x) = Z_\sigma^\tau(x), \quad (1.5.38)$$

where $Z_\sigma^\tau(x)$ is the Krätzel function defined in (1.4.15). In addition, when $\sigma = 1$, both transforms in (1.5.33) and (1.5.35) reduce to the Meijer transform.

### *1.5.1 Mellin Transform of a $\mathcal{P}$-Transform*

First, we obtain the $\mathcal{P}$-transform's Mellin transform. For let us recall the definition of Mellin transform. The Mellin transform of the function $f(y)$ is defined as:

$$(\mathcal{M}f)(s) = \int_0^\infty y^{s-1} f(x)\mathrm{d}y, \ s \in \mathbb{C}, \ x > 0, \quad (1.5.39)$$

whenever $(\mathcal{M}f)(s)$ exists. The following result shows that the Mellin transform of a $\mathcal{P}$-transform of both forms yields a Mellin transform.

**Theorem 1.5.1.** *Assume $f \in \mathbb{L}_{\tau,r}(0,\infty), s, \tau \in \mathbb{C}, \ \beta > 0, x > 0$ such that $\sigma > 0$ in the case of a type-1 $\mathcal{P}$-transform and $\sigma \in \mathbb{R}, \sigma \neq 0$ be the condition for type-2 $\mathcal{P}$-transform. Then for the type-1 $\mathcal{P}$-transform,* *the Mellin transform is given by:*

$$(\mathcal{M}\ \mathcal{P}_{\tau}^{\sigma,\beta,p_-} f)(s) = \frac{\Gamma(s)\Gamma\left(\frac{\tau+\beta s}{\sigma}\right)\Gamma\left(\frac{1}{1-p}+1\right)}{\sigma(1-p)^{\frac{\tau+\beta s}{\sigma}}\Gamma\left(\frac{1}{1-p}+1+\frac{\tau+\beta s}{\sigma}\right)}(\mathcal{M}f)(1-s), \tag{1.5.40}$$

*for* $\mathrm{Re}\left(\frac{\tau+\beta s}{\sigma}\right) > 0$ *and similarly for the type-2* $\mathcal{P}$*-transform, the Mellin transform is given by*

$$(\mathcal{M}\ \mathcal{P}_{\tau}^{\sigma,\beta,p_+} f)(s) = \frac{\Gamma(s)\Gamma\left(\frac{\tau+\beta s}{\sigma}\right)\Gamma\left(\frac{1}{p-1}-\frac{\tau+\beta s}{\sigma}\right)}{|\sigma|(p-1)^{\frac{\tau+\beta s}{\sigma}}\Gamma\left(\frac{1}{p-1}\right)}(\mathcal{M}f)(1-s), \tag{1.5.41}$$

*where* $\mathrm{Re}\left(\frac{\tau+\beta s}{\sigma}\right) > 0$ *and* $\mathrm{Re}\left(\frac{1}{1-p}-\frac{\tau+\beta s}{\sigma}\right) > 0$.

*Proof.* We embark on the proof for a type-1 $\mathcal{P}$-transform, and the result for type-2 $\mathcal{P}$-transform follows suit. Using (1.5.33) and (1.5.34) using the definition of Mellin transform given in (1.5.39), we get

$$(\mathcal{M}\ \mathcal{P}_{\tau}^{\sigma,\beta,p_-} f)(s) = \int_0^\infty x^{s-1}(\mathcal{P}_{\tau}^{\sigma,\beta,p_-} f)(x)\mathrm{d}x,$$

$$= \int_0^\infty x^{s-1}\int_0^\infty Z_{\sigma,\beta}^{\tau,p_-}(xt)f(t)\mathrm{d}t\mathrm{d}x,$$

$$= \int_0^\infty x^{s-1}\int_0^\infty \int_0^{\left[\frac{1}{(1-p)}\right]^{\frac{1}{\sigma}}} y^{\tau-1}[1-(1-p)y^\sigma]^{\frac{1}{1-p}}\mathrm{e}^{-xty^{-\beta}}\mathrm{d}yf(t)\mathrm{d}t\mathrm{d}x.$$

Due to uniform convergence, we can change the order of Integration to obtain

$$= \int_0^\infty f(t)\int_0^{\left[\frac{1}{(1-p)}\right]^{\frac{1}{\sigma}}} y^{\tau-1}[1-(1-p)y^\sigma]^{\frac{1}{1-p}}\int_0^\infty x^{s-1}\mathrm{e}^{-xty^{-\beta}}\mathrm{d}x\mathrm{d}y\mathrm{d}t.$$

By making the substitution $xty^{-\beta} = u$ and simplifying through the gamma function [[8]1.1(1)], we get

$$= \Gamma(s)\int_0^\infty \frac{1}{t^s} f(t) \int_0^{\left[\frac{1}{(1-p)}\right]^{\frac{1}{\sigma}}} y^{\tau+\beta s-1}[1-(1-p)y^\sigma]^{\frac{1}{1-p}} \mathrm{d}y\mathrm{d}t.$$

Again using a substitution $(1-p)y^\sigma = v$ we get

$$\begin{aligned}(\mathcal{M}\ \mathcal{P}_\tau^{\sigma,\beta,p_-} f)(s) &= \frac{\Gamma(s)\Gamma\left(\frac{\tau+\beta s}{\sigma}\right)\Gamma\left(\frac{1}{1-p}+1\right)}{\sigma[(1-p)]^{\frac{\tau+\beta s}{\sigma}}\Gamma\left(\frac{1}{1-p}+1+\frac{\tau+\beta s}{\sigma}\right)}\int_0^\infty t^{-s} f(t)\mathrm{d}t,\\ &= \frac{\Gamma(s)\Gamma\left(\frac{\tau+\beta s}{\sigma}\right)\Gamma\left(\frac{1}{1-p}+1\right)}{\sigma(1-p)^{\frac{\tau+\beta s}{\sigma}}\Gamma\left(\frac{1}{1-p}+1+\frac{\tau+\beta s}{\sigma}\right)}(\mathcal{M}f)(1-s),\end{aligned}$$

where $\mathrm{Re}\,(s) > 0$, $\mathrm{Re}\left(\frac{\tau+\beta s}{\sigma}\right) > 0$. The case of type-2 $\mathcal{P}$-transform follows similarly. We consider the function given in (1.5.36) and through type-2 beta integral, we arrive at the result given in (1.5.41) for $\sigma \in \mathbb{R}$. This completes the proof.

**Corollary 1.5.1.** *If we take the same settings as in Theorem 1.5.1 and if $\beta = 1$, then*

$$\begin{aligned}\lim_{p\to 1^-}(\mathcal{M}\ \mathcal{P}_\tau^{\sigma,1,p_-} f)(s) &= \lim_{p\to 1^+}(\mathcal{M}\ \mathcal{P}_\tau^{\sigma,1,p_+} f)(s)\\ &= (\mathcal{M}\ \mathcal{K}_\tau^\sigma f)(s) = \frac{\Gamma(s)\Gamma\left(\frac{\tau+s}{\sigma}\right)}{|\sigma|}(\mathcal{M}f)(1-s).\end{aligned} \tag{1.5.42}$$

**Corollary 1.5.2.** *Let $\sigma \in \mathbb{R}$, $\sigma \neq 0$, $\tau \in \mathbb{C}, a > 0, \beta > 0, p > 1$ and $s \in \mathbb{C}$, $\mathrm{Re}\,(s) > 0$. Then for $Z_{\sigma,\beta}^{\tau,p_-}(x)$ which is defined in (1.5.34), we have*

$$(\mathcal{M}Z_{\sigma,\beta}^{\tau,p_-})(s) = \frac{1}{\sigma(1-p)^{\frac{\tau+\beta s}{\sigma}}} \frac{\Gamma(s)\Gamma\left(\frac{\tau+\beta s}{\sigma}\right)\Gamma\left(\frac{1}{1-p}+1\right)}{\Gamma\left(\frac{1}{1-p}+\frac{\tau+\beta s}{\sigma}\right)}, \quad (1.5.43)$$

*where* $\sigma > 0$ *and* $\mathrm{Re}\,(\tau + \beta s) > 0$. *For* $Z_{\sigma,\beta}^{\tau,p_+}(x)$ *defined in (1.5.36), we have*

$$(\mathcal{M}Z_{\sigma,\beta}^{\tau,p_+})(s) = \frac{1}{|\sigma|(p-1)^{\frac{\tau+\beta s}{\sigma}}} \frac{\Gamma(s)\Gamma\left(\frac{\tau+\beta s}{\sigma}\right)\Gamma\left(\frac{1}{p-1}-\frac{\tau+\beta s}{\sigma}\right)}{\Gamma\left(\frac{1}{p-1}\right)}, \quad (1.5.44)$$

*where* $\mathrm{Re}\,(\tau + \beta s) > 0$ *and* $\mathrm{Re}\left(\frac{1}{p-1} - \frac{\tau+\beta s}{\sigma}\right) > 0$ *are the conditions in case of* $\sigma > 0$ *and* $\mathrm{Re}\,(\tau + \beta s) < 0$ *and* $\mathrm{Re}\left(\frac{1}{p-1} - \frac{\tau+\beta s}{\sigma}\right) > 0$ *are the conditions in case of* $\sigma < 0$.

*Proof.* The proof proceeds exactly as of Theorem 1.5.1.

**Corollary 1.5.3.** *If the settings of Corollary 1.5.2 are satisfied with* $\beta = 1$ *then*

$$\begin{aligned}\lim_{p\to 1^-}(\mathcal{M}Z_{\sigma,1}^{\tau,p_-})(s) &= \lim_{p\to 1^+}(\mathcal{M}Z_{\sigma,1}^{\tau,p_+})(s)\\ &= (\mathcal{M}Z_\sigma^\tau)(s) = \frac{1}{|\sigma|}\Gamma(s)\Gamma\left(\frac{\tau+s}{\sigma}\right), \sigma \neq 0, \tau \in \mathbb{C}.\end{aligned} \quad (1.5.45)$$

### *1.5.2 H-Function Representations*

Now, we present the $H$-function representation of the functions defined in (1.5.34) and (1.5.36). If $m, n, r, q$ are integers with $1 \leq m \leq q, 0 \leq n \leq r$, for $a_i, b_j \in \mathbb{C}$ along with $\alpha_i, \beta_j \in \mathbb{R}^+ = (0, \infty)$, $i = 1, 2, \cdots, r; j = 1, 2, \cdots, q$, then the Fox's $H$-function is defined as a

Mellin-Barnes type integral as [18, 33]:

$$H_{r,q}^{m,n}(z) = H_{r,q}^{m,n}\left(z\Big|_{(b_1,\beta_1),\cdots,(b_q,\beta_q)}^{(a_1,\alpha_1),\cdots,(a_r,\alpha_r)}\right) = \frac{1}{2\pi i}\int_{\mathcal{L}} h(s)z^{-s}\mathrm{d}s, \quad (1.5.46)$$

with

$$h(s) = \frac{\left\{\prod_{j=1}^{m}\Gamma(b_j+\beta_j s)\right\}\left\{\prod_{j=1}^{n}\Gamma(1-a_j-\alpha_j s)\right\}}{\left\{\prod_{j=m+1}^{q}\Gamma(1-b_j-\beta_j s)\right\}\left\{\prod_{j=n+1}^{p}\Gamma(a_j+\alpha_j s)\right\}}. \quad (1.5.47)$$

Here in (1.5.47), a null product is taken as one if it occurs. $\mathcal{L}$ is the contour that differentiates the poles of numerator gamma functions $\Gamma(b_j+\beta_j s),\ j = 1,\cdots,m$ on the left from those of the numerator gamma functions $\Gamma(1-a_j-\alpha_j s),\ j = 1,\cdots,n$ on the right. For an understanding of $H$-function and detailed convergence conditions, refer to any of the books Mathai [33] or Mathai and Saxena [37] or Srivastava et al. [50] or Kilbas and Saigo [18].

Also from (1.5.46) and (1.5.47), it can be seen that $H_{r,q}^{m,n}(z)$ in (1.5.46) is the Mellin transform inverse of $h(s)$ in (1.5.47).

**Theorem 1.5.2.** *Let $\tau, z \in \mathbb{C},\ a > 0,\ \sigma > 0,\ \beta > 0$ be such that $p < 1$, then*

$$Z_{\sigma,\beta}^{\tau,p-}(z) = \frac{\Gamma\left(\frac{1}{1-p}+1\right)}{\sigma(1-p)^{\frac{\tau}{\sigma}}}H_{1,2}^{2,0}\left[[a(1-p)]^{\frac{\beta}{\sigma}}z\Big|_{(0,1),\ (\frac{\tau}{\sigma},\frac{\beta}{\sigma})}^{\left(\frac{1}{1-p}+1+\frac{\tau}{\sigma},\frac{\beta}{\sigma}\right)}\right], \quad (1.5.48)$$

*where $Z_{\sigma,\beta}^{\tau,p-}(z)$ is as defined in (1.5.34).*

*Proof.* The proof quickly follows by taking the inverse Mellin transform of (1.5.43) in Corollary 1.5.2 and using (1.5.46) and (1.5.47).

**Theorem 1.5.3.** *If we take the conditions $\sigma \in \mathbb{R}, \tau \in \mathbb{C}, a > 0, \beta > 0$ and $p > 1$. If $\sigma > 0$ then we have*

$$Z_{\sigma,\beta}^{\tau,p_+}(z) = \frac{1}{\sigma(p-1)^{\frac{\tau}{\sigma}}\Gamma\left(\frac{1}{p-1}\right)} H_{1,2}^{2,1}\left[(p-1)^{\frac{\beta}{\sigma}} z \middle|_{(0,1),\ (\frac{\tau}{\sigma},\frac{\beta}{\sigma})}^{(1-\frac{1}{p-1}+\frac{\tau}{\sigma},\frac{\beta}{\sigma})}\right]. \tag{1.5.49}$$

*If $\sigma < 0$, then*

$$Z_{\sigma,\beta}^{\tau,p_+}(z) = -\frac{1}{\sigma(p-1)^{\frac{\tau}{\sigma}}\Gamma\left(\frac{1}{q-1}\right)} H_{1,2}^{2,1}\left[(p-1)^{\frac{\beta}{\sigma}} z \middle|_{(0,1),\ (\frac{1}{p-1}-\frac{\tau}{\sigma},-\frac{\beta}{\sigma})}^{(1-\frac{\tau}{\sigma},-\frac{\beta}{\sigma})}\right], \tag{1.5.50}$$

*where $Z_{\sigma,\beta}^{\tau,p}(z)$ is as defined in (1.5.36).*

*Proof.* The proof can be easily seen by taking the inverse Mellin transform of (1.5.44) in Corollary 1.5.2 and using (1.5.46) and (1.5.47).

### *1.5.3 Composition with Fractional Operators*

Further results are derived by employing the left-sided Saigo operator for a power function, which is illustrated in the lemma below:

**Lemma 1.5.1.** *Assuming that $\kappa, \delta, \eta, \lambda \in \mathbb{C}$ and for the condition* $\mathrm{Re}\,(\kappa) > 0$, $\mathrm{Re}\,(\lambda + 1) > 0$ *and* $\mathrm{Re}\,(\lambda + 1 + \eta - \delta) > 0$, *we have*

$$(I_{0+}^{\kappa,\delta,\eta} t^{\lambda})(x) = \frac{\Gamma(\lambda+1)\Gamma(\lambda+1+\eta-\delta)}{\Gamma(\lambda+1-\delta)\Gamma(\lambda+1+\kappa+\eta)} x^{\lambda-\delta}. \tag{1.5.51}$$

*Proof.* Using the formulas (1.2.9) and (1.2.11), we have

$$(I_{0+}^{\kappa,\delta,\eta} t^{\lambda})(x) = \frac{x^{-\kappa-\delta}}{\Gamma(\kappa)} \int_0^x (x-t)^{\kappa-1} \sum_{k=0}^{\infty} \frac{(\kappa+\delta)_k(-\eta)_k}{(\kappa)_k} \frac{\left(1-\frac{t}{x}\right)^k}{k!} t^{\lambda} \mathrm{d}t.$$

The uniform convergence of the integral allows us to change the integration and summation order, and through the substitution $\frac{t}{x} = u$ along with the use of beta integral, we get

$$
\begin{aligned}
(I_{0+}^{\kappa,\delta,\eta} t^{\lambda})(x) &= \frac{x^{-\kappa-\delta}}{\Gamma(\kappa)} \sum_{k=0}^{\infty} \frac{(\kappa+\delta)_k(-\eta)_k}{(\kappa)_k k!} \frac{\Gamma(\kappa+k)\Gamma(\lambda+1)}{\Gamma(\kappa+k+\lambda+1)}, \\
&= \frac{x^{-\kappa-\delta}\Gamma(\lambda+1)}{\Gamma(\kappa+\lambda+1)} {}_2F_1(\kappa+\delta, -\eta; \kappa+\lambda+1; 1).
\end{aligned}
$$

By (1.2.12), we get the result in (1.5.51).

**Corollary 1.5.4.** *If we take the same conditions as in Lemma 1.5.1 with $\delta = 0_-$ we can come up with the left-sided or first-kind Kober fractional operator for a power function as*

$$
I_{0+}^{\kappa,0,\eta} t^{\lambda}(x) = (\mathcal{I}_{\eta,\kappa}^{+} t^{\lambda})x = \frac{\Gamma(\lambda+1+\eta)}{\Gamma(\lambda+1+\kappa+\eta)} x^{\lambda}, \tag{1.5.52}
$$

*and with the condition $\delta = -\kappa$ we get the left-sided or first-kind Riemann-Liouville fractional operator for a power function as*

$$
(I_{0+}^{\kappa,-\kappa,\eta} t^{\lambda})(x) = (\mathcal{I}_{0+}^{\kappa} t^{\lambda})x = \frac{\Gamma(\lambda+1)}{\Gamma(\lambda+1+\kappa)} x^{\lambda+\kappa}. \tag{1.5.53}
$$

Next, we investigate the composition of the left-hand-sided Saigo fractional integral with $\mathcal{P}$-transform and arrive at the following theorem.

**Theorem 1.5.4.** *If the conditions $f \in \mathbb{L}_{\tau,r}(0,\infty)$, $\kappa, \tau, \eta \in \mathbb{C}$, $\beta > 0$, $\mathrm{Re}\,(\kappa) > 0$ are satisfied along with $\sigma > 0$ in case of type-1 $\mathcal{P}$-transform and $\sigma \in \mathbb{R}$ in case of type-2 $\mathcal{P}$-transform, then the Saigo fractional integral for a type-1 $\mathcal{P}$-transform is given by:*

$$(I_{0+}^{\kappa,\delta,\eta}\mathcal{P}_{\tau}^{\sigma,\beta,p-}f)(x)$$
$$=\frac{\Gamma\left(\frac{1}{1-p}+1\right)}{x^{\delta}\sigma(1-p)^{\frac{\tau}{\sigma}}}\int_0^{\infty}H_{3,4}^{2,2}\left[(1-p)^{\frac{\beta}{\sigma}}vx\Big|_{(0,1),\ (\frac{\tau}{\sigma},\frac{\beta}{\sigma}),(\delta,1),(-\kappa-\eta,1)}^{(0,1),(\delta-\eta,1),\left(\frac{1}{1-p}+1+\frac{\tau}{\sigma},\frac{\beta}{\sigma}\right)}\right]f(v)\mathrm{d}v,\tag{1.5.54}$$

*and for a type-2 $\mathcal{P}$-transform, we get:*

$$(I_{0+}^{\kappa,\delta,\eta}\mathcal{P}_{\tau}^{\sigma,\beta,p+}f)(x)=\frac{1}{x^{\delta}\sigma(p-1)^{\frac{\tau}{\sigma}}\Gamma\left(\frac{1}{p-1}\right)}$$
$$\times\int_0^{\infty}H_{3,4}^{2,3}\left[(p-1)^{\frac{\beta}{\sigma}}vx\Big|_{(0,1),\ (\frac{\tau}{\sigma},\frac{\beta}{\sigma}),(\delta,1),(-\kappa-\eta,1)}^{(0,1),(\delta-\eta,1),\left(1-\frac{1}{p-1}+\frac{\tau}{\sigma},\frac{\beta}{\sigma}\right)}\right]f(v)\mathrm{d}v,\tag{1.5.55}$$

*for $\sigma>0$ and*

$$(I_{0+}^{\kappa,\delta,\eta}\mathcal{P}_{\tau}^{\sigma,\beta,p+}f)(x)=-\frac{1}{x^{\delta}\sigma(1-p)^{\frac{\tau}{\sigma}}\Gamma\left(\frac{1}{p-1}\right)}$$
$$\times\int_0^{\infty}H_{3,4}^{2,3}\left[(p-1)^{\frac{\beta}{\sigma}}vx\Big|_{(0,1),\ \left(\frac{1}{p-1}-\frac{\tau}{\sigma},-\frac{\beta}{\sigma}\right),(\delta,1),(-\kappa-\eta,1)}^{(0,1),(\delta-\eta,1),(1-\frac{\tau}{\sigma},-\frac{\beta}{\sigma})}\right]f(v)\mathrm{d}v.\tag{1.5.56}$$

*Proof.* We embark the proof by using (1.2.9) along with the formulas (1.5.33) and (1.5.34) and using the inverse Mellin transform of (1.5.40), we arrive at:

$$
\begin{aligned}
(I_{0+}^{\kappa,\delta,\eta}\mathcal{P}_{\tau}^{\sigma,\beta,p_{-}}f)(x) &= \frac{x^{-\kappa-\delta}}{\Gamma(\kappa)}\int_0^x (x-t)^{\kappa-1} \\
&\times {}_2F_1(\kappa+\delta,-\eta;\kappa;1-\frac{t}{x})\int_0^\infty Z_{\sigma,\beta}^{\tau,p}(tv)f(v)\mathrm{d}v\mathrm{d}t, \\
&= \frac{x^{-\kappa-\delta}\Gamma\left(\frac{1}{1-p}+1\right)}{\Gamma(\kappa)\sigma(1-p)^{\frac{\tau}{\sigma}}}\int_0^x (x-t)^{\kappa-1}{}_2F_1(\kappa+\delta,-\eta;\kappa;1-\frac{t}{x}) \\
&\times \int_0^\infty \frac{1}{2\pi i}\int_{\mathcal{L}} \frac{\Gamma(s)\Gamma\left(\frac{\tau+\beta s}{\sigma}\right)}{\Gamma\left(\frac{1}{1-p}+1+\frac{\tau+\beta s}{\sigma}\right)}\{(1-p)^{\frac{\beta}{\sigma}}vt\}^{-s}\mathrm{d}sf(v)\mathrm{d}v\mathrm{d}t,
\end{aligned}
$$

where $\mathrm{Re}\,(s) > 0$, $\mathrm{Re}\,(\frac{\tau+\beta s}{\sigma}) > 0$. Since the two integrals are uniformly convergent, we change the order of integration, and using Lemma 1.5.1, we get:

$$
\begin{aligned}
(I_{0+}^{\kappa,\delta,\eta}\mathcal{P}_{\tau}^{\sigma,\beta,p_{-}}f)(x) &= \frac{\Gamma\left(\frac{1}{1-p}+1\right)}{x^{\delta}\sigma(1-p)^{\frac{\tau}{\sigma}}}\int_0^\infty \frac{1}{2\pi i}\int_{\mathcal{L}} \frac{\Gamma(s)\Gamma\left(\frac{\tau+\beta s}{\sigma}\right)}{\Gamma\left(\frac{1}{1-p}+1+\frac{\tau+\beta s}{\sigma}\right)} \\
&\times \frac{\Gamma(1-s)\Gamma(1+\eta-\delta-s)}{\Gamma(1-\delta-s)\Gamma(1+\kappa+\eta-s)}\{(1-p)^{\frac{\beta}{\sigma}}vx\}^{-s}\mathrm{d}sf(v)\mathrm{d}v,
\end{aligned}
$$

where $\mathrm{Re}\,(s) > 0$, $\mathrm{Re}\,(1-s) > 0$, $\mathrm{Re}\,(1-\delta-s) > 0$, $\mathrm{Re}\,(\tau+\beta s) > 0$, $\mathrm{Re}\,(1+\eta-\delta-s) > 0$, $\mathrm{Re}\,(1+\kappa+\eta-s) > 0$. This yields the result in (1.5.54) for type-1 $\mathcal{P}$-transform.

**Corollary 1.5.5.** *If we take the same settings of Theorem 1.5.4 with* $\delta = 0_{-}$, *then the left-sided or first-kind Kober fractional operator of a type-1* $\mathcal{P}$*-transform is:*

$$(I_{0+}^{\kappa,-\kappa,\eta}\mathcal{P}_{\tau}^{\sigma,\beta,p_-}f)(x) = (\mathcal{I}_{\eta,\kappa}^{+}\mathcal{P}_{\tau}^{\sigma,\beta,p_-}f)(x),$$
$$= \frac{\Gamma\left(\frac{1}{1-p}+1\right)}{\sigma(1-p)^{\frac{\tau}{\sigma}}}\int_0^\infty H_{2,3}^{2,1}\left[(1-p)^{\frac{\beta}{\sigma}}vx\Big|_{(0,1),\ (\frac{\tau}{\sigma},\frac{\beta}{\sigma}),\ (-\kappa-\eta,1)}^{(-\eta,1),\ \left(\frac{1}{1-p}+1+\frac{\tau}{\sigma},\frac{\beta}{\sigma}\right)}\right]f(v)\mathrm{d}v, \tag{1.5.57}$$

*and for a type-2 $\mathcal{P}$-transform we have*

$$(I_{0+}^{\kappa,\delta,\eta}\mathcal{P}_{\tau}^{\sigma,\beta,p_+}f)(x) = (\mathcal{I}_{\eta,\kappa}^{+}\mathcal{P}_{\tau}^{\sigma,\beta,p_+}f)(x) = \frac{1}{\sigma(p-1)^{\frac{\tau}{\sigma}}\Gamma\left(\frac{1}{p-1}\right)}$$
$$\times\int_0^\infty H_{2,3}^{2,2}\left[(p-1)^{\frac{\beta}{\sigma}}vx\Big|_{(0,1),\ (\frac{\tau}{\sigma},\frac{\beta}{\sigma}),(-\kappa-\eta,1)}^{(-\eta,1),\left(1-\frac{1}{p-1}+\frac{\tau}{\sigma},\frac{\beta}{\sigma}\right)}\right]f(v)\mathrm{d}v, \tag{1.5.58}$$

*for $\sigma > 0$ and*

$$(I_{0+}^{\kappa,\delta,\eta}\mathcal{P}_{\tau}^{\sigma,\beta,p_+}f)(x) = (\mathcal{I}_{\eta,\kappa}^{+}\mathcal{P}_{\tau}^{\sigma,\beta,p_+}f)(x) = -\frac{1}{\sigma(1-p)^{\frac{\tau}{\sigma}}\Gamma\left(\frac{1}{p-1}\right)}$$
$$\times\int_0^\infty H_{2,3}^{2,2}\left[(p-1)^{\frac{\beta}{\sigma}}vx\Big|_{(0,1),\ \left(\frac{1}{p-1}-\frac{\tau}{\sigma},-\frac{\beta}{\sigma}\right),(-\kappa-\eta,1)}^{(-\eta,1),(1-\frac{\tau}{\sigma},-\frac{\beta}{\sigma})}\right]f(v)\mathrm{d}v, \tag{1.5.59}$$

*for $\sigma < 0$.*

**Corollary 1.5.6.** *If the settings are same as that of Theorem 1.5.4 along with $\delta = -\kappa$, then the left-sided or first-kind Riemann-Liouville fractional operator for a type-1 $\mathcal{P}$-transform is:*

$$
(I_{0+}^{\kappa,-\kappa,\eta}\mathcal{P}_{\tau}^{\sigma,\beta,p_-}f)(x) = (\mathcal{I}_{0+}^{\kappa}\mathcal{P}_{\tau}^{\sigma,\beta,p_-}f)(x) = \frac{x^{\kappa}\Gamma\left(\frac{1}{1-p}+1\right)}{\sigma(1-p)^{\frac{\tau}{\sigma}}}
$$

$$
\times \int_0^{\infty} H_{2,3}^{2,1}\left[(1-p)^{\frac{\beta}{\sigma}}vx\middle|\begin{matrix}(0,1),\ \left(\frac{1}{1-p}+1+\frac{\tau}{\sigma},\frac{\beta}{\sigma}\right)\\(0,1),\ (\frac{\tau}{\sigma},\frac{\beta}{\sigma}),\ (-\kappa,1)\end{matrix}\right] f(v)\mathrm{d}v, \tag{1.5.60}
$$

*and for a type-2 $\mathcal{P}$-transform we get*

$$
(I_{0+}^{\kappa,\delta,\eta}\mathcal{P}_{\tau}^{\sigma,\beta,p_+}f)(x) = (\mathcal{I}_{0+}^{\kappa}\mathcal{P}_{\tau}^{\sigma,\beta,p_+}f)(x) = \frac{x^{\kappa}}{\sigma(p-1)^{\frac{\tau}{\sigma}}\Gamma\left(\frac{1}{p-1}\right)}
$$

$$
\times \int_0^{\infty} H_{2,3}^{2,2}\left[(p-1)^{\frac{\beta}{\sigma}}vx\middle|\begin{matrix}(0,1),\left(1-\frac{1}{p-1}+\frac{\tau}{\sigma},\frac{\beta}{\sigma}\right)\\(0,1),\ (\frac{\tau}{\sigma},\frac{\beta}{\sigma}),\ (-\kappa,1)\end{matrix}\right] f(v)\mathrm{d}v, \tag{1.5.61}
$$

*for $\sigma > 0$ and*

$$
(I_{0+}^{\kappa,\delta,\eta}\mathcal{P}_{\tau}^{\sigma,\beta,p_+}f)(x) = (\mathcal{I}_{0+}^{\kappa}\mathcal{P}_{\tau}^{\sigma,\beta,p_+}f)(x) = -\frac{x^{\kappa}}{\sigma(1-p)^{\frac{\tau}{\sigma}}\Gamma\left(\frac{1}{p-1}\right)}
$$

$$
\times \int_0^{\infty} H_{2,3}^{2,2}\left[(p-1)^{\frac{\beta}{\sigma}}vx\middle|\begin{matrix}(0,1),\ (1-\frac{\tau}{\sigma},-\frac{\beta}{\sigma})\\(0,1),\ \left(\frac{1}{p-1}-\frac{\tau}{\sigma},-\frac{\beta}{\sigma}\right),(-\kappa,1)\end{matrix}\right] f(v)\mathrm{d}v, \tag{1.5.62}
$$

*for $\sigma < 0$.*

**Corollary 1.5.7.** *If we take the conditions of Theorem 1.5.4 with $\beta = 1$ then*

$$\lim_{p\to 1_-}(I_{0+}^{\kappa,\delta,\eta}\mathcal{P}_\tau^{\sigma,\beta,p_-}f)(x) = \lim_{p\to 1_+}(I_{0+}^{\kappa,\delta,\eta}\mathcal{P}_\tau^{\sigma,\beta,p_+}f)(x) = (I_{0+}^{\kappa,\delta,\eta}\mathcal{K}_\tau^\sigma f)(x)$$
$$= \frac{1}{x^\delta\sigma}\int_0^\infty H_{2,4}^{2,2}\left[vx\Big|_{(0,1),\ (\frac{\tau}{\sigma},\frac{\beta}{\sigma}),(\delta,1),(-\kappa-\eta,1)}^{(0,1),(\delta-\eta,1)}\right]f(v)\mathrm{d}v, \quad (1.5.63)$$

*for* $\sigma > 0$ *and*

$$\lim_{p\to 1_+}(I_{0+}^{\kappa,\delta,\eta}\mathcal{P}_\tau^{\sigma,\beta,p_+}f)(x) = (I_{0+}^{\kappa,\delta,\eta}\mathcal{K}_\tau^\sigma f)(x)$$
$$= -\frac{1}{x^\delta\sigma}\int_0^\infty H_{3,3}^{1,3}\left[vx\Big|_{(0,1),\ (\delta,1),\ (-\kappa-\eta,1)}^{(0,1),\ (\delta-\eta,1),\ (1-\frac{\tau}{\sigma},-\frac{\beta}{\sigma})}\right]f(v)\mathrm{d}v, \quad (1.5.64)$$

*for* $\sigma < 0$.

**Corollary 1.5.8.** *Taking the conditions of Theorem 1.5.4 with* $\delta = 0_-, \beta = 1$, *we have*

$$\lim_{p\to 1_-}(\mathcal{I}_{\eta,\kappa}^+\mathcal{P}_\tau^{\sigma,\beta,p_-}f)(x) = \lim_{p\to 1_+}(\mathcal{I}_{\eta,\kappa}^+\mathcal{P}_\tau^{\sigma,\beta,p_+}f)(x) = (\mathcal{I}_{\eta,\kappa}^+\mathcal{K}_\tau^\sigma f)(x),$$
$$= \frac{1}{\sigma}\int_0^\infty H_{1,3}^{2,1}\left[vx\Big|_{(0,1),\ (\frac{\tau}{\sigma},\frac{\beta}{\sigma}),(-\kappa-\eta,1)}^{(-\eta,1)}\right]f(v)\mathrm{d}v, \quad (1.5.65)$$

*for* $\sigma > 0$ *and*

$$\lim_{p\to 1_+}(\mathcal{I}_{\eta,\kappa}^+\mathcal{P}_\tau^{\sigma,\beta,p_+}f)(x) = (\mathcal{I}_{\eta,\kappa}^+\mathcal{K}_\tau^\sigma f)(x),$$
$$= -\frac{1}{\sigma}\int_0^\infty H_{2,2}^{1,2}\left[vx\Big|_{(0,1),(-\kappa-\eta,1)}^{(-\eta,1),(1-\frac{\tau}{\sigma},-\frac{\beta}{\sigma})}\right]f(v)\mathrm{d}v, \quad (1.5.66)$$

*for* $\sigma < 0$.

**Corollary 1.5.9.** *Taking the conditions of Theorem 1.5.4 with* $\delta = -\kappa$ *and* $\beta = 1$, *we have*

$$\lim_{p\to 1_-}(\mathcal{I}^\kappa_{0+}\mathcal{P}^{\sigma,\beta,p_-}_\tau f)(x) = \lim_{p\to 1_+}(\mathcal{I}^\kappa_{0+}\mathcal{P}^{\sigma,\beta,p_+}_\tau f)(x) = (\mathcal{I}^\kappa_{0+}\mathcal{K}^\sigma_\tau f)(x),$$
$$= \frac{x^\kappa}{\sigma}\int_0^\infty H^{2,1}_{1,3}\left[vx\Big|^{(0,1)}_{(0,1),\ (\frac{\tau}{\sigma},\frac{\beta}{\sigma}),\ (-\kappa,1)}\right] f(v)\mathrm{d}v, \qquad (1.5.67)$$

*for* $\sigma > 0$ *and*

$$\lim_{p\to 1_+}(\mathcal{I}^\kappa_{0+}\mathcal{P}^{\sigma,\beta,p_+}_\tau f)(x) = (\mathcal{I}^\kappa_{0+}\mathcal{K}^\sigma_\tau f)(x),$$
$$= -\frac{x^\kappa}{\sigma}\int_0^\infty H^{1,2}_{2,2}\left[vx\Big|^{(0,1),\ (1-\frac{\tau}{\sigma},-\frac{\beta}{\sigma})}_{(0,1),\ (-\kappa,1)}\right] f(v)\mathrm{d}v, \qquad (1.5.68)$$

*for* $\sigma < 0$.

To complete the discussion on the applications of fractional operators in $\mathcal{P}$-transform, we present the composition of $\mathcal{P}$-transform with the right-sided or Second-kind Riemann-Liouville fractional integral $\mathcal{I}^\kappa_-$ defined in (1.1.5) and the differential operator $\mathcal{D}^\kappa_-$ of complex order $\kappa \in \mathbb{C}$ defined in (1.2.6.)

**Theorem 1.5.5.** *If* $f \in \mathbb{L}_{\tau,r}(0,\infty), \kappa, \tau \in \mathbb{C}, \beta > 0,\ \mathrm{Re}\,(\kappa) > 0, x > 0$ *are the conditions along with* $\sigma > 0$ *for a type-1* $\mathcal{P}$*-transform and* $\sigma \in \mathbb{R}$ *for a type-2* $\mathcal{P}$*-transform, then there holds the formulae*

$$(\mathcal{I}^\kappa_-\mathcal{P}^{\sigma,\beta,p_-}_\tau f)(x) = (\mathcal{P}^{\sigma,\beta,p_-}_{\tau+\beta\kappa}x^{-\kappa}f)(x), p < 1, \qquad (1.5.69)$$
$$(\mathcal{I}^\kappa_-\mathcal{P}^{\sigma,\beta,p_+}_\tau f)(x) = (\mathcal{P}^{\sigma,\beta,p_+}_{\tau+\beta\kappa}x^{-\kappa}f)(x), p > 1. \qquad (1.5.70)$$

*Proof.* We first prove the result for type-1 $\mathcal{P}$-transform. Using formulas (1.1.5), (1.5.33) and (1.5.34), we have:

$$(\mathcal{I}_{-}^{\kappa}\mathcal{P}_{\tau}^{\sigma,\beta,p_{-}} f)(x) = \frac{1}{\Gamma(\kappa)} \int_{x}^{\infty} (t-x)^{\kappa-1} \int_{0}^{\infty} Z_{\sigma,\beta}^{\tau,p_{-}}(tv) f(v) \mathrm{d}v \mathrm{d}t,$$
$$= \frac{1}{\Gamma(\kappa)} \int_{x}^{\infty} (t-x)^{\kappa-1} \int_{0}^{\infty} f(v)$$
$$\times \int_{0}^{\left(\frac{1}{1-p}\right)^{\frac{1}{\sigma}}} y^{\tau-1}[1-(1-p)y^{\sigma}]^{\frac{1}{1-p}} \mathrm{e}^{-tvy^{-\beta}} \mathrm{d}y \mathrm{d}v \mathrm{d}t.$$

As uniform convergence allows us to change the order of integration, we get:

$$(\mathcal{I}_{-}^{\kappa}\mathcal{P}_{\tau}^{\sigma,\beta,q_{-}} f)(x) = \int_{0}^{\infty} f(u) \int_{0}^{\left(\frac{1}{1-p}\right)^{\frac{1}{\sigma}}} y^{\tau-1}[1-(1-p)y^{\sigma}]^{\frac{1}{1-p}}$$
$$\times \frac{1}{\Gamma(\kappa)} \int_{x}^{\infty} (t-x)^{\kappa-1} \mathrm{e}^{-tvy^{-\beta}} \mathrm{d}t \mathrm{d}y \mathrm{d}v,$$
$$= \int_{0}^{\infty} f(v) \int_{0}^{\left(\frac{1}{1-p}\right)^{\frac{1}{\sigma}}} y^{\tau-1}[1-(1-p)y^{\sigma}]^{\frac{1}{1-p}} (\mathcal{I}_{-}^{\kappa}(\mathrm{e}^{-tvy^{-\beta}}))(x) \mathrm{d}y \mathrm{d}v.$$

By employing the formula of Samko et al. [48]

$$(\mathcal{I}_{-}^{\kappa}\mathrm{e}^{-\lambda t})(x) = \lambda^{-\kappa}\mathrm{e}^{-\lambda x}, \quad \mathrm{Re}\,(\kappa) > 0,\ \mathrm{Re}\,(\lambda) > 0, \tag{1.5.71}$$

we have

$$(\mathcal{I}_{-}^{\kappa}\mathcal{P}_{\tau}^{\sigma,\beta,p_{-}} f)(x)$$
$$= \int_{0}^{\infty} f(v) v^{-\kappa} \int_{0}^{\left(\frac{1}{1-p}\right)^{\frac{1}{\sigma}}} y^{\tau+\beta\kappa-1}[1-(1-p)y^{\sigma}]^{\frac{1}{1-p}} \mathrm{e}^{-xvy^{-\beta}} \mathrm{d}y \mathrm{d}v,$$
$$= \int_{0}^{\infty} f(v) v^{-\kappa} Z_{\sigma,\beta}^{\tau+\beta\kappa,p_{-}}(xv) \mathrm{d}v = (\mathcal{P}_{\tau+\beta\kappa}^{\sigma,\beta,p_{-}} x^{-\kappa} f)(x).$$

According to (1.1.5), the result in (1.5.69) follows. Similarly, the result for type-2 $\mathcal{P}$-transform given in (1.5.70) can be proved.

**Corollary 1.5.10.** *Taking the conditions as in Theorem 1.5.5 with* $\beta = 1$, *we have*

$$\lim_{p\to 1_-} (\mathcal{I}_-^\kappa \mathcal{P}_\tau^{\sigma,1,p_-} f)(x) = \lim_{p\to 1_+} (\mathcal{I}_-^\kappa \mathcal{P}_\tau^{\sigma,1,p_+} f)(x)$$
$$= (\mathcal{I}_-^\kappa \mathcal{K}_\tau^\sigma f)(x) = (\mathcal{K}_{\tau+\kappa}^\sigma x^{-\kappa} f)(x). \quad (1.5.72)$$

The next theorem gives the composition formula for fractional derivative given in (1.2.6) with the $\mathcal{P}$-transforms of both types.

**Theorem 1.5.6.** *If* $\kappa, \tau \in \mathbb{C}$, $\operatorname{Re}(\kappa) > 0, \sigma > 0$ *and* $\beta > 0$ *are the conditions, then*

$$(\mathcal{D}_-^\kappa \mathcal{P}_\tau^{\sigma,\beta,p_-} f)(x) = (\mathcal{P}_{\tau-\beta\kappa}^{\sigma,\beta,p_-} x^\kappa f)(x), p < 1, \quad (1.5.73)$$
$$(\mathcal{D}_-^\kappa \mathcal{P}_\tau^{\sigma,\beta,p_+} f)(x) = (\mathcal{P}_{\tau-\beta\kappa}^{\sigma,\beta,p_+} x^\kappa f)(x), p > 1. \quad (1.5.74)$$

*Proof.* If $p < 1$ and $n = [\operatorname{Re}(\kappa)] + 1$ then using (1.2.6), (1.5.33) and (1.5.69), with $\kappa$ substituted by $n - \kappa$, we obtain

$$(\mathcal{D}_-^\kappa \mathcal{P}_\tau^{\sigma,\beta,p_-} f)(x) = \left(-\frac{\mathrm{d}}{\mathrm{d}x}\right)^n (\mathcal{P}_{\tau+\beta n-\beta\kappa}^{\sigma,\beta,p_-} x^{-(n-\kappa)} f)(x),$$
$$= \left(-\frac{\mathrm{d}}{\mathrm{d}x}\right)^n \int_0^\infty f(v) v^{-(n-\kappa)}$$
$$\times \int_0^{\left(\frac{1}{(1-p)}\right)^{\frac{1}{\sigma}}} y^{\tau+\beta n-\beta\kappa-1} [1-(1-p)y^\sigma]^{\frac{1}{1-p}} \mathrm{e}^{-xvy^{-\beta}} \mathrm{d}y\mathrm{d}v.$$

Due to the uniform convergence, by taking the differentiation inside the integral, we get:

$$
\begin{aligned}
&(\mathcal{D}_{-}^{\kappa}\mathcal{P}_{\tau}^{\sigma,\beta,p}f)(x)\\
&\quad = \int_0^{\infty} f(v)v^{\kappa}\int_0^{\left(\frac{1}{1-p}\right)^{\frac{1}{\sigma}}} y^{\tau-\beta\kappa-1}[1-(1-p)y^{\sigma}]^{\frac{1}{1-p}}\mathrm{e}^{-xvy^{-\beta}}\mathrm{d}y\mathrm{d}v,\\
&\quad = \int_0^{\infty} Z_{\sigma,\beta}^{\tau-\beta\kappa,p_-}(xv)v^{\kappa}f(v)\mathrm{d}v.
\end{aligned}
$$

In view of (1.5.33), we get the result in (1.5.73). The corresponding result in (1.5.74) for $p > 1$ can be proved similarly.

**Corollary 1.5.11.** *If we take the same conditions as in Theorem 1.5.6 along with* $\beta = 1$*, we have*

$$
\begin{aligned}
\lim_{p\to 1_-}(\mathcal{D}_{-}^{\kappa}\mathcal{P}_{\tau}^{\sigma,1,p_-}f)(x) &= \lim_{p\to 1_+}(\mathcal{D}_{-}^{\kappa}\mathcal{P}_{\tau}^{\sigma,1,p_+}f)(x)\\
&= (\mathcal{D}_{-}^{\kappa}\mathcal{K}_{\tau}^{\sigma}f)(x) = (\mathcal{K}_{\tau-\kappa}^{\sigma}x^{\kappa}f)(x). \quad (1.5.75)
\end{aligned}
$$

**Remark 1.5.1.** *Lemma 1.5.1 and Theorem 1.5.4 to Theorem 1.5.6 and their corollaries were established in Kumar & Kilbas [22].*

---

## 1.6 Conclusion

Although fractional calculus is as old as classical calculus, it lay dormant for a while for unknown reasons. Fractional calculus has received much attention since the topic was brought back into the spotlight in the 20th century. Numerous novel differential and fractional integral operators have been found recently. It has been used in several fields and subjects, increasing its interdisciplinary scope. The Generalized krätzel function discussed in this chapter covers a wider family of functions and creates a trajectory of functions from standard Krätzel to generalized Krätzel function. With the right convergence conditions, the composition of fractional operators and $\mathcal{P}$-transform mentioned in this chapter may be applied to other fractional operators. This chapter's major goal is to pique readers' enthusiasm for the potential applications of fractional calculus in other disciplines.

## Acknowledgment

The author would like to thank the University of Kerala for providing financial support under project No. 1622/2021/UOK to complete this research work.

## Interest Statement

The author declares that they have no conflict of interest.

# References


[1] Kumar, D., (2011). *P-transforms, Integral Transforms and Special Functions, 22*(8), 603–616.

[2] Kumar, D., & Kilbas, A. A. (2010). *Fractional Calculus of P-transforms, Fractional Calculus and Applied Analysis, 13*(3), 309–328.

[3] Abel, N. H., (1881). Solution de quelques problémes á l'aide d'intégrales définies, *Qeuvres Complétes Christianian (Grondah1) 1*, 16–18.

[4] Bromwich, T. J. I., (1919). Examples of operational methods in mathematical physics, *Philosophical Magazine 37*, 407–419.

[5] Center, W., (1848). On the value of $\left(\frac{\mathrm{d}}{\mathrm{d}x}\right)^{\theta} x^0$ when $\theta$ is a positive proper fraction, *Cambridge & Dublin Math. Journal 3*, 163–169.

[6] Devis, H. T., (1936). *The theory of linear operators*, Bloomington, Ind., Principia Press.

[7] De Morgan, A., (1840). *The differential and integral calculus combining differentiation, integration, development, differential equations, differences, summation, calculus of variations ... with applications to algebra, plane, and solid geometry*, London (Baldwin and Craddock)' Published in 25 parts under the superintendence of the Society for the Diffusion of Useful Knowledge, 597–599.

[8] Erdélyi, A., Magnus, W., Oberhettinger, F. & Tricomi, F. G., (1981). *Higher Transcendental Functions*, Vol I, McGraw-Hill, New York-Toronto-London (1953); Reprinted: Krieger, Melbourne, Florida.

[9] Euler, L., (1738). On transcendental progressions that is, those whose general terms cannot be given algebraically, *Commentarii Academiae Scientiarum Imperialis Scientiarum Petropolitanae 5*, 55.

[10] Fourier, J. B. J., (1822). Theorie Analytique de la Chaleur, *Oeuvres de Fourier Paris (Firmin Didot) 1*, 508.

[11] Greer, H. R., (1858). On fractional differentiation, *Quarterly Journal of Math. 3*, 327–330.

[12] Gregory, D. F., (1841). *Examples of the processes of the differential and integral calculus*, first edition, 350, second edition, 354, Cambridge, Deighton.

[13] Hadamard, J., (1892). Essai sur l' étude des fonctions dennées par leur developpment de Taylor, *Journal Math. Pures. Appl. 8*, 101–186.

[14] Hargreave, C. J., (1848). On the solution of linear differential equations, *Philosophical Transactions 148*, 31–34

[15] Heaviside, O., (1892). *Electrical Papers*, MacMillan, London.

[16] Kelland, P., (1839). On general differentiation, *Trans. Roy. Soc. Edinburgh 14*, 567–618.

[17] Kilbas, A. A., & Kumar, D., (2009). On Generalized Krätzel Function, *Integral Transforms Spec. Funct. 20*(11), 835–846.

[18] Kilbas, A. A., & Saigo, M., (2004). *H-Transforms. Theory and Applications.*, Chapman and Hall/CRC, Boca Raton, Florida.

[19] Krätzel, E., (1975). *Integral transformations of Bessel type*, In: Generalized Functions and Operational Calculus (Proc. Conf. Varna 1975). Bulg. Acad. Sci., Sofia (1979), pp. 148–155.

[20] Krug, A., (1890). Theorie der Derivationen Wien Denkenschriften, *Mathematisch-Naturwissenschaftliche Classe 57*, 151–228.

[21] Kumar, D., (2011). $\mathcal{P}$-transforms, *Integral Transforms and Special Functions 22*(8), 603–616.

[22] Kumar, D., & Kilbas, A. A., (2010). Fractional Calculus of $\mathcal{P}$-transforms, *Fractional Calculus and Applied Analysis 13* (3), 309–328.

[23] Lacroix, S. F., (1819). Thraité du Calcul Différentiel et du Calcul Intégral, *Paris Courcier 3*, second edition, 409–410.

[24] Lagrange, J. L., (1849). Sur un nouvelle espsce de calcul relatif B la differentiation et B l'integration des quantites variables Oeuvres de Lagrange, *Paris (Gauthier-Villars) 3*, 441–476.

[25] Laplace, P. S., (1820). Théorie Analytique des Probabilités, *Paris Courcier.*

[26] Laurent, H., (1884). Sur le calcul des dérivdes á indices quelconques Nouv, *Annales de Mathématiques 3*(3), 240–252.

[27] Leibniz, G. W., (1695a). Letter from Hanover, Germany to G. F. A. L'Hospital, September 30, 1695, *Mathematische Schriften 1849, reprinted 1962, Hildesheim, Germany (Olns Verlag) 2*, 301–302.

[28] Leibniz, G. W., (1695b). Letter from Hanover, Germany to Johann Bernoulli December 28, *ibid. 3*, 1, 226.

[29] Leibniz, G. W., (1697). Letter from Hanover, Germany to John Wallis, May 28, *ibid. 4*, 25.

[30] Letnikov, A. V., (1872). An explanation of the concepts of the theory of differentiation of arbitrary index, *Moskow Matem. Sbornik 6*, 413–445.

[31] Liouville, J., (1832). Mémoire sur l'intégration de l'équation $(mx^2 + nx + p)\frac{\mathrm{d}^2 y}{\mathrm{d}x^2} + (qx + r)\frac{\mathrm{d}y}{\mathrm{d}x} + sy = 0$ á l'aide des diffgrentielles a indices quelconques, *Journal d l'Ecole Polytechnique 13, cahier 21*, 163–186.

[32] Liouville, J., (1834). Mémoire sur le théoréme des fonctions conpliémentaires Journal fü die reine und angewandte Mathematik, *Crelle's Journal 11*, 1–19.

[33] Mathai, A. M., (2003). *A Handbook of Generalized Special Functions for Statistical and Physical Sciences*, Clarendon Press, Oxford.

[34] Mathai, A. M., (2005). A Pathway to matrix-variate gamma and normal densities, *Linear Algebra Appl. 396*, 317–328.

[35] Mathai, A. M., & Haubold, H. J., (2007). Pathway model, superstatistics, Tsallis statistics and a generalized measure of entropy, *Physica A 375*, 110–122.

[36] Mathai, A. M., & Haubold, H. J., (2017). *An Introduction to Fractional Calculus*, Nova Publishers, NewYork.

[37] Mathai, A. M., & Saxena, R. K., (1978). *The H-function with Applications in Statistics and Other Disciplines*, Halsted Press[John Wiley & Sons], New York.

[38] McBrid, A. C., & Roach, G. F. (editors), (1985). *Fractional Calculus*, University of Stratchclyde, August 1984, (Pitman Advanced Publishing Program), Glasgow.

[39] Miller, K. S., & Ross, B., (1993). *An introduction to fractional calculus and fractional differential equations*, Wiley Interscience, New York.

[40] Nekrassov, P. A., (1888). General differentiation, *Moskow Matem. Sbornik 14*, 45–168.

[41] Nishimoto, K., (1990). *Proceedings of the International Conference on Fractional Calculus and its Applications*, College of Engineering, Nihon University, Tokyo, May-June 1989.

[42] Oldham, K., & Spanier, J., (1974). *The fractional calculus*, Academic Press, New York.

[43] Peacock, G., (1833). Report on the recent progress and present state of affairs of certain branches of analysis, in *Report to the British Association for the Advancement of Science*, 185–352.

[44] Riemann, B. G., (1876). Versuch einer Auffassung der Integration und Differentiation, *Gesannnelte Werke*, 331–344.

[45] Ross, B., (1974). Fractional calculus and its Applications (Springer-Verlag), *Lecture Notes in Mathematics 457*.

[46] Ross, B., (1977). The development of Fractional Calculus 1965–1900, *Historia Mathematica 4*, 75–89.

[47] Saigo, M., (1978). A remark on integral operators involving the Gauss hypergeometric functions, *Math. Rep. College of General Edu. Kyushu University 11*, 135–143.

[48] Samko, S. G., Kilbas, A. A., & Marichev, O. I., (1993). *Fractional Integrals and Derivatives, Theory and applications*, Gordon and Breach, New York.

[49] Sonin, N. Y., (1869). On differentiation with arbitrary index, *Moscow Matem, Sbornik 6* (1), 1–38.

[50] Srivastava, H. M., Gupta K. C., & Goyal, S. P., (1982). *The H-functions of One and Two Variables with Applications*, South Asian Publishers, New Delhi, Madras, India.

# Chapter 2

## Laplace Transformation of Fractional Integrals and Derivatives

**Min-Jie Luo[1] and Ravinder Krishna Raina[2]**

[1]*Department of Mathematics, School of Mathematics and Statistics, Donghua University, Shanghai, People's Republic of China*

[2]*M.P. University of Agriculture and Technology, Udaipur, Rajasthan, India*

**Abstract**

The present chapter first provides the necessary definitions of the Laplace transformation and its various important properties. The interrelationships between the Laplace transform and fractional integrals and derivatives are then explored. In subsequent sections, the Laplace transformations of Caputo's fractional derivatives, Hilfer's fractional derivatives, and Prabhakar's fractional calculus operators are discussed. The Laplace transformations of multiple Erdélyi-Kober integral operators and the multidimensional analogs of the defined fractional integral operators are also examined. The article includes several results and their applications. Additionally, some important and useful remarks (including updated references) are presented in this chapter. It is believed that the observations specifically highlighted may benefit the interested reader for further investigations on the subject.

## 2.1 Fundamentals of Laplace Transformation

### *2.1.1 Laplace Transformation*

Let $\mathbb{R} := (-\infty, \infty)$ and $\mathbb{R}_+ := (0, \infty)$. Suppose $L^1(A)$, with $A \subset \mathbb{R}$, be the space of all Lebesgue measurable, complex-valued functions $f : A \to \mathbb{C}$ with finite norm given by:

$$\|f\|_1 := \int_A |f(t)| \mathrm{d}t.$$

Let $\mathbb{E}_c$ ($c \in \mathbb{R}$) be the set of all functions $f : \mathbb{R}_+ \to \mathbb{C}$ such that $f(t)\mathrm{e}^{-ct} \in L^1(\mathbb{R}_+)$, and let $L^1_{\mathrm{loc}}(A)$ denote the space of all *locally integrable functions* on $A \subset \mathbb{R}$. In other words, the function $f \in L^1_{\mathrm{loc}}(A)$ if and only if:

$$\int_K |f(t)| \mathrm{d}t < \infty, \text{ for all compact set } K \text{ in } A,$$

and it may be noted that $L^1(\mathbb{R}_+) \subset L^1_{\mathrm{loc}}(\mathbb{R}_+)$.

**Definition 2.1.1** (Laplace transform)**.** *Let $f \in \mathbb{E}_c$ for some $c \in \mathbb{R}$. The* Laplace transform *of $f$ is defined by:*

$$(\mathcal{L}f)(s) = \mathcal{L}[f](s) := \int_{\mathbb{R}_+} \mathrm{e}^{-st} f(t) \mathrm{d}t, \tag{2.1.1}$$

*where $s := c + \mathrm{i}t \in \mathbb{C}$.*

**Remark 2.1.1.** *The relation (2.1.1) also defines an integral operator which has the function $\mathrm{e}^{-st}$ as its kernel and is usually denoted by $\mathcal{L}$ and called* Laplace transformation.

**Remark 2.1.2.** *Instead of $\mathbb{E}_c$, we may alternatively consider the set $\widetilde{\mathbb{E}}_c$ consisting of all functions $f \in L^1_{\mathrm{loc}}(\mathbb{R}_+)$ with $f(t) = \mathcal{O}(\mathrm{e}^{ct})$ for some $c \in \mathbb{R}$. It is clear that if $f \in \widetilde{\mathbb{E}}_c$, then $f \in \widetilde{\mathbb{E}}_{c+\varepsilon}$, where $\varepsilon > 0$ is arbitrary. In fact, if $f \in \widetilde{\mathbb{E}}_c$, then there exist $T, M > 0$ such that for arbitrary $\varepsilon > 0$:*

$$\begin{aligned}\int_{\mathbb{R}_+} \mathrm{e}^{-(c+\varepsilon)t}|f(t)|\mathrm{d}t &= \left(\int_0^T + \int_T^\infty\right) \mathrm{e}^{-(c+\varepsilon)t}|f(t)|\mathrm{d}t \\ &\leq \int_0^T |f(t)|\mathrm{d}t + M\int_T^\infty \mathrm{e}^{-\varepsilon t}\mathrm{d}t < \infty.\end{aligned}$$

**Remark 2.1.3.** *For $f(t)\mathrm{e}^{-ct} \in L^1(\mathbb{R}_+) \cap L^2(\mathbb{R}_+)$ for some $c \geq 0$, the function $f$ is then called a* Laplace-Pego function of order $c$ *(see [22, p. 992]).*

If $f \in \mathbb{E}_\sigma$ for one $\sigma \in \mathbb{R}$, then the Laplace integral in (2.1.1) converges absolutely for all $s = c + it$, $c \geq \sigma$, $t \in \mathbb{R}$, and one can observe that

$$|\mathcal{L}[f](s)| \leq \int_{\mathbb{R}_+} \mathrm{e}^{-ct}|f(t)|\mathrm{d}t \leq \int_{\mathbb{R}_+} \mathrm{e}^{-\sigma t}|f(t)|\mathrm{d}t = \|\mathrm{e}^{-ct}f(t)\|_1 < \infty.$$

For a given function $f$, the index

$$\sigma_{ac}(f) := \inf\left\{c : \|\mathrm{e}^{-ct}f(t)\|_1 < \infty\right\} \in \mathbb{R} \qquad (2.1.2)$$

is called the *abscissa of absolute convergence* (see [51, p. 97]).

Let $L_\sigma$ denote the set of all functions with $\sigma$ as the abscissa of absolute convergence of their Laplace transforms, i.e., $L_\sigma := \{f : \sigma_{ac}(f) = \sigma\}$. Clearly, if $f \in L_\sigma$, then $f \in \mathbb{E}_{\sigma+\varepsilon}$ for $\varepsilon$ arbitrary.

**Example 2.1.1.** Consider the function $f(t) = 1/(1+t^2)$. Obviously, $f \in L_0$ and $\mathcal{L}[f](s)$ exists for $\mathrm{Re}\,(s) \geq 0$.

**Example 2.1.2.** Let the function $f(t) = 1$, then evidently $f \in L_0$ and $\mathcal{L}[f](s) = 1/s$, $\operatorname{Re}(s) > 0$.

**Example 2.1.3.** For $\operatorname{Re}(z) > 0$, $t^{z-1} \in L_0$. Thus for $\min\{\operatorname{Re}(z), \operatorname{Re}(s)\} > 0$, we have

$$\mathcal{L}[t^{z-1}](s) = \int_{\mathbb{R}_+} t^{z-1} \mathrm{e}^{-st} \mathrm{d}t = \frac{\Gamma(z)}{s^z}, \tag{2.1.3}$$

where $\Gamma(z)$ represents the Gamma function defined by:

$$\Gamma(z) := \int_{\mathbb{R}_+} t^{z-1} \mathrm{e}^{-t} \mathrm{d}t, \ \operatorname{Re}(z) > 0.$$

There are several valuable monographs on the subject of Laplace transform see, [5], [50], and [51]. Also, many books have chapters devoted to the Laplace transform (see, for instance, [13]).

The following result shows that the Laplace transformation is a linear operation ([13, p. 576, Theorem 9.1]).

**Theorem 2.1.1** (Linearity). *Let $\alpha, \beta \in \mathbb{C}$. If $f \in L_{\sigma_1}$ and $g \in L_{\sigma_2}$, then*

$$\mathcal{L}[\alpha f + \beta g](s) = \alpha \mathcal{L}[f](s) + \beta \mathcal{L}[g](s), \ \operatorname{Re}(s) > \max\{\sigma_1, \sigma_2\}.$$

The usual convolution of $f$ and $g$ is defined by:

$$(f * g)(t) := \int_0^t f(\tau) g(t - \tau) \mathrm{d}\tau,$$

and it is not difficult to establish the following result (see [13, p. 581, Theorem 9.8]).

**Theorem 2.1.2** (Convolution). *If $f \in L_{\sigma_1}$ and $g \in L_{\sigma_2}$, then*

$$\mathcal{L}[f * g](s) = \mathcal{L}[f](s) \mathcal{L}[g](s), \ \operatorname{Re}(s) > \max\{\sigma_1, \sigma_2\}.$$

In addition, $\sigma_{ac}(f * g) \le \max\{\sigma_1, \sigma_2\}$, where $\sigma_{ac}$ is defined by (2.1.2) (see [28, p. 266]).

**Theorem 2.1.3** (Integral). *Let $f \in L_\sigma$ and*

$$\varphi(t) := \int_0^t f(u)\mathrm{d}u.$$

*Then*

$$\mathcal{L}[\varphi](s) = \frac{1}{s}\mathcal{L}[f](s), \quad \mathrm{Re}\,(s) > \max\{\sigma, 0\}.$$

*Proof.* Since $\varphi = 1 * f$ and $1 \in L_0$, the use of Theorem 2.1.2 gives

$$\mathcal{L}[\varphi](s) = \mathcal{L}[1*f](s) = \mathcal{L}[1](s)\mathcal{L}[f](s) = \frac{1}{s}\mathcal{L}[f](s), \quad \mathrm{Re}\,(s) > \max\{\sigma, 0\}.$$

Let $D := \dfrac{\mathrm{d}}{\mathrm{d}t}$, we have the following result involving the derivative.

**Theorem 2.1.4** (Derivative). *Let $f \in L_{\sigma_0}$, $Df \in L_{\sigma_1}$ and $f(0^+) := \lim_{x \to 0^+} f(x)$ exist. Then*

$$\mathcal{L}[Df](s) = s\mathcal{L}[f](s) - f(0^+), \quad \mathrm{Re}\,(s) > \max\{\sigma_0, \sigma_1, 0\}. \tag{2.1.4}$$

*In general, if $D^j f \in L_{\sigma_j}$ $(j = 0, \cdots, k)$ and $(D^j f)(0^+)$ exists for $j = 0, \cdots, k-1$, then*

$$\begin{gathered}\mathcal{L}[D^k f](s) = s^k\mathcal{L}[f](s) - \sum_{j=0}^{k-1} s^{k-j-1}(D^j f)(0^+), \\ \mathrm{Re}\,(s) > \max\{\sigma_0, \cdots, \sigma_k, 0\}.\end{gathered} \tag{2.1.5}$$

*Proof.* Since

$$f(t) - f(0^+) = \int_0^t Df(u)\mathrm{d}u,$$

we can use Theorem 2.1.3 to obtain

$$\mathcal{L}[f(t) - f(0^+)](s) = \mathcal{L}\left[\int_0^t Df(u)\mathrm{d}u\right](s) = \frac{1}{s}\mathcal{L}[Df](s),$$

from which (2.1.4) follows.

To prove (2.1.5), we use the mathematical induction. For $k = 2$, let $F := Df$, then $F \in L_{\sigma_1}$, $DF \in L_{\sigma_2}$ and thus

$$\begin{aligned}\mathcal{L}[D^2 f](s) &= \mathcal{L}[DF](s)\\ &= s\mathcal{L}[F](s) - F(0^+)\ (\mathrm{Re}\,(s) > \max\{\sigma_1, \sigma_2, 0\})\\ &= s^2\mathcal{L}[f](s) - sf(0^+) - (Df)(0^+)\ (\mathrm{Re}\,(s) > \max\{\sigma_0, \sigma_1, \sigma_2, 0\}).\end{aligned}$$

Assume that (2.1.5) holds for $k = m$, then

$$\begin{aligned}\mathcal{L}[D^{m+1} f](s) &= \mathcal{L}[D(D^m)f](s)\\ &= s\mathcal{L}[D^m f](s) - (D^m f)(0^+)\ (\mathrm{Re}\,(s) > \max\{\sigma_m, \sigma_{m+1}, 0\})\\ &= s^{m+1}\mathcal{L}[f](s) - \sum_{j=0}^{m-1} s^{m-j}(D^j f)(0^+) - (D^m f)(0^+)\\ &= s^{m+1}\mathcal{L}[f](s) - \sum_{j=0}^{m} s^{m-j}(D^j f)(0^+)\\ &\quad (\mathrm{Re}\,(s) > \max\{\sigma_0, \cdots, \sigma_{m+1}, 0\}).\end{aligned}$$

The proof is now complete.

### *2.1.2 n-Dimensional Laplace Transformation*

Let $\mathbb{R}^n := \mathbb{R} \times \cdots \times \mathbb{R}$ and similarly $\mathbb{R}^n_+ := \mathbb{R}_+ \times \cdots \times \mathbb{R}_+$. For $\mathbf{t} := (t_1, \cdots, t_n) \in \mathbb{C}^n$, $\mathbf{s} := (s_1, \cdots, s_n) \in \mathbb{C}^n$ and $\boldsymbol{\alpha} := (\alpha_1, \cdots, \alpha_n) \in \mathbb{C}^n$, we shall write:

$$\begin{aligned}
\mathrm{Re}\,(\mathbf{s}) &:= (\,\mathrm{Re}\,(s_1), \cdots, \mathrm{Re}\,(s_n)) \in \mathbb{R}^n, \\
\mathbf{s} \pm \mathbf{t} &:= (s_1 \pm t_1, \cdots, s_n \pm t_n), \\
\mathbf{st} &:= s_1 t_1 + \cdots + s_n t_n, \\
\mathbf{s}^{\boldsymbol{\alpha}} &:= s_1^{\alpha_1} \cdots s_n^{\alpha_n}, \\
\Gamma(\boldsymbol{\alpha}) &:= \Gamma(\alpha_1) \cdots \Gamma(\alpha_n)
\end{aligned}$$

and

$$\int_{\mathbb{R}^n_+} f(\mathbf{t}) \mathrm{d}\mathbf{t} := \int_{\mathbb{R}_+} \cdots \int_{\mathbb{R}_+} f(t_1, \cdots, t_n) \mathrm{d}t_1 \cdots \mathrm{d}t_n.$$

In addition, $\mathbf{s} > (\geq)\mathbf{t}$ stands for $s_j > (\geq) t_j$ $(j = 1, \cdots, n)$.

For given $\mathbf{c} \in \mathbb{R}^n$, denote by $\mathbb{E}_{\mathbf{c}}$ the set of all functions $f : \mathbb{R}^n_+ \to \mathbb{C}$ such that $f(\mathbf{t})\mathrm{e}^{-\mathbf{ct}} \in L^1(\mathbb{R}^n_+)$.

**Definition 2.1.2** ($n$-dimensional Laplace transformation [2, p. 82]). *Let $f \in \mathbb{E}_{\mathbf{c}}$ for some $\mathbf{c} \in \mathbb{R}^n$. The* $n$-dimensional Laplace transformation *of $f$ is defined by:*

$$\mathcal{L}_n[f](\mathbf{s}) := \int_{\mathbb{R}^n_+} \mathrm{e}^{-\mathbf{st}} f(\mathbf{t}) \mathrm{d}\mathbf{t}$$

*and the multidimensional convolution of $f$ and $g$ is defined by:*

$$(f * g)(\mathbf{t}) := \int_{\mathbb{R}^n_+(\mathbf{t})} f(\boldsymbol{\tau}) g(\mathbf{t} - \boldsymbol{\tau}) \mathrm{d}\boldsymbol{\tau},$$

*where* $\mathbb{R}^n_+(\mathbf{t}) := \{\mathbf{x} : \mathbf{x} \in \mathbb{R}^n, \mathbf{0} < \mathbf{x} < \mathbf{t}\} \quad (\mathbf{t} \in \mathbb{R}^n_+)$.

The following theorem will be used in Section 2.8.

**Theorem 2.1.5** ([2, p. 114]). *Let $f, g \in \mathbb{E}_{\mathbf{c}}$. Then $f * g \in \mathbb{E}_{\mathbf{c}}$ and*

$$\mathcal{L}_n[f * g](\mathbf{s}) = \mathcal{L}_n[f](\mathbf{s})\mathcal{L}_n[f](\mathbf{s}), \quad \mathrm{Re}\,(\mathbf{s}) \geq \mathbf{c}.$$

### *2.1.3 Some Advanced Examples*

**Example 2.1.4.** The two-parametric Mittag-Leffler function $E_{\alpha,\beta}(z)$ is defined by (see [12, p. 56]):

$$E_{\alpha,\beta}(z) := \sum_{k=0}^{\infty} \frac{z^k}{\Gamma(\alpha k + \beta)}, \tag{2.1.6}$$

where $z \in \mathbb{C}$, $\alpha \in \mathbb{C}$ ( $\mathrm{Re}\,(\alpha) > 0$) and $\beta \in \mathbb{C}$. It is well-known that (see [12, p. 84, (4.10.1)])

$$\mathcal{L}[t^{\beta-1}E_{\alpha,\beta}(\lambda t^{\alpha})](s) = \frac{s^{\alpha-\beta}}{s^{\alpha}-\lambda}, \quad \alpha > 0, \ \mathrm{Re}\,(\beta) > 0, \ \lambda \in \mathbb{C}, \ \mathrm{Re}\,(s) > |\lambda|^{1/\alpha}. \tag{2.1.7}$$

When $\beta = 1$, (2.1.6) reduces to the classical Mittag-Leffler function

$$E_{\alpha}(z) := \sum_{k=0}^{\infty} \frac{z^k}{\Gamma(\alpha k + 1)}, \tag{2.1.8}$$

and (2.1.7) becomes

$$\mathcal{L}[E_{\alpha}(\lambda t^{\alpha})](s) = \frac{s^{\alpha-1}}{s^{\alpha}-\lambda}, \quad \alpha > 0, \ \lambda \in \mathbb{C}, \ \mathrm{Re}\,(s) > |\lambda|^{1/\alpha}. \tag{2.1.9}$$

To prove (2.1.7), we need the asymptotic formulas of $E_{\alpha,\beta}(z)$ at infinity. If $0 < \alpha < 2$, $\beta \in \mathbb{C}$, $m \in \mathbb{N}$ and $\mu$ is a real number such that

$$\frac{\pi\alpha}{2} < \mu < \min\{\pi, \pi\alpha\}, \tag{2.1.10}$$

then as $|z| \to \infty$ the following expansions holds (see [29, pp. 32–34, Theorems 1.3 and 1.4]; see also [12] and [18])

$$E_{\alpha,\beta}(z) = \begin{cases} \dfrac{1}{\alpha} z^{(1-\beta)/\alpha} e^{z^{1/\alpha}} - \displaystyle\sum_{k=1}^{m} \frac{z^{-k}}{\Gamma(\beta - \alpha k)} + \mathcal{O}\left(\frac{1}{|z|^{m+1}}\right), & |\arg(z)| \le \mu; \\ -\displaystyle\sum_{k=1}^{m} \frac{z^{-k}}{\Gamma(\beta - \alpha k)} + \mathcal{O}\left(\frac{1}{|z|^{m+1}}\right), & \mu \le |\arg(z)| \le \pi. \end{cases}$$

If $\alpha \ge 2$, $\beta \in \mathbb{C}$, $m \in \mathbb{N}$ and $\mu$ is a real number satisfying (2.1.10), then we have ([29, p. 35, Theorem 1.7]; see also [12] and [18])

$$E_{\alpha,\beta}(z) = \frac{1}{\alpha} \sum_{|\arg(z)+2n| \le \frac{\pi\alpha}{2}} \left(z^{1/\alpha} e^{2\pi i n/\alpha}\right)^{1-\beta} \exp\left(z^{1/\alpha} e^{2\pi i n}/\alpha\right)$$
$$- \sum_{k=1}^{m} \frac{z^{-k}}{\Gamma(\beta - \alpha k)} + \mathcal{O}\left(\frac{1}{|z|^{m+1}}\right), \quad |z| \to \infty, \ |\arg(z)| \le \frac{\pi\alpha}{2}.$$

Thus for $\alpha > 0$, $\mathrm{Re}\,(\beta) > 0$, $\lambda \in \mathbb{C}$, the function $t^{\beta-1} E_\alpha(\lambda t^\alpha) \in L_{|\lambda|^{1/\alpha}}$ and its Laplace transform can be evaluated as follows:

$$\mathcal{L}[t^{\beta-1} E_{\alpha,\beta}(\lambda t^\alpha)](s) = \sum_{k=0}^{\infty} \frac{\lambda^k}{\Gamma(\alpha k + \beta)} \mathcal{L}[t^{\alpha k + \beta - 1}](s) \ (\min\{\mathrm{Re}\,(s), \mathrm{Re}\,(\beta)\} > 0)$$
$$\overset{\text{Eq. (2.1.3)}}{=\!=\!=} \sum_{k=0}^{\infty} \frac{\lambda^k}{s^{\alpha k+\beta}} = \frac{s^{\alpha-\beta}}{s^\alpha - \lambda}, \ \alpha > 0, \ \mathrm{Re}\,(\beta) > 0, \ \mathrm{Re}\,(s) > |\lambda|^{1/\alpha}.$$

Note that the condition $|\lambda/s^\alpha| < 1$ can be removed. This completes the proof of (2.1.7).

As the *Queen Function of the Fractional Calculus*, a large family of special functions can easily be derived from the classical Mittag-Leffler function.

**Example 2.1.5.** The Prabhakar (three-parametric Mittag-Leffler) function $E^\gamma_{\alpha,\beta}(z)$ is defined by (see [12, p. 97]):

$$E^\gamma_{\alpha,\beta}(z) := \sum_{k=0}^{\infty} \frac{(\gamma)_k}{\Gamma(\alpha k + \beta)} \frac{z^k}{k!}, \tag{2.1.11}$$

where $z \in \mathbb{C}$, $\alpha \in \mathbb{C}$ $(\operatorname{Re}(\alpha) > 0)$, $\gamma, \beta \in \mathbb{C}$ and $(\gamma)_k$ is the Pochhammer symbol defined by:

$$(\gamma)_k := \frac{\Gamma(\gamma+k)}{\Gamma(\gamma)} = \begin{cases} 1, & (k = 0; \gamma \in \mathbb{C} \setminus \{0\}); \\ \gamma(\gamma+1)\cdots(\gamma+k-1), & (k \in \mathbb{N}; \gamma \in \mathbb{C}). \end{cases}$$

It is well-known that [12, p. 102, Eq. (5.1.26)]

$$\mathcal{L}[t^{\beta-1}E^{\gamma}_{\alpha,\beta}(\lambda t^{\alpha})](s) = \frac{s^{\alpha\gamma-\beta}}{(s^{\alpha}-\lambda)^{\gamma}}, \ \alpha > 0, \ \operatorname{Re}(\beta) > 0, \ \lambda, \gamma \in \mathbb{C}, \ \operatorname{Re}(s) > |\lambda|^{1/\alpha}. \tag{2.1.12}$$

In fact, under the conditions given in (2.1.12), we have

$$\begin{aligned} \mathcal{L}[t^{\beta-1}E^{\gamma}_{\alpha,\beta}(\lambda t^{\alpha})](s) &= \sum_{k=0}^{\infty} \frac{(\gamma)_k}{\Gamma(\alpha k+\beta)} \frac{\lambda^k}{k!} \mathcal{L}[t^{\alpha k+\beta-1}](s) \\ &\overset{\text{Eq. (2.1.3)}}{=\!=\!=} s^{-\beta} \sum_{k=0}^{\infty} \frac{(\gamma)_k}{k!} \frac{\lambda^k}{s^{\alpha k}} = s^{-\beta}\left(1 - \frac{\lambda}{s^{\alpha}}\right)^{-\gamma}, \end{aligned}$$

which is equivalent to (2.1.12).

**Example 2.1.6.** In 2005, Raina [33] introduced a noteworthy generalization of the Prabhakar function (2.1.11), which is defined by:

$$\mathcal{F}^{\boldsymbol{\sigma}}_{\rho,\lambda}(z) := \sum_{k=0}^{\infty} \frac{\sigma(k)}{\Gamma(\rho k+\lambda)} z^k, \tag{2.1.13}$$

where $\lambda \in \mathbb{C}$, $\operatorname{Re}(\rho) > 0$ and $\boldsymbol{\sigma} = \{\sigma(k)\}_{k=0}^{\infty}$ represents any complex-valued sequence that makes the above series converge absolutely in $|z| < R \leq \infty$. Let

$$\mathcal{F}(z) := \sum_{k=0}^{\infty} \sigma(k) z^k, \ |z| < R_0.$$

Following [23, p. 2080], it is easy to see that

$$\mathcal{L}[t^{\lambda-1}\mathcal{F}^{\sigma}_{\rho,\lambda}(\omega t^{\rho})](s) = s^{-\lambda}\mathcal{F}(\omega s^{-\rho}),\ \rho > 0,\ \operatorname{Re}(\lambda) > 0,\ \operatorname{Re}(s) > R_0^{-1/\rho}|\omega|^{1/\rho}. \tag{2.1.14}$$

If we set $\sigma(k) = (\gamma)_k/k!$ in (2.1.13), then $R_0 = 1$ and we obtain from (2.1.14) the corresponding result (2.1.12) for the Prabhakar function (2.1.11).

Next, we examine various Laplace transforms involving the hypergeometric functions of several variables.

**Example 2.1.7.** Let us consider the Laplace transform of the function $g$ defined by:

$$g(t) := \frac{t^{c-1}}{\Gamma(c)}\Phi_2^{(n)}\left[b_1, \cdots, b_n; c; \lambda_1 t, \cdots, \lambda_n t\right],$$

where $\Phi_2^{(n)}$ denotes the confluent hypergeometric function defined by [45, p. 202, Eq. (2.2)]:

$$\begin{aligned}&\Phi_2^{(n)}\left[b_1, \cdots, b_n; c; z_1, \cdots, z_n\right]\\ &\quad := \sum_{m_1,\cdots,m_n=0}^{\infty} \frac{(b_1)_{m_1}\cdots(b_n)_{m_n}}{(c)_{m_1+\cdots+m_n}}\frac{z_1^{m_1}}{m_1!}\cdots\frac{z_n^{m_n}}{m_n!} \quad (\mathbf{z} \in \mathbb{C}^n).\end{aligned} \tag{2.1.15}$$

By interchanging the order of summation and integration, we have

$$\mathcal{L}[g](s) = \sum_{m_1,\cdots,m_n=0}^{\infty} \frac{(b_1)_{m_1}\cdots(b_n)_{m_n}}{(c)_{m_1+\cdots+m_n}}\frac{\lambda_1^{m_1}}{m_1!}\cdots\frac{\lambda_n^{m_n}}{m_n!}\frac{1}{\Gamma(c)}\mathcal{L}[t^{c+m_1+\cdots+m_n-1}](s)dt,$$

and using the formula that

$$\frac{1}{\Gamma(c)}\mathcal{L}[t^{c+m_1+\cdots+m_n-1}](s) = \frac{(c)_{m_1+\cdots+m_n}}{s^{c+m_1+\cdots+m_n}},\ \operatorname{Re}(c) > 0,\ \operatorname{Re}(s) > 0,$$

it easily follows that [45, p. 204, Eq. (2.11)]

$$\mathcal{L}[g](s) = s^{-c} \prod_{j=1}^{n} \left(1 - \frac{\lambda_j}{s}\right)^{-b_j}, \tag{2.1.16}$$

where $\operatorname{Re}(c) > 0$ and $\operatorname{Re}(s) > \max\{0, \operatorname{Re}(\lambda_1), \cdots, \operatorname{Re}(\lambda_n)\}$. The formula (2.1.16) is useful in solving certain Volterra-type fractional integro-differential equations with $\Phi_2^{(n)}$ as their kernels [45]. A generalization of (2.1.16) can be found in [40].

**Example 2.1.8.** Consider the function

$$g(t_1, \cdots, t_n) := \frac{t_1^{a_1-1}}{\Gamma(a_1)} \cdots \frac{t_n^{a_n-1}}{\Gamma(a_n)} \Phi_2^{(n)} \left[b_1, \cdots, b_n; c; \lambda_1 t_1, \cdots, \lambda_n t_n\right],$$

where $\Phi_2^{(n)}$ is defined by (2.1.15). Then

$$\begin{aligned} \mathcal{L}_n[g](s_1, \cdots, s_n) &= \sum_{m_1, \cdots, m_n = 0}^{\infty} \frac{(b_1)_{m_1} \cdots (b_n)_{m_n}}{(c)_{m_1+\cdots+m_n}} \frac{\lambda_1^{m_1}}{m_1!} \cdots \frac{\lambda_n^{m_n}}{m_n!} \\ &\quad \cdot \prod_{j=1}^{n} \frac{1}{\Gamma(a_j)} \mathcal{L}[t_j^{a_j+m_j-1}](s) \\ &= \sum_{m_1, \cdots, m_n = 0}^{\infty} \frac{(b_1)_{m_1} \cdots (b_n)_{m_n}}{(c)_{m_1+\cdots+m_n}} \frac{\lambda_1^{m_1}}{m_1!} \cdots \frac{\lambda_n^{m_n}}{m_n!} \prod_{j=1}^{n} \frac{(a_j)_{m_j}}{s_j^{a_j+m_j}} \\ &= s_1^{-a_1} \cdots s_n^{-a_n} F_B \left[a_1, \cdots, a_n, b_1, \cdots, b_n; c; \frac{\lambda_1}{s_1}, \cdots, \frac{\lambda_n}{s_n}\right], \end{aligned}$$

where $\min\{\operatorname{Re}(a_1), \cdots, \operatorname{Re}(a_n)\} > 0$, $\operatorname{Re}(s_j) > \max\{0, \operatorname{Re}(\lambda_j)\}$ $(j = 1, \cdots, n)$ and $F_B$ denotes the Lauricella hypergeometric function defined by:

$$\begin{aligned} &F_B\left[a_1, \cdots, a_n, b_1, \cdots, b_n; c; z_1, \cdots, z_n\right] \\ &= \sum_{m_1, \cdots, m_n = 0}^{\infty} \frac{(a_1)_{m_1} \cdots (a_n)_{m_n} (b_1)_{m_1} \cdots (b_n)_{m_n}}{(c)_{m_1+\cdots+m_n}} \frac{z_1^{m_1}}{m_1!} \cdots \frac{z_n^{m_n}}{m_n!} \\ &\quad (\max\{|z_1|, \cdots, |z_n|\} < 1). \end{aligned}$$

## 2.2 Laplace Transformation of the Riemann-Liouville Fractional Integrals and Derivatives

The classical Riemann-Liouville fractional integrals defined on a finite interval $[a, b]$ can be naturally extended to the half-axis $\mathbb{R}_+$, and the resulting fractional integrals are usually called the *Liouville fractional integrals.*

**Definition 2.2.1** ([18, p. 79]). *Let* $\mathrm{Re}\,(\alpha) > 0$ *and* $x \in \mathbb{R}_+$. *The Liouville fractional integrals* $I^\alpha_{0+}f$ *and* $I^\alpha_{-}f$ *of order* $\alpha \in \mathbb{C}$ *are defined by:*

$$(I^\alpha_{0+}f)(x) := \frac{1}{\Gamma(\alpha)} \int_0^x (x-t)^{\alpha-1} f(t)\mathrm{d}t \tag{2.2. 1}$$

*and*

$$(I^\alpha_{-}f)(x) := \frac{1}{\Gamma(\alpha)} \int_x^\infty (t-x)^{\alpha-1} f(t)\mathrm{d}t, \tag{2.2. 2}$$

*respectively.*

The operator $I^\alpha_-$ is also known as the Weyl fractional integral operator.

**Definition 2.2.2** ([18, p. 80]). *Let* $\mathrm{Re}\,(\alpha) \geq 0$, $n = [\mathrm{Re}\,(\alpha)] + 1$ *and* $x \in \mathbb{R}_+$. *The Liouville fractional derivatives* $D^\alpha_{0+}f$ *and* $D^\alpha_{-}f$ *of order* $\alpha \in \mathbb{C}$ *are defined by:*

$$\begin{aligned}(D^\alpha_{0+}f)(x) &:= D^n(I^{n-\alpha}_{0+}f)(x) \\ &= \frac{1}{\Gamma(n-\alpha)} \left(\frac{\mathrm{d}}{\mathrm{d}x}\right)^n \int_0^x (x-t)^{n-\alpha-1} f(t)\mathrm{d}t\end{aligned}$$

*and*

$$\begin{aligned}(D_{-}^{\alpha}f)(x) &:= (-D)^n(I_{-}^{n-\alpha}f)(x)\\ &= \frac{1}{\Gamma(n-\alpha)}\left(-\frac{\mathrm{d}}{\mathrm{d}x}\right)^n \int_x^{\infty}(t-x)^{n-\alpha-1}f(t)\mathrm{d}t,\end{aligned}$$

*respectively.*

### *2.2.1 Main Results*

We first consider the Laplace transforms of $I_{0+}^{\alpha}f$ and $D_{0+}^{\alpha}f$.

**Theorem 2.2.1.** *Let* $\mathrm{Re}\,(\alpha) > 0$ *and* $f \in L_{\sigma}$. *Then*

$$(\mathcal{L}I_{0+}^{\alpha}f)(s) = s^{-\alpha}\mathcal{L}[f](s), \quad \mathrm{Re}\,(s) > \max\{\sigma, 0\}. \tag{2.2. 3}$$

*Proof.* We note that

$$(I_{0+}^{\alpha}f)(t) = \frac{t^{\alpha-1}}{\Gamma(\alpha)} * f(t),$$

where $f \in L_{\sigma}$ and $t^{\alpha-1} \in L_0$. The desired result (2.2. 3) follows on using Theorem 2.1.2.

**Theorem 2.2.2.** *Let* $\mathrm{Re}\,(\alpha) > 0$ *and* $n = [\mathrm{Re}\,(\alpha)] + 1$. *Also, let* $f \in L_{\sigma}$, $D^jI_{0+}^{n-\alpha}f \in L_{\sigma_j}$ $(j = 0, \cdots, n)$ *and* $(D^jI_{0+}^{n-\alpha}f)(0^+)$ *exist for* $j = 0, \cdots, n-1$. *Then*

$$(\mathcal{L}D_{0+}^{\alpha}f)(s) = s^{\alpha}\mathcal{L}[f](s) - \sum_{j=0}^{n-1} s^{n-j-1}(D^jI^{n-\alpha}f)(0^+), \tag{2.2. 4}$$

*where* $\mathrm{Re}\,(s) > \max\{\sigma, \sigma_0, \cdots, \sigma_n, 0\}$.

*Proof.* From Definition 2.2.2 and Theorem 2.1.4, we have

$$\begin{aligned}(\mathcal{L}D^{\alpha}_{0+}f)(s) &= (\mathcal{L}D^{n}I^{n-\alpha}_{0+}f)(s)\\ &= s^{n}(\mathcal{L}I^{n-\alpha}_{0+}f)(s) - \sum_{j=0}^{n-1} s^{n-k-1}(D^{j}I^{n-\alpha}f)(0^{+}).\end{aligned} \tag{2.2. 5}$$

Since $\mathrm{Re}\,(n-\alpha) > 0$ and $f \in L_{\sigma}$, we can use Theorem 2.2.1 to get

$$(\mathcal{L}I^{n-\alpha}_{0}f)(s) = s^{\alpha-n}\mathcal{L}[f](s), \;\; \mathrm{Re}\,(s) > \max\{\sigma, 0\}. \tag{2.2. 6}$$

Combining (2.2. 5) and (2.2. 6), we obtain the result (2.2. 4).

**Remark 2.2.1.** *In some versions of the above theorem (e.g., [18, p. 84]), we need the spaces*

$$AC[a,b] := \{f : [a,b] \to \mathbb{C} : f \textit{ is absolutely continous}\}$$

*and*

$$AC^{n}[a,b] := \left\{f : [a,b] \to \mathbb{C} : (D^{n-1}f) \in AC[a,b]\right\}.$$

Unlike the cases of $\mathcal{L}I^{\alpha}_{0+}f$ and $\mathcal{L}D^{\alpha}_{0+}f$, the results for $\mathcal{L}I^{\alpha}_{-}f$ and $\mathcal{L}D^{\alpha}_{-}f$ are much more complicated. Here we just prove a result for $\mathcal{L}I^{\alpha}_{-}f$.

**Theorem 2.2.3.** *Let* $\mathrm{Re}\,(\alpha) > 0$ *and let* $I^{\alpha}_{-}f$ *exist on* $\mathbb{R}_{+}$ *and* $I^{\alpha}_{-}f \in L_{\sigma}$. *Also, let* $g_{\alpha}(t) := t^{\alpha}f(t)$ *and assume that* $g_{\alpha} \in L_{\sigma}$ *if* $\sigma < 0$ *and* $g_{\alpha} \in L^{1}(\mathbb{R}_{+})$ *if* $\sigma \geq 0$. *Then*

$$(\mathcal{L}I^{\alpha}_{-}f)(s) = \frac{1}{\Gamma(\alpha+1)}\int_{0}^{\infty} {}_{1}F_{1}\left[\begin{matrix}1\\ \alpha+1\end{matrix}; -su\right] g_{\alpha}(u)\mathrm{d}u, \;\; \mathrm{Re}\,(s) > \sigma, \tag{2.2. 7}$$

*where* ${}_1F_1$ *denotes the confluent hypergeometric function (see* [18, p. 29, Eq. (1.6.14)]*).*

*Proof.* For $\operatorname{Re}(s) > \sigma$, the formula can be formally established by observing that

$$(\mathcal{L}I_{-}^{\alpha}f)(s) = \frac{1}{\Gamma(\alpha)}\int_0^{\infty} \mathrm{e}^{-st}\left[\int_t^{\infty}(u-t)^{\alpha-1}f(u)\mathrm{d}u\right]\mathrm{d}t$$
$$= \int_0^{\infty} k(s,u)f(u)\mathrm{d}u.$$

The function $k(s,u)$ is given by

$$k(s,u) := \frac{1}{\Gamma(\alpha)}\int_0^{u} \mathrm{e}^{-st}(u-t)^{\alpha-1}\mathrm{d}t \xlongequal{t\to ut} \frac{u^{\alpha}}{\Gamma(\alpha)}\int_0^{1} \mathrm{e}^{-sut}(1-t)^{\alpha-1}\mathrm{d}t$$
$$= \frac{u^{\alpha}}{\Gamma(\alpha+1)}{}_1F_1\left[\begin{matrix}1\\ \alpha+1\end{matrix};-su\right] \quad (\operatorname{Re}(\alpha)>0),$$

wherein, we have used the integral representation for the function ${}_1F_1$ (see [18, p. 29, Eq. (1.6.15)]). To see that (2.2. 7) does hold for a certain class of functions characterized by the theorem, we first recall that the confluent hypergeometric function ${}_1F_1$ has the following asymptotic behavior (see [18, p. 29]):

$${}_1F_1\left[\begin{matrix}b\\ c\end{matrix};z\right] \sim \begin{cases}\dfrac{\Gamma(c)}{\Gamma(b)}\mathrm{e}^{z}z^{b-c}, & \operatorname{Re}(z)\to\infty;\\[2mm] \dfrac{\Gamma(c)\mathrm{e}^{-\mathrm{i}\pi b}}{\Gamma(c-b)}, & \operatorname{Re}(z)\to-\infty.\end{cases}$$

If $\sigma > 0$, then for $\operatorname{Re}(s) > \sigma$,

$$\frac{1}{\Gamma(\alpha+1)}{}_1F_1\left[\begin{matrix}1\\ \alpha+1\end{matrix};-su\right] \sim \frac{1}{\Gamma(\alpha)}s^{-1}u^{-1}, \quad u\to\infty.$$

If $\sigma < 0$, then for $\sigma < \operatorname{Re}(s) < 0$,

$$\frac{1}{\Gamma(\alpha+1)}{}_1F_1\left[\begin{matrix}1\\ \alpha+1\end{matrix};-su\right] \sim (-s)^{-\alpha}u^{-\alpha}\mathrm{e}^{-su},\ u\to\infty.$$

In addition, for $\operatorname{Re}(s)=0$, we have

$$\left|\frac{1}{\Gamma(\alpha+1)}{}_1F_1\left[\begin{matrix}1\\ \alpha+1\end{matrix};-su\right]\right| \le \frac{|\alpha|}{\operatorname{Re}(\alpha)},\quad u\in\mathbb{R}.$$

Finally, we evaluate now $I_-^\alpha\mathcal{L}f$ and $D_-^\alpha\mathcal{L}f$. The following two theorems were first established by Conlan and Koh in their attempt to solve a certain problem in system analysis (see [4] and [20]). Later, Raina and Koul [34] proved equivalent forms of these results and pointed out some applications in finding new expansions.

**Theorem 2.2.4.** *Let* $\operatorname{Re}(\alpha)>0$ *and let* $f\in L_\sigma$ *so that* $(I_-^\alpha\mathcal{L}f)(s)$ *exists on* $\mathbb{R}_+$. *Also, let* $g_\alpha(t):=t^\alpha f(t)$ *and* $g_{-\alpha}\in L_\sigma$. *Then*

$$(I_-^\alpha\mathcal{L}f)(s)=\mathcal{L}[g_{-\alpha}](s),\ \operatorname{Re}(s)>\max\{\sigma,0\}. \tag{2.2. 8}$$

*Proof.* Note that

$$\begin{aligned}(I_-^\alpha\mathcal{L}f)(s) &= \frac{1}{\Gamma(\alpha)}\int_s^\infty (t-s)^{\alpha-1}\left[\int_{\mathbb{R}_+}\mathrm{e}^{-tu}f(u)\mathrm{d}u\right]\mathrm{d}t\\ &= \frac{1}{\Gamma(\alpha)}\int_{\mathbb{R}_+}\left[\int_s^\infty (t-s)^{\alpha-1}\mathrm{e}^{-tu}\mathrm{d}t\right]f(u)\mathrm{d}u,\end{aligned}$$

where

$$\int_s^\infty (t-s)^{\alpha-1}\mathrm{e}^{-tu}\mathrm{d}t \overset{t\to t+s}{=\!=\!=} \mathrm{e}^{-su}\int_{\mathbb{R}_+} t^{\alpha-1}\mathrm{e}^{-tu}\mathrm{d}t = \mathrm{e}^{-su}u^{-\alpha}\Gamma(\alpha),\ u,s\in\mathbb{R}_+.$$

Thus, we have

$$
\begin{aligned}
(I_-^\alpha \mathcal{L} f)(s) &= \int_{\mathbb{R}_+} \mathrm{e}^{-su} u^{-\alpha} f(u) \mathrm{d}u \\
&= \int_{\mathbb{R}_+} \mathrm{e}^{-su} g_{-\alpha}(u) \mathrm{d}u = \mathcal{L}[g_{-\alpha}](s), \quad \mathrm{Re}\,(s) > \max\{\sigma, 0\}.
\end{aligned}
$$

**Theorem 2.2.5.** *Let* $\mathrm{Re}\,(\alpha) > 0$ *and let* $f \in L_\sigma$ *so that* $(D_-^\alpha \mathcal{L} f)(s)$ *exists on* $\mathbb{R}_+$. *Also, let* $g_\alpha(t) := t^\alpha f(t)$ *and* $g_\alpha \in L_\sigma$. *Then*

$$
(D_-^\alpha \mathcal{L} f)(s) = \mathcal{L}[g_\alpha](s), \quad \mathrm{Re}\,(s) > \max\{\sigma, 0\}.
$$

*Proof.* By Definition 2.2.2 and Theorem 2.2.4, we obtain

$$
\begin{aligned}
(D_-^\alpha \mathcal{L} f)(s) = (-D)^n (I_-^{n-\alpha} \mathcal{L} f)(s) &= (-D)^n \mathcal{L}[g_{\alpha-n}](s) \\
&= \mathcal{L}[t^n g_{\alpha-n}(t)](s) = \mathcal{L}[g_\alpha](s),
\end{aligned}
$$

where $n = [\,\mathrm{Re}\,(\alpha)] + 1$.

Multidimensional analogs of the above results are also considered in [31].

### 2.2.2 Applications

**Example 2.2.1.** Let $0 < \alpha < 1$ and $x \in \mathbb{R}_+$. Consider the fractional differential equation

$$
(D_{0+}^\alpha f)(x) = \lambda f(x) \ (x > 0)
$$

together with the initial condition $(I_{0+}^{1-\alpha} f)(0^+) = \lim_{x \to 0^+} (I_{0+}^{1-\alpha} f)(x) = c_0$. This problem is also known as the eigenvalue problem (see [41]). Applying the Laplace transform and Theorem 2.2.2 gives

$$s^\alpha \mathcal{L}[f](s) - (I_{0+}^{1-\alpha} f)(0^+) = \lambda \mathcal{L}[f](s) \quad \Rightarrow \quad \mathcal{L}[f](s) = \frac{c_0}{s^\alpha - \lambda}.$$

In view of (2.1.7), we obtain $f(x) = c_0 x^{\alpha-1} E_{\alpha,\alpha}(\lambda x^\alpha)$, where $E_{\alpha,\beta}(z)$ is the two-parametric Mittag-Leffler function defined by (2.1.6).

A function $f : \mathbb{R}_+ \to \mathbb{R}$ is completely monotone if $f \in C^\infty$ and $(-1)^n f^{(n)}(x) \geq 0$. The set of all completely functions on $\mathbb{R}_+$ is denoted by $CM(\mathbb{R}_+)$.

**Proposition 2.2.1.** *Let $\alpha > 0$ and $f \geq 0$.*

*(i) If $f \in L_0$, then $(\mathcal{L} I_{0+}^\alpha f)(s) \in CM(\mathbb{R}_+)$.*

*(ii) If $f \in L_0$ so that $(D_-^\alpha f)(s)$ exists on $\mathbb{R}_+$, then*

$$(D_-^\alpha \mathcal{L} f)(x) \in CM(\mathbb{R}_+).$$

*Proof.* The theorem follows immediately from the fact that the Laplace transform of a non-negative function is completely monotone (see [27, p. 396, Theorem 7]).

Let $\alpha \geq 0$. A function defined on $\mathbb{R}_+$ is said to be *completely monotone of order* $\alpha$ if $x^\alpha f(x) \in CM(\mathbb{R}_+)$. We write $f \in CM_\alpha(\mathbb{R}_+)$ if $x^\alpha f(x) \in CM(\mathbb{R}_+)$. Koumandos and Pedersen [21] found a sufficient and necessary condition for a function $f \in CM_\alpha(\mathbb{R}_+)$. The right-hand side Liouville fractional integral of a measure $\mu$ on $[0, \infty)$ is defined by:

$$\left(I_{0+}^\alpha(\mu)\right)(x) := \frac{1}{\Gamma(\alpha)} \int_0^x (x-t)^{\alpha-1} \mathrm{d}\mu(t), \quad \alpha > 0.$$

In view of the above details relating to the completely monotone functions and results of Proposition 2.2.1, we mention the following:

**Theorem 2.2.6** ([21, p. 35, Remark 1.6]). *A function $f \in CM_\alpha(\mathbb{R}_+)$ if and only if $f$ is the Laplace transform of a fractional integral of a positive Radon measure on $[0, \infty)$, that is $f(x) = (\mathcal{L}I^\alpha_{0+}(\mu))(x)$.*

We would also like here to mention Vojta's interesting paper [48], in which he not only gives further relations of the complete monotonicity concerning the Liouville fractional integral $I^\alpha_-$ and the Laplace transform, but also relates the right-hand side of (2.2. 8) to the Lambert $W$-function.

---

## 2.3 Laplace Transformation of the Caputo fractional Derivatives

In order to overcome some shortcomings of the Riemann-Liouville fractional derivatives, Caputo [3] proposed new definitions of fractional derivatives, which are defined as follows:

**Definition 2.3.1** ([18, p. 97]). *Let $\mathrm{Re}(\alpha) > 0$ and $x \in \mathbb{R}_+$. The Caputo fractional derivatives ${}^C D^\alpha_{0+} f$ and ${}^C D^\alpha_- f$ of order $\alpha$ are defined by:*

$$({}^C D^\alpha_{0+} f)(x) := \frac{1}{\Gamma(n-\alpha)} \int_0^x (x-t)^{n-\alpha-1} f^{(n)}(t)\mathrm{d}t \qquad (2.3.1)$$

*and*

$$({}^C D^\alpha_{-} f)(x) := \frac{1}{\Gamma(n-\alpha)} \int_x^\infty (t-x)^{n-\alpha-1} f^{(n)}(t)\mathrm{d}t, \qquad (2.3.2)$$

*respectively.*

### 2.3.1 Main Results

**Theorem 2.3.1.** *Let* $\mathrm{Re}\,(\alpha) > 0$ *and* $n = [\mathrm{Re}\,(\alpha)] + 1$. *If* $D^j f \in L_{\sigma_j}$ $(j = 0, \cdots, n)$ *and* $(D^j f)(0^+)$ *exists for* $j = 0, \cdots, n-1$, *then*

$$(\mathcal{L}\,{}^C D^\alpha_{0+} f)(s) = s^\alpha \mathcal{L}[f](s) - \sum_{j=0}^{n-1} s^{\alpha-j-1}(D^j f)(0^+),$$

*where* $\mathrm{Re}\,(s) > \max\{\sigma_0, \cdots, \sigma_n, 0\}$.

*Proof.* Observe that

$$({}^C D^\alpha_{0+} f)(x) = \frac{t^{n-\alpha-1}}{\Gamma(n-\alpha)} * f^{(n)}(t),$$

where $f^{(n)} \in L_{\sigma_n}$ and $t^{n-\alpha-1} \in L_0$. Using Theorems 2.1.2 and 2.1.4, we obtain

$$\begin{aligned}(\mathcal{L}\,{}^C D^\alpha_{0+} f)(s) &= \mathcal{L}\left[\frac{t^{n-\alpha-1}}{\Gamma(n-\alpha)}\right] \mathcal{L}\left[f^{(n)}\right](s) \quad (\mathrm{Re}\,(s) > \max\{0, \sigma_n\}) \\ &= s^{\alpha-n}\left[s^n \mathcal{L}[f](s) - \sum_{j=0}^{n-1} s^{n-j-1}(D^j f)(0^+)\right] \\ &= s^\alpha \mathcal{L}[f](s) - \sum_{j=0}^{n-1} s^{\alpha-j-1}(D^j f)(0^+).\end{aligned}$$

### 2.3.2 Applications

**Example 2.3.1.** For $0 < \alpha < 1$ and $x \in \mathbb{R}_+$, we consider the fractional differential equation

$$({}^C D_{0+}^{\alpha} f)(x) = \lambda f(x) \ (x > 0)$$

together with the initial condition $f(0^+) = \lim_{x \to 0^+} f(x) = c_0$. Applying the Laplace transform and Theorem 2.2.2, we get

$$s^{\alpha} \mathcal{L}[f](s) - s^{\alpha-1} f(0^+) = \lambda \mathcal{L}[f](s) \quad \Rightarrow \quad \mathcal{L}[f](s) = \frac{c_0 s^{\alpha-1}}{s^{\alpha} - \lambda}.$$

In view of (2.1.9), we obtain $f(x) = c_0 E_{\alpha}(\lambda x^{\alpha})$, where $E_{\alpha}(z)$ is the classical Mittag-Leffler function defined by (2.1.8).

**Example 2.3.2.** Let $0 < \nu \leq 1$. Consider the fractional recursive differential equation

$$({}^C D_{0+}^{\nu} p_k)(t) = -\lambda \left(p_k(t) - p_{k-1}(t)\right), \ k \geq 0,$$

where $p_{-1}(t) = 0$, subject to the initial conditions

$$p_k(0) = \begin{cases} 1, & k = 0, \\ 0, & k \geq 1. \end{cases}$$

This equation is proposed by Beghim and Orsingher while studying Poisson-type processes (see [1, p. 688, Theorem 2.1]). When $k = 0$, the equation reduces to $({}^C D_{0+}^{\nu} p_0)(t) = -\lambda p_0(t)$, from which the use of Laplace transform and Theorem 2.2.2 gives

$$s^{\nu} \mathcal{L}[p_0](s) - s^{\nu-1} = -\lambda \mathcal{L}[p_0](s) \quad \Rightarrow \quad \mathcal{L}[p_0](s) = \frac{s^{\nu-1}}{s^{\nu} - \lambda}.$$

For $k \geq 1$, we have

$$s^{\nu} \mathcal{L}[p_k](s) = -\lambda \left(\mathcal{L}[p_k](s) - \mathcal{L}[p_{k-1}](s)\right),$$

and thus

$$\mathcal{L}[p_k](s) = \frac{\lambda}{s^\nu + \lambda}\mathcal{L}[p_{k-1}](s) = \frac{\lambda^2}{(s^\nu + \lambda)^2}\mathcal{L}[p_{k-2}](s)$$
$$= \cdots = \frac{\lambda^k}{(s^\nu + \lambda)^k}\mathcal{L}[p_0](s) = \frac{\lambda^k s^{\nu-1}}{(s^\nu + \lambda)^{k+1}}.$$

In view of (2.1.12), we obtain

$$p_k^\nu(t) = (\lambda t^k)^k E_{\nu,\nu k+1}^{k+1}(-\lambda t^\nu),\ k \geq 0,\ t \in \mathbb{R}_+.$$

---

## 2.4 Laplace Transformation of Hifler's Fractional Derivatives

In 2000, Hilfer [14] introduced the following fractional derivatives.

**Definition 2.4.1.** *Let* $\mathrm{Re}\,(\alpha) \geq 0,\ 0 < \alpha < 1$ *and* $0 \leq \beta \leq 1$. *The right-sided and left-sided Hilfer fractional derivatives* $D_{0+}^{\alpha,\beta} f$ *and* $D_{-}^{\alpha,\beta} f$ *of order* $\alpha$ *and type* $\beta$ *are defined by:*

$$(D_{0+}^{\alpha,\beta} f)(x) := \left( I_{0+}^{\beta(1-\alpha)} \frac{\mathrm{d}}{\mathrm{d}x} (I_{0+}^{(1-\beta)(1-\alpha)} f) \right)(x) \tag{2.4.1}$$

*and*

$$(D_{-}^{\alpha,\beta} f)(x) := \left( I_{-}^{\beta(1-\alpha)} \frac{\mathrm{d}}{\mathrm{d}x} (I_{-}^{(1-\beta)(1-\alpha)} f) \right)(x), \tag{2.4.2}$$

*respectively.*

**Remark 2.4.1.** *When* $\beta = 0$, (2.4.1) *and* (2.4.2) *reduce to the Liouville fractional derivatives* (2.2. 1) *and* (2.2. 2), *respectively.*

*Moreover, for* $\beta = 1$, (2.4.1) *and* (2.4.2) *give the Caputo fractional derivatives* (2.3.1) *and* (2.3.2), *respectively.*

Later in [15], Hilfer et al. considered the following extended form of (2.4.1):

$$\begin{aligned}(D_{0+}^{\alpha,\beta} f)(x) &:= \left( I_{0+}^{\beta(1-\alpha)} \frac{\mathrm{d}}{\mathrm{d}x} (I_{0+}^{(1-\beta)(1-\alpha)} f) \right)(x) \\ &= (I_{0+}^{\beta(n-\alpha)} D_{0+}^{\alpha+\beta n-\alpha\beta} f)(x), \end{aligned} \tag{2.4.3}$$

where $n-1 < \alpha \leq n \in \mathbb{N}$ and $0 \leq \beta \leq 1$.

### 2.4.1 Main Results

The following result is due to Tomovski.

**Theorem 2.4.1** ([46, p. 3380]). *Let* $n-1 < \alpha \leq n$ $(n \in \mathbb{N})$ *and* $0 \leq \beta \leq 1$. *Let* $f$ *be a "sufficiently good" function with* $\sigma$ *its abscissa of absolute convergence. Then we have*

$$\mathcal{L}[D_{0+}^{\alpha,\beta} f](s) = s^{\alpha}\mathcal{L}[f](s) - \sum_{j=0}^{n-1} s^{n-j-1-\beta(n-\alpha)} (D^j I_{0+}^{(1-\beta)(n-\alpha)} f)(0^+), \tag{2.4.4}$$

*where* $\mathrm{Re}\,(s) > \max\{0, \sigma\}$.

For the sake of brevity, above and in what follows, we use the term "sufficiently good" function to mean that the function has sufficiently many derivatives and converges fast enough at infinity so that the required fractional operations of such a function (e.g., $I_{0+}^{\alpha} I_{0+}^{\beta} f$ and $I_{0+}^{\alpha} D_{0+}^{\alpha,\beta} f$) are valid and that the Laplace transform of the associated results exist for $\mathrm{Re}\,(s) > \max\{0, \sigma\}$.

*Proof.* Applying the integral operator $I_{0+}^{\alpha}$ to each side of (2.4.3) and then using the compositional properties ([18, 73, Lemma 2.3] and [18, 74, Lemma 2.5])

$$(I_{0+}^{\alpha} I_{0+}^{\beta} \varphi)(x) = (I_{0+}^{\alpha+\beta} \varphi)(x) \ (\varphi \in L^1(\mathbb{R}_+)),$$

$$(I_{0+}^{\alpha} D_{0+}^{\alpha} \varphi)(x) = \varphi(x) - \sum_{j=0}^{n-1} (D^j I_{0+}^{n-\alpha} \varphi)(0^+) \frac{x^{\alpha-n+j}}{\Gamma(\alpha - n + j + 1)}$$
$$(\varphi \in L^1(\mathbb{R}_+),\ I_{0+}^{n-\alpha} \varphi \in AC^n[0, b], \forall b > 0),$$

we obtain

$$\begin{aligned}
&(I_{0+}^{\alpha} D_{0+}^{\alpha,\beta} f)(x) \\
&= (I_{0+}^{\alpha} I_{0+}^{\beta(n-\alpha)} D_{0+}^{\alpha+\beta n-\alpha\beta} f)(x) = (I_{0+}^{\alpha+\beta n-\alpha\beta} D_{0+}^{\alpha+\beta n-\alpha\beta} f)(x) \\
&= f(x) - \sum_{j=0}^{n-1} (D^j I_{0+}^{(n-\alpha)(1-\beta)} f)(0^+) \frac{x^{j-(n-\alpha)(1-\beta)}}{\Gamma(j - (n-\alpha)(1-\beta) + 1)}.
\end{aligned} \tag{2.4.5}$$

Next, applying the Laplace transform on each side of (2.4.5), we get

$$\begin{aligned}
&\mathcal{L}[I_{0+}^{\alpha} D_{0+}^{\alpha,\beta} f](s) \\
&= \mathcal{L}[f](s) - \sum_{j=0}^{n-1} (D^j I_{0+}^{(n-\alpha)(1-\beta)} f)(0^+) \frac{\mathcal{L}[x^{j-(n-\alpha)(1-\beta)}](s)}{\Gamma(j - (n-\alpha)(1-\beta) + 1)},
\end{aligned}$$

which after a little simplification yields

$$s^{-\alpha} \mathcal{L}[D_{0+}^{\alpha,\beta} f](s) = \mathcal{L}[f](s) - \sum_{j=0}^{n-1} s^{(n-\alpha)(1-\beta)-j-1} (D^j I_{0+}^{(n-\alpha)(1-\beta)} f)(0^+),$$

and the result (2.4.4) follows.

### *2.4.2 Applications*

**Example 2.4.1.** Let $0 < \alpha < 1$, $0 \le \beta \le 1$ and $x \in \mathbb{R}_+$. Consider the fractional differential equation

$$(D_{0+}^{\alpha,\beta} f)(x) = \lambda f(x)$$

under the initial condition $(I_{0+}^{(1-\beta)(1-\alpha)} f)(0^+) = \lim_{x \to 0^+} (I_{0+}^{(1-\beta)(1-\alpha)} f)(x) = c_0$. Such an initial problem was first proposed and solved by Hilfer (see [14, p. 115]; see also [41]). Applying the Laplace transforms and Theorem 2.4.1, we have

$$s^\alpha \mathcal{L}[f](s) - c_0 s^{\beta(\alpha-1)} = \lambda \mathcal{L}[f](s) \quad \Rightarrow \quad \mathcal{L}[f](s) = \frac{c_0 s^{\beta(\alpha-1)}}{s^\alpha - \lambda}.$$

From (2.1.7), we obtain

$$f(x) = x^{(1-\beta)(\alpha-1)} E_{\alpha,\alpha+\beta(1-\alpha)}(\lambda x^\alpha),$$

where $E_{\alpha,\beta}(z)$ is defined by (2.1.6).

**Example 2.4.2.** Let $\mathrm{Re}(\nu) > 0$, $0 < \alpha < 1$, $0 \le \beta \le 1$ and $x \in \mathbb{R}_+$. Consider the Volterra-type fractional differintegral equation

$$(D_{0+}^{\alpha,\beta} f)(x) + \frac{\lambda}{\Gamma(\nu)} \int_0^x (x-t)^{\nu-1} f(t) \mathrm{d}t = g(x)$$

in $L^1(\mathbb{R}_+)$ with the initial condition $(I_{0+}^{(1-\beta)(1-\alpha)} f)(0+) = c_0$. This kind of equation was proposed and solved by Tomovski et al. (see [47, p. 808, Theorem 8]). Application of the Laplace transform and Theorem 2.4.1 gives

$$s^\alpha \mathcal{L}[f](s) - c_0 s^{\beta(\alpha-1)} + \lambda s^{-\nu} \mathcal{L}[f](s) = \mathcal{L}[g](s),$$

from which

$$\mathcal{L}[f](s) = c_0 \frac{s^{\beta(\alpha-1)+\nu}}{s^{\alpha+\nu} + \lambda} + \frac{s^\nu}{s^{\alpha+\nu} + \lambda} \mathcal{L}[g](s). \tag{2.4.6}$$

In view of (2.1.16), the relation (2.4.6) becomes

$$\mathcal{L}[f](s) = c_0 \mathcal{L}[t^{(1-\beta)(\alpha-1)} E_{\alpha+\nu,\alpha-\beta(\alpha-1)}(-\lambda t^{\alpha+\nu})](s) \\ + \mathcal{L}[t^{\alpha-1} E_{\alpha+\nu,\alpha}(-\lambda t^{\alpha+\nu}) * g(t)](s).$$

Taking now the inverse Laplace transform, we then obtain

$$f(t) = c_0 t^{(1-\beta)(\alpha-1)} E_{\alpha+\nu,\alpha-\beta(\alpha-1)}(-\lambda t^{\alpha+\nu}) + t^{\alpha-1} E_{\alpha+\nu,\alpha}(-\lambda t^{\alpha+\nu}) * g(t).$$

---

## 2.5 Laplace Transformation of the Prabhaker Fractional Integrals and Derivatives

In 1971, Prabhakar [30] studied a class of weakly-singular linear Volterra operators whose kernels involves the three-parameter Mittag-Leffler function $E^{\gamma}_{\alpha,\beta}(z)$. The fractional operators derived from Prabhakar's work have stimulated a large number of research results and have led to an important theory known as the *Prabhakar Fractional Calculus* (see [9], [10], [11], and [17]).

**Definition 2.5.1** ([11, p. 25]). *Let* $\alpha > 0$, $\mathrm{Re}\,(\beta) > 0$ *and* $x \in \mathbb{R}_+$. *The Prabhakar fractional integral* $\mathcal{J}^{\gamma}_{\alpha,\beta,\lambda;0}f$ *is defined by:*

$$(\mathcal{J}^{\gamma}_{\alpha,\beta,\lambda;0}f)(s) := \int_0^x (x-t)^{\beta-1} E^{\gamma}_{\alpha,\beta}[\lambda(x-t)^{\alpha}] f(t)\mathrm{d}t,$$

*where* $E^{\gamma}_{\alpha,\beta}(z)$ *is defined by* (2.1.11).

**Definition 2.5.2** ([11, p. 26]). *Let* $\alpha > 0$, $\mathrm{Re}\,(\beta) > 0$, $m = [\mathrm{Re}\,(\beta)] + 1$ *and* $x \in \mathbb{R}_+$. *The Prabhakar derivative of Riemann-Liouville type* ${}^{RL}\mathcal{D}^{\gamma}_{\alpha,\beta,\lambda;0}f$ *is defined by:*

$$({}^{RL}\mathcal{D}^{\gamma}_{\alpha,\beta,\lambda;0}f)(x) := (D^m \mathcal{J}^{-\gamma}_{\alpha,m-\beta,\lambda;0}f)(x).$$

**Definition 2.5.3** ([11, p. 26]). *Let* $\alpha > 0$, $\mathrm{Re}\,(\beta) > 0$, $m = [\mathrm{Re}\,(\beta)] + 1$ *and* $x \in \mathbb{R}_+$. *The Prabhakar derivative of Caputo type* ${}^{C}\mathcal{D}^{\gamma}_{\alpha,\beta,\lambda;0}f$ *is defined by:*

$$({}^{C}\mathcal{D}^{\gamma}_{\alpha,\beta,\lambda;0}f)(x) := (\mathcal{J}^{-\gamma}_{\alpha,m-\beta,\lambda;0}D^m f)(x).$$

As mentioned in Example 2.1.6, Raina's function $\mathcal{F}^{\sigma}_{\rho,\lambda}(z)$ is a very general Mittag-Leffler type function. With the help of this function, Raina [33, p. 194, Eq. (2.2)] also introduced a generalized fractional integral operators which provide a useful generalization of the Prabhakar fractional integral (see also [23]).

### *2.5.1 Main Results*

**Theorem 2.5.1.** *Let* $\alpha > 0$, $\mathrm{Re}\,(\beta) > 0$ *and* $f \in L_\sigma$. *Then*

$$(\mathcal{L}\mathcal{J}^{\gamma}_{\alpha,\beta,\lambda;0}f)(s) = \frac{s^{\alpha\gamma-\beta}}{(s^\alpha - \lambda)^\gamma}\mathcal{L}[f](s), \quad \mathrm{Re}\,(s) > \max\{\sigma, |\lambda|^{1/\alpha}\}. \tag{2.5.1. 1}$$

*Proof.* Note that

$$(\mathcal{J}^{\gamma}_{\alpha,\beta,\lambda;0}f)(t) = t^{\beta-1}E^{\gamma}_{\alpha,\beta}(\lambda t^\alpha) * f(t),$$

where $f \in L_\sigma$ and $t^{\beta-1}E^{\gamma}_{\alpha,\beta}(\lambda t^\alpha) \in L_{|\lambda|^{1/\alpha}}$. Using Theorem 2.1.2, we obtain

$$(\mathcal{L}\mathcal{J}^{\gamma}_{\alpha,\beta,\lambda;0}f)(s) = \mathcal{L}[t^{\beta-1}E^{\gamma}_{\alpha,\beta}(\lambda t^\alpha)](s)\mathcal{L}[f](s) = \frac{s^{\alpha\gamma-\beta}}{(s^\alpha - \lambda)^\gamma}\mathcal{L}[f](s),$$

provided that $\operatorname{Re}(s) > \max\{\sigma, |\lambda|^{1/\alpha}\}$.

**Remark 2.5.1.** *There is an alternative way to derive* (2.5.1. 1). *Recall that [11, p. 25, Eq. (5.5)]*

$$(\mathcal{J}^{\gamma}_{\alpha,\beta,\lambda;0}f)(t) = \sum_{k=0}^{\infty} \frac{(\gamma)_k}{k!}\lambda^k (I_{0+}^{\alpha k+\beta} f)(t),$$

*then in view of Theorem 2.2.1, we have*

$$\begin{aligned}(\mathcal{L}\mathcal{J}^{\gamma}_{\alpha,\beta,\lambda;0}f)(s) &= \sum_{k=0}^{\infty} \frac{(\gamma)_k}{k!}\lambda^k (\mathcal{L}I_{0+}^{\alpha k+\beta} f)(s)\\ &= \left(\sum_{k=0}^{\infty} \frac{(\gamma)_k}{k!}\lambda^k s^{-\alpha k-\beta}\right)\mathcal{L}[f](s) = s^{-\beta}\left(1-\frac{\lambda}{s^{\alpha}}\right)^{-\gamma}\mathcal{L}[f](s),\end{aligned}$$

*which is evidently* (2.5.1. 1).

**Theorem 2.5.2.** *Let* $\alpha, \beta > 0$ *and* $m = [\beta]+1$. *Let* $f$ *be a "sufficiently good" function with* $\sigma$ *its abscissa of absolute convergence. Then we have*

$$\mathcal{L}[{}^{RL}\mathcal{D}^{\gamma}_{\alpha,\beta,\lambda;0}f](s) = s^{\beta-\alpha\gamma}(s^{\alpha}-\lambda)^{\gamma}\mathcal{L}[f](s) - \sum_{j=0}^{m-1} s^{m-j-1}(\mathcal{J}^{-\gamma}_{\alpha,m-\beta-j,\lambda;0}f)(0^+),$$

*where* $\operatorname{Re}(s) > \max\{\sigma, |\lambda|^{1/\alpha}\}$.

*Proof.* Using Definition 2.5.2 and Theorem 2.1.4, we have

$$\mathcal{L}[{}^{RL}\mathcal{D}^{\gamma}_{\alpha,\beta,\lambda;0}f](s) = s^{m}\mathcal{L}[\mathcal{J}^{-\gamma}_{\alpha,n-\beta,\lambda;0}f](s) - \sum_{j=0}^{m-1} s^{m-j-1}(D^{j}\mathcal{J}^{-\gamma}_{\alpha,n-\beta,\lambda;0}f)(0^+).$$

The result then follows by using (2.5.1. 1) and the relation (see [17, p. 44, Eq. (5.6)])

$$D^j \mathcal{J}^{\gamma}_{\alpha,\beta,\lambda;0} f = \mathcal{J}^{\gamma}_{\alpha,\beta-j,\lambda;0} f \ (f \in C[0,b], \ \forall b > 0).$$

**Theorem 2.5.3.** *Let* $\alpha, \beta > 0$ *and* $m = [\beta]+1$. *Let* $f$ *be a "sufficiently good" function with* $\sigma$ *its abscissa of absolute convergence. Then we have*

$$\mathcal{L}[{}^C\mathcal{D}^{\gamma}_{\alpha,\beta,\lambda;0} f](s) = s^{\beta-\alpha\gamma}(s^{\alpha}-\lambda)^{\gamma} \left[ \mathcal{L}[f](s) - \sum_{j=0}^{m-1} s^{-j-1}(D^j f)(0^+) \right],$$

*where* $\operatorname{Re}(s) > \max\{\sigma, |\lambda|^{1/\alpha}\}$.

*Proof.* The proof is simply based on Definition 2.5.3 and Theorems 2.5.1 and 2.1.4. We omit further details. The reader may also refer to [6, p. 124].

### *2.5.2 Applications*

**Example 2.5.1.** Let $\alpha > 0$ and $0 < \beta \leq 1$. Consider the Cauchy problem

$$\begin{cases} {}^{RL}\mathcal{D}^{\gamma}_{\alpha,\beta,\lambda;0} f(x) = Af(x), \\ (\mathcal{J}^{-\gamma}_{\alpha,1-\beta,\lambda;0} f)(0^+) = c_0. \end{cases}$$

Applying the Laplace transforms and Theorem 2.5.3, we get

$$s^{\beta-\alpha\gamma}(s^{\alpha}-\lambda)^{\gamma}\mathcal{L}[f](s) - c_0 = A\mathcal{L}[f](s),$$

from which

$$\mathcal{L}[f](s) = \frac{c_0}{s^{\beta-\alpha\gamma}(s^\alpha - \lambda)^\gamma - A}$$
$$= c_0 \frac{s^{\alpha\gamma-\beta}}{(s^\alpha - \lambda)^\gamma} \frac{1}{1 - \frac{As^{\alpha\gamma-\beta}}{(s^\alpha-\lambda)^\gamma}} = c_0 \sum_{k=0}^{\infty} A^k \frac{s^{(\alpha\gamma-\beta)(k+1)}}{(s^\alpha - \lambda)^{\gamma(k+1)}},$$

where $|A||s^{\alpha\gamma-\beta}| < |(s^\alpha - \lambda)^\gamma|$. In view of (2.1.12), we obtain

$$f(t) = c_0 \sum_{k=0}^{\infty} A^k t^{\beta(k+1)-1} E^{\gamma(k+1)}_{\alpha,\beta(k+1)}(\lambda t^\alpha).$$

---

## 2.6 Laplace Transformation of the Saigo Fractional Integrals

In 1978, Saigo [36] introduced a pair of fractional integral operators involving the Gauss hypergeometric function ${}_2F_1$. During the past decades, various properties and applications have been discovered for Saigo's operators (see, for example, [37] and [38]; see also [42]).

**Definition 2.6.1.** *Let* $\mathrm{Re}(\alpha) > 0$, $\beta, \eta \in \mathbb{C}$ *and* $x \in \mathbb{R}_+$. *The Saigo fractional integrals* $I^{\alpha,\beta,\eta}_{0+} f$ *and* $J^{\alpha,\beta,\eta}_{-} f$ *are defined by*

$$I^{\alpha,\beta,\eta}_{0+} f(x) := \frac{x^{-\alpha-\beta}}{\Gamma(\alpha)} \int_0^x (x-t)^{\alpha-1} {}_2F_1 \left[ \begin{matrix} \alpha+\beta, -\eta \\ \alpha \end{matrix}; 1 - \frac{t}{x} \right] f(t)\mathrm{d}t \tag{2.6.0.1}$$

*and*

$$J_{-}^{\alpha,\beta,\eta} f(x) := \frac{1}{\Gamma(\alpha)} \int_x^\infty (t-x)^{\alpha-1} {}_2F_1 \left[ \begin{matrix} \alpha+\beta, -\eta \\ \alpha \end{matrix} ; 1 - \frac{x}{t} \right] f(t) t^{-\beta-\alpha} \mathrm{d}t, \tag{2.6.0.2}$$

*respectively.*

Alternatively, (2.6.0.1) and (2.6.0.2) can be written as

$$I_{0+}^{\alpha,\beta,\eta} f(x) := \frac{x^{-\beta}}{\Gamma(\alpha)} \int_0^1 (1-\sigma)^{\alpha-1} {}_2F_1 \left[ \begin{matrix} \alpha+\beta, -\eta \\ \alpha \end{matrix} ; 1 - \sigma \right] f(x\sigma) \mathrm{d}\sigma \tag{2.6.0.3}$$

and

$$J_{-}^{\alpha,\beta,\eta} f(x) := \frac{x^{-\beta}}{\Gamma(\alpha)} \int_1^\infty \sigma^{-\beta-\alpha} (\sigma-1)^{\alpha-1} {}_2F_1 \left[ \begin{matrix} \alpha+\beta, -\eta \\ \alpha \end{matrix} ; 1 - \frac{1}{\sigma} \right] f(x\sigma) \mathrm{d}\sigma. \tag{2.6.0.4}$$

In the sequel, where we show the relationship between the Saigo fractional integrals and the (multiple) Erdélyi-Kober fractional integrals (see Section 2.7), the above two expressions will be more convenient.

### *2.6.1 Main Results*

The Laplace transformation of the Saigo fractional integrals has been well studied by Srivastava, Saigo, and Raina [42]. Some typical results are presented here.

**Theorem 2.6.1** ([42, p. 6]). *Let* $\lambda \in \mathbb{C}$ *and* $x^\lambda I_{0+}^{\alpha,\beta,\eta} f \in L_\sigma$. *Then*

$$(\mathcal{L} x^\lambda I_{0+}^{\alpha,\beta,\eta} f)(s) = \int_0^\infty S(\alpha, \beta, \eta, \lambda; s, x) f(x) \mathrm{d}x,$$

*where*

$$S(\alpha,\beta,\eta,\lambda;s,x)$$
$$:= \frac{\Gamma(\beta-\lambda)\Gamma(\eta-\lambda)}{\Gamma(-\lambda)\Gamma(\alpha+\beta+\eta-\lambda)} x^{\lambda-\beta} {}_2F_2\left[\begin{matrix}\lambda+1, \lambda-\alpha-\beta-\eta+1\\ \lambda-\beta+1, \lambda-\eta+1\end{matrix}; -sx\right]$$
$$+ \frac{\Gamma(\eta-\beta)\Gamma(\lambda-\beta)}{\Gamma(\alpha+\eta)\Gamma(-\beta)} s^{\beta-\lambda} {}_2F_2\left[\begin{matrix}1-\alpha-\eta, \beta+1\\ \beta-\eta+1, \beta-\lambda+1\end{matrix}; -sx\right]$$
$$+ \frac{\Gamma(\beta-\eta)\Gamma(\lambda-\eta)}{\Gamma(\alpha+\beta)\Gamma(-\eta)} s^{\eta-\lambda} x^{\eta-\beta} {}_2F_2\left[\begin{matrix}1-\alpha-\beta, \eta+1\\ \eta-\beta+1, \eta-\lambda+1\end{matrix}; -sx\right]. \tag{2.6.1.5}$$

*Proof.* We begin by recalling the following generalized fractional integration by parts for Saigo's fractional integral operators (see [42, p. 5, Eq. (2.14)]):

$$\int_0^\infty g(x) I_{0+}^{\alpha,\beta,\eta} f(x) \mathrm{d}x = \int_0^\infty f(x) J_{-}^{\alpha,\beta,\eta} g(x) \mathrm{d}x. \tag{2.6.1.6}$$

Setting $g(x) = x^\lambda \mathrm{e}^{-sx}$ in (2.6.1.6) gives

$$(\mathcal{L} x^\lambda I_{0+}^{\alpha,\beta,\eta} f)(s) = \int_0^\infty f(x) J_{-}^{\alpha,\beta,\eta} x^\lambda \mathrm{e}^{-sx} \mathrm{d}x.$$

The proof is now complete since we have (see, for example, [25, p. 638] and [42, p. 7])

$$J_{-}^{\alpha,\beta,\eta} x^\lambda \mathrm{e}^{-sx} = S(\alpha,\beta,\eta,\lambda;s,x). \tag{2.6.1.7}$$

For the sake of completeness, we provide here an outline of the proof of (2.6.1.7). We shall use the following well-known result [39, p. 294, Eq. (3.7)]:

$$J_{-}^{\alpha,\beta,\eta}x^{\lambda} = \frac{\Gamma(\beta-\lambda)\Gamma(\eta-\lambda)}{\Gamma(-\lambda)\Gamma(\alpha+\beta+\eta-\lambda)}x^{\lambda-\beta} \tag{2.6.1.8}$$

$$(\operatorname{Re}(\alpha) > 0,\ \operatorname{Re}(\lambda) < \min\{\operatorname{Re}(\beta), \operatorname{Re}(\eta)\}).$$

By replacing $e^{-sx}$ with its Mellin-Barnes contour integral representation, we obtain

$$J_{-}^{\alpha,\beta,\eta}x^{\lambda}e^{-sx} = \frac{1}{2\pi i}\int_L \Gamma(z)s^{-z}J_{-}^{\alpha,\beta,\eta}x^{\lambda-z}dz.$$

With the help of (2.6.1.8), we have

$$J_{-}^{\alpha,\beta,\eta}x^{\lambda}e^{-sx} = \frac{x^{\lambda-\beta}}{2\pi i}\int_L \frac{\Gamma(\beta-\lambda+z)\Gamma(\eta-\lambda+z)\Gamma(z)}{\Gamma(z-\lambda)\Gamma(\alpha+\beta+\eta-\lambda+z)}(sx)^{-z}dz,$$

where $\operatorname{Re}(\lambda) - \max\{\operatorname{Re}(\beta), \operatorname{Re}(\eta)\} < \operatorname{Re}(z)$. Let

$$\Phi(z) := \frac{\Gamma(\beta-\lambda+z)\Gamma(\eta-\lambda+z)\Gamma(z)}{\Gamma(z-\lambda)\Gamma(\alpha+\beta+\eta-\lambda+z)}.$$

Then, by the residue theorem, we have

$$J_{-}^{\alpha,\beta,\eta}x^{\lambda}e^{-sx} = x^{\lambda-\beta}\sum_{n=0}^{\infty}\operatorname*{Res}_{z=-n}\Phi(z)(sx)^{-z} + x^{\lambda-\beta}\sum_{n=0}^{\infty}\operatorname*{Res}_{z=\lambda-\beta-n}\Phi(z)(sx)^{-z}$$
$$+ x^{\lambda-\beta}\sum_{n=0}^{\infty}\operatorname*{Res}_{z=\lambda-\eta-n}\Phi(z)(sx)^{-z},$$

which after a little simplification gives the relation (2.6.1.7). We recommend the reader to refer to [38] for a more general case.

**Theorem 2.6.2** ([42, p. 6]). *Let $\mu \in \mathbb{C}$ and $x^{\mu}J_{-}^{\alpha,\beta,\eta}f \in L_{\sigma}$. Then*

$$(\mathcal{L}x^{\mu}J_{-}^{\alpha,\beta,\eta}f)(s) = \int_0^{\infty} T(\alpha,\beta,\eta,\mu;s,x)f(x)dx,$$

*where*

$$T(\alpha,\beta,\eta,\mu;s,x) := \frac{\Gamma(\mu+1)\Gamma(\mu-\beta+\eta+1)}{\Gamma(\mu-\beta+1)\Gamma(\mu+\alpha+\eta+1)} \cdot x^{\mu-\beta}{}_2F_2\left[\begin{matrix}\mu+1,\mu-\beta+\eta+1\\ \mu-\beta+1,\mu+\alpha+\eta+1\end{matrix};-sx\right]. \tag{2.6.1.9}$$

*Proof.* The proof is similar to the proof of Theorem 2.6.1. We find that the use of (2.6.1.6) readily gives

$$(\mathcal{L}x^{\mu}J_{-}^{\alpha,\beta,\eta}f)(s) = \int_0^{\infty} f(x) I_{0+}^{\alpha,\beta,\eta} x^{\mu} \mathrm{e}^{-sx} \mathrm{d}x.$$

It remains to show that $I_{0+}^{\alpha,\beta,\eta}x^{\mu}\mathrm{e}^{-sx} = T(\alpha,\beta,\eta,\mu;s,x)$. Here we adopt a different approach. Since ([39, p. 295, Eq. (3.8)])

$$I_{0+}^{\alpha,\beta,\eta}x^{\mu} = \frac{\Gamma(\mu+1)\Gamma(\mu-\beta+\eta+1)}{\Gamma(\mu-\beta+1)\Gamma(\mu+\alpha+\eta+1)}x^{\mu-\beta}$$

$$(\mathrm{Re}\,(\alpha)>0,\ \ \mathrm{Re}\,(\mu)>\max\{0,\ \mathrm{Re}\,(\beta-\eta)\}-1),$$

we have immediately

$$\begin{aligned}
I_{0+}^{\alpha,\beta,\eta}x^{\mu}\mathrm{e}^{-sx} &= \sum_{n=0}^{\infty}\frac{(-s)^n}{n!}I_{0+}^{\alpha,\beta,\eta}x^{\mu+n}\\
&= \sum_{n=0}^{\infty}\frac{(-s)^n}{n!}\frac{\Gamma(\mu+1+n)\Gamma(\mu-\beta+\eta+1+n)}{\Gamma(\mu-\beta+1+n)\Gamma(\mu+\alpha+\eta+1+n)}x^{\mu-\beta+n}\\
&= \frac{\Gamma(\mu+1)\Gamma(\mu-\beta+\eta+1)}{\Gamma(\mu-\beta+1)\Gamma(\mu+\alpha+\eta+1)}\\
&\quad\cdot x^{\mu-\beta}{}_2F_2\left[\begin{matrix}\mu+1,\mu-\beta+\eta+1\\ \mu-\beta+1,\mu+\alpha+\eta+1\end{matrix};-sx\right],
\end{aligned}$$

which completes the proof.

### *2.6.2 Remarks*

(i) In [42, pp. 8–9], Srivastava, Saigo, and Raina have also established the formulas for

$$I_{0+}^{\alpha,\beta,\eta} s^{\lambda} \mathcal{L}[f](s) \quad \text{and} \quad J_{-}^{\alpha,\beta,\eta} s^{\mu} \mathcal{L}[f](s).$$

Some applications of these formulas are also mentioned. These results are not given here. The interested reader may refer to [42] for further details.

(ii) Let $\mathrm{Re}\,(\alpha) > 0$ and $n = [\mathrm{Re}\,(\alpha)] + 1$. The Saigo fractional derivatives $D_{0+}^{\alpha,\beta,\eta} f$ and $D_{-}^{\alpha,\beta,\eta} f$ are defined by (see [38, p. 92, Eqs. (2.11) and (2.12)]):

$$(D_{0+}^{\alpha,\beta,\eta} f)(x) := D^n (I_{0+}^{-\alpha+n,-\beta-n,\alpha+\eta-n} f)(x)$$

and

$$(D_{-}^{\alpha,\beta,\eta} f)(x) := (-D)^n (I_{-}^{-\alpha+n,-\beta-n,\alpha+\eta} f)(x),$$

respectively. However, to the best knowledge of the authors of the present article, the Laplace transformation of the operators $(D_{0+}^{\alpha,\beta,\eta} f)(x)$ and $(D_{-}^{\alpha,\beta,\eta} f)(x)$ do not seem to have been studied so far. In addition, it may be worthwhile also to study the Laplace transformation of the Caputo-type Saigo fractional derivatives introduced recently by Rao et al. [35, p. 23, Eqs. (34) and (35)].

(iii) In [24], Luo and Raina introduced a pair of fractional integral operators whose kernels involve a very special class of generalized hypergeometric functions, $r+2Fr+1$. Many properties of these operators have been discovered over the past few years (see [25] and [26]). It may be, therefore, of interest to establish corresponding results in light of Theorems 2.6.1 and 2.6.2.

---

## 2.7 Laplace Transformation of the (Multiple) Erdélyi-Kober Fractional Integrals

In 1986, Kiryakova introduced a pair of generalized fractional integrals involving Meijer's $G$-function which have now become the fundamental concepts of the *Generalized Fractional Calculus* (see [19]). Meijer's $G$-function is defined by:

$$G_{p,q}^{m,n}\left[\sigma \left| \begin{matrix} (a_i)_{1,p} \\ (b_j)_{1,q} \end{matrix} \right.\right] = \frac{1}{2\pi \mathrm{i}} \int_L \mathcal{G}_{p,q}^{m,n}(s)\sigma^{-s}\mathrm{d}s,$$

where

$$\mathcal{G}_{p,q}^{m,n}(s) := \frac{\prod_{j=1}^{m} \Gamma(b_j+s) \prod_{i=1}^{n} \Gamma(1-a_i-s)}{\prod_{i=n+1}^{p} \Gamma(a_i+s) \prod_{j=m+1}^{q} \Gamma(1-b_j-s)}$$

and $L$ is a suitable contour (see [19, p. 313]).

**Definition 2.7.1** ([19, p. 11]). *Let $m \geq 1$ be an integer, $\beta > 0$, $\gamma_1, \cdots, \gamma_m$ and $\delta_1 > 0, \cdots, \delta_m > 0$ be arbitrary real numbers. Consider the set $\gamma = (\gamma_1, \cdots, \gamma_m)$ as a multiweight and $\delta = (\delta_1, \cdots, \delta_m)$ as a positive multiorder of integration. The multiple*

*Erdélyi-Kober fractional integrals* $I_{\beta,m}^{(\gamma_k),(\delta_k)} f$ *and* $W_{\beta,m}^{(\gamma_k),(\delta_k)} f$ *are defined by:*

$$I_{\beta,m}^{(\gamma_k),(\delta_k)} f(x) = \int_0^1 G_{m,m}^{m,0}\left[\sigma \left| \begin{matrix} (\gamma_k+\delta_k)_1^m \\ (\gamma_k)_1^m \end{matrix} \right.\right] f(x\sigma^{1/\beta})\mathrm{d}\sigma \tag{2.7.1}$$

*and*

$$W_{\beta,m}^{(\gamma_k),(\delta_k)} f(x) = \int_1^\infty G_{m,m}^{m,0}\left[\frac{1}{\sigma} \left| \begin{matrix} (\gamma_k+\delta_k+1)_1^m \\ (\gamma_k+1)_1^m \end{matrix} \right.\right] f(x\sigma^{1/\beta})\mathrm{d}\sigma, \tag{2.7.2}$$

*respectively.*

Recall that [19, p. 18, Eq. (1.1.18)]

$$G_{2,2}^{2,0}\left[\sigma \left| \begin{matrix} \gamma_1+\delta_1, \gamma_2+\delta_2 \\ \gamma_1, \gamma_2 \end{matrix} \right.\right] = \begin{cases} \dfrac{\sigma^{\gamma_2}(1-\sigma)^{\delta_1+\delta_2-1}}{\Gamma(\delta_1+\delta_2)} {}_2F_1\left[\begin{matrix} \gamma_2+\delta_2-\gamma_1, \delta_1 \\ \delta_1+\delta_2 \end{matrix}; 1-\sigma\right], & \sigma < 1, \\ 0, & \sigma > 1. \end{cases}$$

So by comparing (2.7.1) and (2.7.2) with (2.6.0.3) and (2.6.0.4), respectively, we obtain the following relations:

$$I_{0+}^{\alpha,\beta,\eta} = x^{-\beta} I_{1,2}^{(\eta-\beta,0),(-\eta,\alpha+\eta)} \tag{2.7.3}$$

and

$$J_{-}^{\alpha,\beta,\eta} = x^{-\beta} W_{1,2}^{(\eta,\beta),(-\eta,\alpha+\eta)}. \tag{2.7.4}$$

### *2.7.1 Main Results*

The generalized fractional integration by parts for the (multiple) Erdélyi-Kober fractional integrals can be found in [19, p. 40, Lemma 1.3.4]. However, the result in [19] is not quite applicable to our current situation. So let us derive some slightly different versions of the generalized fractional integration by parts. First, we have

$$\begin{aligned}
&\int_0^\infty g(x) I_{\beta,m}^{(\gamma_k),(\delta_k)} f(x)\mathrm{d}x \\
&= \int_0^\infty \int_0^1 G_{m,m}^{m,0}\left[\sigma \left| \begin{matrix} (\gamma_k+\delta_k)_1^m \\ (\gamma_k)_1^m \end{matrix}\right.\right] f(x\sigma^{1/\beta}) g(x)\mathrm{d}\sigma\mathrm{d}x \\
&\overset{x\to x\sigma^{-1/\beta}}{=\!=\!=} \int_0^\infty \int_0^1 G_{m,m}^{m,0}\left[\sigma \left| \begin{matrix} (\gamma_k+\delta_k)_1^m \\ (\gamma_k)_1^m \end{matrix}\right.\right] f(x) g(x\sigma^{-1/\beta})\sigma^{-1/\beta}\mathrm{d}\sigma\mathrm{d}x \\
&= \int_0^\infty f(x)\left[\int_0^1 G_{m,m}^{m,0}\left[\sigma \left| \begin{matrix} (\gamma_k+\delta_k)_1^m \\ (\gamma_k)_1^m \end{matrix}\right.\right] g(x\sigma^{-1/\beta})\sigma^{-1/\beta}\mathrm{d}\sigma\right]\mathrm{d}x \\
&\overset{\sigma\to\sigma^{-1}}{=\!=\!=} \int_0^\infty f(x)\left[\int_1^\infty G_{m,m}^{m,0}\left[\frac{1}{\sigma} \left| \begin{matrix} (\gamma_k+\delta_k)_1^m \\ (\gamma_k)_1^m \end{matrix}\right.\right] g(x\sigma^{1/\beta})\sigma^{-2+1/\beta}\mathrm{d}\sigma\right]\mathrm{d}x.
\end{aligned}$$

Using the identity [19, p. 315, Eq. (A.14)]

$$z^\sigma G_{p,q}^{m,n}\left[z \left| \begin{matrix} (a_j) \\ (b_k) \end{matrix}\right.\right] = G_{p,q}^{m,n}\left[z \left| \begin{matrix} (a_j+\sigma) \\ (b_k+\sigma) \end{matrix}\right.\right] \quad (\sigma \in \mathbb{C}),$$

we have

$$\int_0^\infty g(x) I_{\beta,m}^{(\gamma_k),(\delta_k)} f(x)\mathrm{d}x = \int_0^\infty f(x) W_{\beta,m}^{(\gamma_k-1/\beta+1),(\delta_k)} g(x)\mathrm{d}x, \tag{2.7.5}$$

provided that each integral exists.

By letting $\gamma_k \to \gamma_k + 1/\beta - 1$ in (2.7.5) and then setting $f \leftrightarrow g$ in the resulting equation, we get

$$\int_0^\infty g(x) W_{\beta,m}^{(\gamma_k),(\delta_k)} f(x)\mathrm{d}x = \int_0^\infty f(x) I_{\beta,m}^{(\gamma_k+1/\beta-1),(\delta_k)} g(x)\mathrm{d}x, \tag{2.7.6}$$

provided that each integral exists.

In addition, it is worth mentioning that the above derivation can also be used to find the adjoint operators of $I_{\beta,m}^{(\gamma_k),(\delta_k)}$ and $W_{\beta,m}^{(\gamma_k),(\delta_k)}$ (see [16, p. 668]).

In order to express our result, we need Fox's $H$-function which is defined by:

$$H_{p,q}^{m,n}\left[\sigma \left| \begin{matrix} (a_i, \alpha_i)_{1,p} \\ (b_j, \beta_j)_{1,q} \end{matrix} \right.\right] = \frac{1}{2\pi \mathrm{i}} \int_L \mathcal{H}_{p,q}^{m,n}(s)\sigma^{-s}\mathrm{d}s, \tag{2.7.7}$$

where

$$\mathcal{H}_{p,q}^{m,n}(s) := \frac{\prod_{j=1}^{m} \Gamma(b_j + \beta_j s) \prod_{i=1}^{n} \Gamma(1 - a_i - \alpha_i s)}{\prod_{i=n+1}^{p} \Gamma(a_i + \alpha_i s) \prod_{j=m+1}^{q} \Gamma(1 - b_j - \beta_j s)},$$

and $L$ is a suitable contour (see [19, p. 343]). When $\alpha_i = 1$ $(i = 1, \cdots, p)$ and $\beta_j = 1$ $(j = 1, \cdots, p)$, the $H$-function reduces to the $G$-function.

**Theorem 2.7.1.** *Let $\lambda \in \mathbb{C}$ and $x^\lambda I_{\beta,m}^{(\gamma_k),(\delta_k)} f \in L_\sigma$. We have*

$$(\mathcal{L}x^\lambda I_{\beta,m}^{(\gamma_k),(\delta_k)} f)(s)$$
$$= \int_0^\infty x^\lambda H_{m,m+1}^{m+1,0}\left[ sx \middle| \begin{matrix} (\gamma_k + \delta_k - \frac{1+\lambda}{\beta} + 1, \frac{1}{\beta})_1^m \\ (0,1), (\gamma_k - \frac{1+\lambda}{\beta} + 1, \frac{1}{\beta})_1^m \end{matrix} \right] f(x)\mathrm{d}x, \tag{2.7.8}$$

*where the H-function is defined by* (2.7.7).

*Proof.* Letting $g(x) = \mathrm{e}^{-sx}x^\lambda$ in (2.7.5), we obtain

$$(\mathcal{L}x^\lambda I_{\beta,m}^{(\gamma_k),(\delta_k)} f)(s) = \int_0^\infty f(x) W_{\beta,m}^{(\gamma_k - 1/\beta + 1),(\delta_k)} \mathrm{e}^{-sx} x^\lambda \mathrm{d}x.$$

As in the proof of Theorem 2.6.1, we replace $\mathrm{e}^{-sx}$ with its Mellin-Barnes contour integral representation to obtain

$$W_{\beta,m}^{(\gamma_k - 1/\beta + 1),(\delta_k)} \mathrm{e}^{-sx} x^\lambda = \frac{1}{2\pi \mathrm{i}} \int_L \Gamma(z) s^{-z} W_{\beta,m}^{(\gamma_k - 1/\beta + 1),(\delta_k)} x^{\lambda - z} \mathrm{d}z.$$

Recall that [19, p. 46, Lemma 1.4.1]

$$W_{\beta,m}^{(\gamma_k),(\delta_k)} x^q = d_q x^q, \text{ where } d_q := \prod_{k=1}^m \frac{\Gamma(\gamma_k - \frac{q}{\beta})}{\Gamma(\gamma_k + \delta_k - \frac{q}{\beta})}.$$

We have

$$W_{\beta,m}^{(\gamma_k - 1/\beta + 1),(\delta_k)} \mathrm{e}^{-sx} x^\lambda$$
$$= \frac{x^\lambda}{2\pi \mathrm{i}} \int_L \Gamma(z) \prod_{k=1}^m \frac{\Gamma(\gamma_k - \frac{1+\lambda}{\beta} + 1 + \frac{z}{\beta})}{\Gamma(\gamma_k + \delta_k - \frac{1+\lambda}{\beta} + 1 + \frac{z}{\beta})} (sx)^{-z} \mathrm{d}z.$$

The result (2.7.8) follows now immediately by interpreting the contour integral in terms of the $H$-function.

It may be pointed out that Theorem 2.7.1 contains Theorem 2.6.1 as a special case. To see this, we first let $\beta = 1$ and $m = 2$ in (2.7.8) to get

$$(\mathcal{L}x^{\lambda} I_{1,2}^{(\gamma_1,\gamma_2),(\delta_1,\delta_2)} f)(s) = \int_0^{\infty} x^{\lambda} G_{2,3}^{3,0}\left[sx \,\middle|\, \begin{matrix} \gamma_1+\delta_1-\lambda, \gamma_2+\delta_2-\lambda \\ 0, \gamma_1-\lambda, \gamma_2-\lambda \end{matrix}\right] f(x)\mathrm{d}x. \quad (2.7.9)$$

Then, setting $(\gamma_1, \gamma_2) = (\eta-\beta, 0)$ and $(\delta_1, \delta_2) = (-\eta, \alpha+\eta)$ in (2.7.9) and taking (2.7.3) into account, we obtain

$$(\mathcal{L}x^{\lambda+\beta} I_{0+}^{\alpha,\beta,\eta} f)(s) = \int_0^{\infty} x^{\lambda} G_{2,3}^{3,0}\left[sx \,\middle|\, \begin{matrix} -\beta-\lambda, \alpha+\eta-\lambda \\ 0, \eta-\beta-\lambda, -\lambda \end{matrix}\right] f(x)\mathrm{d}x. \quad (2.7.10)$$

Finally, by letting $\lambda \to \lambda - \beta$ in (2.7.10), we find that

$$(\mathcal{L}x^{\lambda} I_{0+}^{\alpha,\beta,\eta} f)(s) = \int_0^{\infty} x^{\lambda-\beta} G_{2,3}^{3,0}\left[sx \,\middle|\, \begin{matrix} -\lambda, \alpha+\eta-\lambda+\beta \\ 0, \eta-\lambda, \beta-\lambda \end{matrix}\right] f(x)\mathrm{d}x. \quad (2.7.11)$$

Using an identity connecting the $G$-function with the generalized hypergeometric functions (see, for example, [25, p. 629, Eq. (56)]), it is not difficult to see that

$$x^{\lambda-\beta} G_{2,3}^{3,0}\left[sx \,\middle|\, \begin{matrix} -\lambda, \alpha+\eta-\lambda+\beta \\ 0, \eta-\lambda, \beta-\lambda \end{matrix}\right] = S(\alpha, \beta, \eta, \lambda; s, x),$$

where $S(\alpha, \beta, \eta, \lambda; s, x)$ is defined by (2.6.1.5).

**Theorem 2.7.2.** *Let $\mu \in \mathbb{C}$ and $x^\mu W_{\beta,m}^{(\gamma_k),(\delta_k)} f \in L_\sigma$. We have*

$$(\mathcal{L}x^\mu W_{\beta,m}^{(\gamma_k),(\delta_k)} f)(s) = \int_0^\infty x^\mu {}_m\Psi_m \left[ \begin{matrix} \left(\gamma_k + \frac{\mu+1}{\beta}, \frac{1}{\beta}\right)_{1,m} \\ \left(\gamma_k + \delta_k + \frac{\mu+1}{\beta}, \frac{1}{\beta}\right)_{1,m} \end{matrix} ; -sx \right] f(x)\mathrm{d}x, \tag{2.7.12}$$

*where ${}_p\Psi_q$ denotes the Fox-Wright generalized hypergeometric function defined by:*

$${}_p\Psi_q \left[ \begin{matrix} (a_i, \alpha_i)_{1,p} \\ (b_j, \beta_j)_{1,q} \end{matrix} ; z \right] := \sum_{n=0}^\infty \frac{\prod_{i=1}^p \Gamma(a_i + \alpha_i n)}{\prod_{j=1}^q \Gamma(b_j + \beta_j n)} \frac{z^n}{n!}.$$

*Proof.* Letting $g(x) = \mathrm{e}^{-sx} x^\mu$ in (2.7.6), we obtain

$$(\mathcal{L}x^\mu W_{\beta,m}^{(\gamma_k),(\delta_k)} f)(s) = \int_0^\infty f(x) I_{\beta,m}^{(\gamma_k+1/\beta-1),(\delta_k)} \mathrm{e}^{-sx} x^\mu \mathrm{d}x.$$

Recall that [19, p. 22, Lemma 1.2.1]

$$I_{\beta,m}^{(\gamma_k),(\delta_k)} x^p = c_p x^p, \text{ where } c_p := \prod_{k=1}^m \frac{\Gamma\left(\gamma_k + \frac{p}{\beta} + 1\right)}{\Gamma\left(\gamma_k + \delta_k + \frac{p}{\beta} + 1\right)}.$$

So we have

$$\begin{aligned} I_{\beta,m}^{(\gamma_k+1/\beta-1),(\delta_k)} \mathrm{e}^{-sx} x^\mu &= \sum_{n=0}^\infty \frac{(-s)^n}{n!} I_{\beta,m}^{(\gamma_k+1/\beta-1),(\delta_k)} x^{\mu+n} \\ &= x^\mu \sum_{n=0}^\infty \frac{(-sx)^n}{n!} \prod_{k=1}^m \frac{\Gamma\left(\gamma_k + \frac{\mu+1}{\beta} + \frac{n}{\beta}\right)}{\Gamma\left(\gamma_k + \delta_k + \frac{\mu+1}{\beta} + \frac{n}{\beta}\right)}. \end{aligned}$$

By interpreting the series as the Fox-Wright generalized hypergeometric function, we obtain the result (2.7.12).

We also point out that Theorem 2.7.2 contains Theorem 2.6.2 as a special case. Letting $\beta = 1$ and $m = 2$ in (2.7.12) we have

$$
\begin{aligned}
&(\mathcal{L}x^{\mu}W_{1,2}^{(\gamma_1,\gamma_2),(\delta_1,\delta_2)}f)(s)\\
&\quad = \int_0^{\infty} x^{\mu}{}_2\Psi_2\left[\begin{matrix}(\gamma_1+\mu+1,1),(\gamma_2+\mu+1,1)\\(\gamma_1+\delta_1+\mu+1,1),(\gamma_2+\delta_2+\mu+1,1)\end{matrix};-sx\right]f(x)\mathrm{d}x.
\end{aligned}
\tag{2.7.13}
$$

Then, setting $(\gamma_1,\gamma_2)=(\eta,\beta)$ and $(\delta_1,\delta_2)=(-\eta,\alpha+\eta)$ in (2.7.13) and taking (2.7.4) into account, we obtain

$$
\begin{aligned}
&(\mathcal{L}x^{\mu+\beta}I_{-}^{\alpha,\beta,\eta}f)(s)\\
&\quad = \int_0^{\infty} x^{\mu}{}_2\Psi_2\left[\begin{matrix}(\eta+\mu+1,1),(\beta+\mu+1,1)\\(\mu+1,1),(\alpha+\beta+\eta+\mu+1,1)\end{matrix};-sx\right]f(x)\mathrm{d}x.
\end{aligned}
\tag{2.7.14}
$$

Finally, by letting $\mu \to \mu-\beta$ in (2.7.14) we find that

$$
\begin{aligned}
&(\mathcal{L}x^{\mu}I_{-}^{\alpha,\beta,\eta}f)(s)\\
&\quad = \int_0^{\infty} x^{\mu-\beta}{}_2\Psi_2\left[\begin{matrix}(\eta+\mu-\beta+1,1),(\mu+1,1)\\(\mu-\beta+1,1),(\alpha+\eta+\mu+1,1)\end{matrix};-sx\right]f(x)\mathrm{d}x\\
&\quad = \int_0^{\infty} T(\alpha,\beta,\eta,\mu;s,x)f(x)\mathrm{d}x,
\end{aligned}
$$

where $T(\alpha,\beta,\eta,\mu;s,x)$ is defined in (2.6.1.9).

### *2.7.2 Remarks*

(i) As generalizations of (2.7.1) and (2.7.2), Galué et al. ([7], and [8]) have introduced the multiple Erdélyi-Kober fractional integral operators involving Fox's $H$-function (see [44, p. 20]). However, results analog to Theorems 2.7.1 and 2.7.2 for these fractional integrals have not so far been established. It may be interesting, therefore, to investigate such results.

(ii) Let $I^\alpha$ be an arbitrary fractional integral operator. When $I^\alpha$ is the Liouville fractional integral (see Definition 2.2.1) or a type of a Prabhakar fractional integral (see Definition 2.5.1), we have a relation of the form

$$(\mathcal{L}I^\alpha f)(s) = K(s)\mathcal{L}[f](s), \tag{2.7.15}$$

where the function $K(s)$ is usually a function in $s$ which corresponds to the special function involving various parameters in the kernel defining the given fractional integral operator. However, for the Saigo and the multiple Erdélyi-Kober fractional integrals (see (2.7.8) and (2.7.11)), the relationship (2.7.15) is apparently quite different. We may point out here a special case for which we can obtain the result similar to (2.7.15). The hyper-Bessel integral operator $L$ is a special case of the multiple Erdélyi-Kober fractional integral. Compared to the Laplace transformation, the *Obrechkoff transformation* $\mathcal{O}$ is more suitable when building the Operational Calculus for the hyper-Bessel integral operator (see [19, p. 178]). In fact, we have [19, p. 180, Eq. (3.9.27)]

$$(\mathcal{O}Lf)(s) = \frac{1}{\beta^m s^\beta}\mathcal{O}[f](s),$$

which is an analog of relation (2.7.15).

## 2.8 Multidimensional Fractional Integrals

There are several useful multidimensional fractional integrals in the literature (see, for example, [2], [18], [32], [44]). Here we consider only one of them and leave the others to the interested reader.

**Definition 2.8.1** ([2, p. 244]). *Let* $\mathrm{Re}(\boldsymbol{\alpha}) > \mathbf{0}$ *and* $\mathbf{x} \in \mathbb{R}^n_+$. *Let* $f : \mathbb{R}^n_+ \to \mathbb{C}$. *The multidimensional Liouville fractional integral* $\mathbf{I}^{\boldsymbol{\alpha}}_{\mathbf{0}+} f$ *is defined by:*

$$(\mathbf{I}^{\boldsymbol{\alpha}}_{\mathbf{0}+} f)(\mathbf{x}) := \frac{1}{\Gamma(\boldsymbol{\alpha})} \int_{\mathbb{R}^n_+(\mathbf{x})} (\mathbf{x} - \mathbf{t})^{\boldsymbol{\alpha}-\mathbf{1}} f(\mathbf{t}) d\mathbf{t}. \tag{2.8.0.1}$$

### *2.8.1 The Main Result*

The following result is well-known and particularly useful (see [2, p. 120] and [43, p. 514]).

**Theorem 2.8.1** ([2, p. 120]). *Let* $\mathrm{Re}(\boldsymbol{\alpha}) > \mathbf{0}$ *and* $f \in \mathbb{E}_{\mathbf{c}}$. *Then*

$$(\mathcal{L}_n \mathbf{I}^{\boldsymbol{\alpha}}_{\mathbf{0}+} f)(\mathbf{s}) = \mathbf{s}^{-\boldsymbol{\alpha}} \mathcal{L}_n[f](\mathbf{s}), \ \ \mathrm{Re}(\mathbf{s}) \geq \mathbf{c}. \tag{2.8.1.2}$$

*Proof.* Observe that

$$(\mathbf{I}^{\boldsymbol{\alpha}}_{\mathbf{0}+} f)(\mathbf{t}) = \frac{\mathbf{t}^{\boldsymbol{\alpha}-\mathbf{1}}}{\Gamma(\boldsymbol{\alpha})} * f(\mathbf{t}).$$

Using Theorem 2.1.5, we get the desired result.

### *2.8.2 Applications*

**Example 2.8.1.** Let $f, g \in \mathbb{E}_{\mathbf{c}}$ $(\mathbf{c} < \mathbf{1})$. Consider the multidimensional integral equation

$$g(\mathbf{x}) = \int_{\mathbb{R}^n} (\mathbf{I}_{0+}^{\boldsymbol{\mu}} f)(\mathbf{x}) h(\boldsymbol{\mu}) \mathrm{d}\boldsymbol{\mu}, \tag{2.8.2.3}$$

where $h$ is an unknown function and $\mathbf{I}_{0+}^{\boldsymbol{\mu}}$ is defined by (2.8.0.1). The equation (2.8.2.3), as a multidimensional generalization of Wainwright's equation [49] and was proposed and solved (formally) by Srivastava and Raina [43]. Applying $\mathcal{L}_n$ to (2.8.2.3) and using (2.8.1.2), we find that

$$\int_{\mathbb{R}^n} \mathbf{s}^{-\boldsymbol{\mu}} h(\boldsymbol{\mu}) \mathrm{d}\boldsymbol{\mu} = \frac{(\mathcal{L}_n g)(\mathbf{s})}{(\mathcal{L}_n f)(\mathbf{s})}, \tag{2.8.2.4}$$

provided that $(\mathcal{L}_n f)(\mathbf{s}) \neq 0$ and the integral converges absolutely. Setting $\mathbf{s} = \mathrm{e}^{\mathrm{i}\boldsymbol{\omega}}$ $(\boldsymbol{\omega} \in \mathbb{R}^n)$ in (2.8.2.4), we get

$$\int_{\mathbb{R}^n} \mathrm{e}^{-\mathrm{i}\boldsymbol{\omega}\boldsymbol{\mu}} h(\boldsymbol{\mu}) \mathrm{d}\boldsymbol{\mu} = \frac{(\mathcal{L}_n g)(\mathrm{e}^{\mathrm{i}\boldsymbol{\omega}})}{(\mathcal{L}_n f)(\mathrm{e}^{\mathrm{i}\boldsymbol{\omega}})}.$$

By using the multidimensional Fourier inversion formula [43, p. 513, Eq. (1.5)], we formally obtain

$$h(\boldsymbol{\mu}) = \frac{1}{(2\pi)^n} \int_{\mathbb{R}^n} \mathrm{e}^{\mathrm{i}\boldsymbol{\omega}\boldsymbol{\mu}} \frac{(\mathcal{L}_n g)(\mathrm{e}^{\mathrm{i}\boldsymbol{\omega}})}{(\mathcal{L}_n f)(\mathrm{e}^{\mathrm{i}\boldsymbol{\omega}})} \mathrm{d}\boldsymbol{\omega}.$$

It may also be mentioned here that an illustrative example relating to the classical Laguerre function is also given in [43].

---

## Acknowledgment

The work of the first author is supported by the National Natural Science Foundation of China (No. 12001095).

# *References*


[1] Beghin, L., & Orsingher, E., (2010). Poisson-type processes governed by fractional and higher-order recursive differential equations. *Electron. J. Probab. 15, Paper No. 22*, 684–709.

[2] Brychkov, Yu. A., Glaeske, H.-J., Prudnikov, A. P., & Tuan, V. K., (1992). *Multidimensional integral transformations.* Gordon and Breach, New York.

[3] Caputo, M., (1967). Linear models of dissipation whose $q$ is almost frequency independent - II. *Geophys. J. R. Astr. Soc.*, 529–539.

[4] Conlan, J., & Koh, E. L., (1975). A fractional differentiation theorem for the Laplace transform. *Can. Math. Bull. 18,* 605–606.

[5] Doetsch, G., (1943). *The Laplace Transformation.* New York.

[6] Eshaghi, S., Ansari, A., Ghaziani, R.K., & Darani, M. A., (2017). Fractional Black-Scholes model with regularized Prabhakar derivative. *Publ. Inst. Math., Nouv. Sér. 102*(116), 121–132.

[7] Galué, L., Kalla, S. L., & Srivastava, H. M., (1993). Further results on an $H$-function generalized fractional calculus. *J. Fractional Calculus. 4*, 89–102.

[8] Galué, L., Kiryakova, V. S., & Kalla, S. L., (1993). Solution of dual integral equations by fractional calculus. *Math. Balk., New Ser. 7*(1), 53–72.

[9] Giusti, A., (2018). A comment on some new definitions of fractional derivative. *Nonlinear Dyn. 93*(3), 1757–1763.

[10] Giusti, A., (2020). General fractional calculus and Prabhakar's theory. *Commun. Nonlinear Sci. Numer. Simul. 83, Article ID 105114*, 7p.

[11] Giusti, A., Colombaro, I., Garra, R., Garrappa, R., Polito, F., Popolizio, M., & Mainardi, F., (2020). A practical guide to Prabhakar fractional calculus. *Fract. Calc. Appl. Anal. 23*(1), 9–54.

[12] Gorenflo, R., Kilbas, A. A., Mainardi, F., & Rogosin, S., (2014). *Mittag-Leffler Functions, Related Topics and Applications.* Springer Monographs in Mathematics, Springer, Heidelberg.

[13] Hansen, E. W., (2014). *Fourier transforms. Principles and applications.* Hoboken, NJ: John Wiley & Sons (ISBN 978-1-118-47914-8/hbk; 978-1-118-90179-3/ebook). xv, 755p.

[14] Hilfer, R., (2000). *Application of Fractional Calculus in Physics.* World Scientific, Singapore.

[15] Hilfer, R., Luchko, Y., & Tomovski, Ž., (2009). Operational method for the solution of fractional differential equations with generalized Riemann-Liouville fractional derivatives. *Fract. Calc. Appl. Anal. 12*(3), 299–318.

[16] Ho, K.-P., (2020). Erdélyi-Kober fractional integrals on Hardy space and BMO. *Proyecciones 39*(3), 663–677.

[17] Kilbas, A. A., Saigo, M., & Saxena, R. K., (2004). Generalized Mittag-Leffler function and generalized fractional calculus operators. *Integr. Transf. Spec. Funct. 15*(1), 31–49.

[18] Kilbas, A. A., Srivastava, H. M., & Trujillo, J. J., (2006). *Theory and Applications of Fractional Differential Equations.* North-Holland Mathematics Studies, vol. 204, Elsevier Science: B.V., Amsterdam.

[19] Kiryakova, V., (1994). *Generalized fractional calculus and applications.* Pitman Res. Notes Math. Ser., vol. 301, Longman/Wiley, New York.

[20] Koh, E. L., & Conlan, J., (1976). Fractional derivatives Laplace transforms and association of variables. *Int. J. Systems Sci. 7*, 591–596.

[21] Koumandos, S., & Pedersen, H. L., (2009). Completely monotonic functions of positive order and asymptotic expansions of the logarithm of Barnes double gamma function and Euler's gamma function. *J. Math. Anal. Appl. 355*, 33–40.

[22] Krukowski, M., (2020). Characterizing compact families via the Laplace transform. *Ann. Acad. Sci. Fenn., Math. 45*(2), 991–1002.

[23] Luo, M.-J., & Raina, R. K., (2014). On certain classes of fractional kinetic equations. *Filomat, 28*(10), 2077–2090.

[24] Luo, M.-J., & Raina, R. K., (2017). Fractional integral operators characterized by some new hypergeometric summation formulas. *Frac. Calc. Appl. Anal. 20*(2), 422–446.

[25] Luo, M.-J., & Raina, R. K., (2019). The decompositional structure of certain fractional integral operators. *Hokkaido Math. J. 48*, 611–650.

[26] Luo, M.-J., & Raina, R. K., (2022). On the composition structures of certain fractional integral operators. *Symmetry 14*(9), 1845.

[27] Miller, K. S., & Samko, S. G., (2001). Completely monotonic functions. *Integral Transforms Spec. Funct. 12*(4), 389–402.

[28] Myers, D. E., (1972). Extensions of Laplace transform spaces and their topologies. *Ann. Pol. Math. 25*, 265–272.

[29] Podlubny, I., (1999). *Fractional differential equations. An introduction to fractional derivatives, fractional differential equations, to methods of their solution and some of their applications.* Academic Press, San Diego.

[30] Prabhakar, T. R., (1971). A singular integral equation with a generalized Mittag-Leffler function in the kernel. *Yokohama Math. J. 19* (1), 7–15.

[31] Raina, R. K., (1991). A note on the multidimensional Weyl fractional operator. *Proc. Indian Acad. Sci., Math. Sci. 101*(3), 179–181.

[32] Raina, R. K., (2006). On weighted inequalities for certain fractional integral operators. *Int. J. Math. Math. Sci. 2006*(13), Article ID 79196.

[33] Raina, R. K., (2005). On generalized Wright's hypergeometric functions and fractional calculus operators. *East Asian Math. J., 21*(2), 191–203.

[34] Raina, R. K., & Koul, C. L., (1979). On Weyl fractional calculus. *Proc. Am. Math. Soc. 73*, 188–192.

[35] Rao, A., Garg, M., & Kalla, S., (2010). Caputo-type fractional derivative of a hypergeometric integral operator. *Kuwait J. Sci. Eng. 37*(1A), 15–29.

[36] Saigo, M., (1978). A remark on integral operators involving the Gauss hypergeometric functions. *Math. Rep. Coll. Gen. Educ. Kyushu Univ. 11*, 135–143.

[37] Saigo, M., (1980). On the Hölder continuity of the generalized fractional integrals and derivatives. *Math. Rep. Coll. Gen. Educ. Kyushu Univ.* 12, 55–62.

[38] Saigo, M., & Kilbas, A. A., (1999). Generalized Fractional Calculus of the $H$-Function. *Fukuoka Univ. Sci. Rep. 29*(1), 31–45.

[39] Saigo, M., Saxena, R. K., & Ram, J., (1992). Certain properties of operators of fractional integration associated with Mellin and Laplace transformations, in: Srivastava, H.M., & Owa S. (Eds.), *Current Topics in Analytic Function Theory*, World Scientific, Singapore, 291–304.

[40] Saxena, R. K., & Kalla, S. L., (2005). Solution of Volterra-type integro-differential equations with a generalized Lauricella confluent hypergeometric function in the kernels. *Int. J. Math. Math. Sci. 2005*(8), 1155–1170.

[41] Srivastava, H. M., (2017). Remarks on some families of fractional-order differential equations. *Integral Transforms Spec. Funct. 28*(7), 560–564.

[42] Srivastava, H. M., Saigo, M., & Raina, R. K., (1993). Some existence and connection theorems associated with the Laplace transform and a certain class of integral operators. *J. Math. Anal. Appl. 172*(1), 1–10.

[43] Srivastava, H. M., & Raina, R. K., (1989). Solution of a certain multiple integral equation. *J. Math. Anal. Appl. 143*(2), 512–516.

[44] Srivastava, H. M., & Saxena, R. K., (2001). Operators of fractional integration and their applications. *Appl. Math. Comput. 118*(1), 1–52.

[45] Srivastava, H. M., & Saxena, R. K., (2005). Some Volterra-type fractional integro-differential equations with a multivariable confluent hypergeometric function in their kernel. *J. Integral Equations Appl. 17*(2), 199–217.

[46] Tomovski, Ž., (2012). Generalized Cauchy type problems for nonlinear fractional differential equations with composite fractional derivative operator. *Nonlinear Analysis: Theory, Methods and Applications 75*, 3364–3384.

[47] Tomovski, Ž., Hilfer, R., & Srivastava, H.M., (2010). Fractional and operational calculus with generalized fractional derivative operators and Mittag-Leffler type functions. *Integral Transf. Spec. Funct. 21*(11-12), 797–814.

[48] Vojta, V., (2014). Fractional calculus and Lambert function I. Liouville-Weyl fractional integral. *Acta Polytechnica 54*(4), 305–319.

[49] Wainwright, W. L., (1975). A functional relation, in "*Fractional Calculus and Its Applications*" (B. Ross, Ed.), pp. 298–305, Springer-Verlag, Berlin/Heidelberg/New York.

[50] Widder, D. V., (1941). *The Laplace Transform.* Princeton University Press, New York.

[51] Widder, D. V., (1971). *An Introduction to Transform Theory.* Academic Press, New York.

# Chapter 3

## *Estimating the Order of Derivation in a Fractional Differential Equation*

**Silvia Seminara,[1] María Inés Troparevsky,[1] and Marcela Fabio[2]**

[1]*Department of Mathematics, Faculty of Engineering, University of Buenos Aires, Argentina*

[2]*ITECA, UNSAM-CONICET, ECyT, CEDEMA, Buenos Aires, Argentina*

**Abstract**

Fractional derivatives have been the subject of great interest and extensive research in recent decades, as they provide useful models for numerous physical problems from diverse areas of application. When a fractional model is proposed, the order of the fractional derivative can be considered as a parameter to be determined and should be chosen to best fit experimental data. In this work, we propose a numerical scheme to estimate, from experimental measurements, the order of a general nonlinear fractional model described by the Caputo derivative. We take advantage of the equivalent integral formulation of the problem and propose an estimation procedure based on an approximate solution to it. We show several examples of very good performance.

## Introduction

Fractional derivatives are integral operators with different properties depending on their kernels, but they share the same characteristic that deeply differentiates them from natural order derivatives: they have *memory*. This non-locality property makes them suitable to describe models where not only local information is required to properly explain the underlying phenomena.

In the last decades, models described by fractional differential equations have been used in different areas of science: chemistry, fluid mechanics, materials engineering, biology, economics, etc. Applications of fractional calculus in diverse areas are cited, for example, in [2], [3], and [27]. Numerous results concerning solutions to this type of equation, involving Riemann-Liouville, Caputo, Caputo-Fabrizio, Atangana-Baleanu and other fractional derivatives were stated (see [16], [21], [23], and [24] for a traditional bibliography) and different explicit and numerical solutions to them were developed (see [12], [14], [18], [20], and [22]).

In general, the order of derivation in a given fractional model is a parameter to be determined and it is an important issue to be explored. Actually, accuracy in the fractional order can improve our understanding of the system behavior and the possibility to properly use it to model real physical problems. The importance of the identification of the fractional order is reflected in numerous recent publications referring to this topic. For example, theoretical results concerning the existence and uniqueness of the fractional order are stated in [5] and [15] for a 1D parabolic equation, and identification in the case of transport processes in heterogeneous materials appears in [28].

In real-world situations, only measured data of a system and its response are available to determine the fractional order of the differential equations that describe the model. Consequently, a procedure to determine the value that best fits the experimental data is highly desirable. The problem of determining the orders of the fractional derivatives from real data in anomalous diffusion models, where materials exhibit sub-diffusion or super-diffusion properties, is studied in [6], [11], [13], and [19]. The inverse boundary value problem for determining the fractional order of non-autonomous semi-linear sub-diffusion equations with memory terms from short-time observations is analyzed in [17]. In [8], the identification of the fractional order for a certain class of systems based on nonlinear function optimization techniques is proposed. Fractional-order system identification using genetic algorithms is discussed in [29]. Regarding the sensitivity of solutions to fractional models with respect to the order of derivation, there are recent works exploring this problem (see, for example, [10] and [15]).

We have made a first approach to the problem in [26], where we identified the order of derivation for the case of fractional linear systems. In the present work, we propose a numerical scheme to estimate, from experimental measurements, the order of a general nonlinear fractional model described by the Caputo derivative. We take advantage of the known equivalent integral formulation of the problem and propose an estimation procedure based on an approximate solution to it. Simulating experimental data, we show several examples of very good performance to illustrate the method.

## 3.1 The Problem

We want to determine the order of derivation, $\alpha$, in a fractional initial value problem (FIVP):

$$\begin{cases} D^\alpha y(t) & = g(t,y), \quad t \in [0,T] \\ y(0) & = y_0 \end{cases} \tag{3.1.1}$$

which best fits an experimental situation. Here $D^\alpha$ denotes the Caputo fractional derivative of order $\alpha \in (0,1)$,

$$D^\alpha y(t) = \frac{1}{\Gamma(1-\alpha)} \int_0^t y'(\tau)(t-\tau)^{-\alpha} d\tau, \ \ t \geq 0,$$

$y(t)$ is the system response, $g(t,y)$ is a known external source and $y_0$ is the known initial condition. In order to apply the Caputo operator to $y(t)$ we suppose that $y(t)$ is an absolute continuous function defined on $[0,T], T > 0$. Note that $D^\alpha y(t)$ can be expressed as a convolution,

$$D^\alpha y(t) = (y' * m_\alpha)(t),$$

where

$$m_\alpha(t) = \begin{cases} 0 & t < 0 \\ \frac{t^{-\alpha}}{\Gamma(1-\alpha)} & t \geq 0. \end{cases}$$

**Remark 3.1.1.** *The Riemann-Liouville's integral operator*

$$I^\alpha y(t) = \frac{1}{\Gamma(\alpha)} \int_0^t y(\tau)(t-\tau)^{\alpha-1} d\tau, \ \ t \geq 0,$$

*defined for $y(t)$ integrable in $[0,T]$, satisfies*

$$I^{\alpha}D^{\alpha}y = y$$

*if $y(0) = 0$, and*

$$D^{\alpha}I^{\alpha}y = y$$

*if $y(t)$ is continuous in $[0,T]$ (see Theorem 3.8 and Theorem 3.7 in [7]).*

### 3.1.1 Uniqueness of $\alpha$

Suppose there are two possible fractional orders $\alpha, \beta \in (0,1)$ satisfying

$$\begin{cases} D^{\alpha}y(t) &= g(t,y), \quad t \in [0,T] \\ y(0) &= 0 \end{cases} \tag{3.1.2}$$

and

$$\begin{cases} D^{\beta}y(t) &= g(t,y), \quad t \in [0,T] \\ y(0) &= 0 \end{cases} \tag{3.1.3}$$

for the same source $g(t,y)$ and the same response $y(t)$ of the system, with $y(0) = 0$.

Suppose $\beta > \alpha$. Then $\gamma = \beta - \alpha \in (0,1)$. By *Remark* 3.1.1, and Lemma 3.13 in [7],

$$y = I^{\alpha}D^{\alpha}y = I^{\alpha}D^{\beta}y = I^{\alpha}(D^{\alpha}D^{\beta-\alpha})y = (I^{\alpha}D^{\alpha})D^{\beta-\alpha}y$$

and, if $D^{\beta-\alpha}y(0) = 0$ (for which it is enough to assume some smoothness condition of $y$ in 0, for example, $y \in C^2[0,\varepsilon))$, we have

$$y = D^\gamma y.$$

In consequence,

$$y(t) = \frac{1}{\Gamma(1-\gamma)} \int_0^t y'(\tau)(t-\tau)^{-\gamma} d\tau, \ \ t \geq 0,$$

and

$$y(t) = (y' * m_\gamma)(t).$$

Supposing we can apply Laplace Transform,

$$L(y)(s) = L(y')(s) \cdot L(m_\gamma)(s)$$

and

$$L(y)(s) = s^{-\gamma+1} \cdot L(y)(s)$$

so it must be $\gamma = 1$, which is not possible.

Then $\alpha = \beta$ in (3.1.2) and (3.1.3).

Note that, if $y(0) = y_0 \neq 0$, a similar reasoning can be applied to $y(t) - y_0$.

## 3.2 Determination of $\alpha$

### *3.2.1 A Linear Case*

When $g(t,y) = \lambda y(t) + f(t)$, the problem reads

$$\begin{cases} D^\alpha y(t) & = \lambda y(t) + f(t), \quad t \in [0,T] \\ y(0) & = 0. \end{cases} \tag{3.2.1}$$

If $0 < \alpha < 1$ and $\lambda < 0$, it is a mathematical model for a relaxing oscillator with a power-law attenuation (see [4]).

There is an explicit formula for the solution $y(t)$ expressed in terms of Mittag-Leffler functions

$$E_\alpha(z) = \sum_{k=0}^{+\infty} \frac{z^k}{\Gamma(\alpha k + 1)}$$

and

$$E_{\alpha,\beta}(z) = \sum_{k=0}^{+\infty} \frac{z^k}{\Gamma(\alpha k + \beta)}.$$

The solution to (3.2.1) is

$$y(t) = y_0 E_\alpha(\lambda t^\alpha) + \int_0^t f(t-\tau)\tau^{\alpha-1} E_{\alpha,\alpha}(\lambda \tau^\alpha) d\tau \tag{3.2.2}$$

(see [7], Remark 7.1).

The Mittag-Leffler functions $E_\alpha$ and $E_{\alpha,\beta}$ are entire functions that play an important role in the context of fractional differential equations since the eigenfunctions of the Caputo differential operator can be written in terms of them.

*3.2.1.1 Two Theoretical Results*

- If we can apply Laplace Transform both to $y, y'$ and $f$ in (3.2.1), we obtain

$$L(y')(s) \cdot L(m_\alpha)(s) = \lambda L(y)(s) + L(f)(s)$$

Consequently,

$$[sL(y)(s) - y(0)]s^{\alpha-1} = \lambda L(y)(s) + L(f)(s)$$

and

$$s^\alpha = \frac{s\lambda L(y)(s) + sL(f)(s)}{sL(y)(s) - y(0)} \tag{3.2.3}$$

It follows that, when the parameter $\lambda$ is known, an external source $f(t)$ is given and we compute the consequent system response by means of (3.2.2), the order $\alpha$ is uniquely determined by (3.2.3).

- Based on previous works, where the order of derivation in the case of homogeneous equations is related to the limit of $y$ or $y'$ for $t \to 0$ or $t \to \infty$ (see [11]) we analyze the evolution of $y'$ trying to relate its behavior to the value of $\alpha$.

  We make the following assumptions: $f$ is derivable and $f'$ is integrable in $[0, T]$.

Recalling (3.2.2), we obtain

$$\begin{aligned} y'(t) = & \ [\lambda y_0 + f(0)]t^{\alpha-1}E_{\alpha,\alpha}(\lambda t^\alpha)] \\ + & \ \int_0^t f'(t-\tau)\tau^{\alpha-1}E_{\alpha,\alpha}(\lambda\tau^\alpha)d\tau. \end{aligned} \tag{3.2.4}$$

**Remark 3.2.1.** *If $|f'(t)| \leq K$ for $t \in [0, T]$, the second term in (3.2.4) vanishes as $t \to 0^+$,*

$$\lim_{t\to 0^+} \int_0^t f'(t-\tau)\tau^{\alpha-1}E_{\alpha,\alpha}(\lambda\tau^\alpha)\,d\tau = 0.$$

*To prove this assertion, we use the following result that can be found in [11],*

$$E_{\alpha,\alpha}(\lambda\tau^\alpha) = \frac{1}{\Gamma(\alpha+1)} + \tau^\alpha r_\lambda(\tau)$$

*where $r_\lambda(\tau)$ is a continuous and bounded function that has a finite limit for $\tau \to 0^+$.*

*Let's consider $|r_\lambda(\tau)| \leq M$ for $\tau \in [0, T]$. Then,*

$$\begin{aligned} &|\textstyle\int_0^t f'(t-\tau)\tau^{\alpha-1} \quad E_{\alpha,\alpha}(\lambda\tau^\alpha)\,d\tau| \\ &\qquad\qquad \leq \tfrac{K}{\Gamma(\alpha+1)}t^\alpha\left[1+\tfrac{t^\alpha}{2}M\Gamma(\alpha)\right] \longrightarrow 0 \end{aligned}$$

*as $t \to 0^+$.*

Then, based on these observations, we can determine the behavior of $\lim_{t\to 0^+} t^\beta y'(t)$ for different values of $\beta$:

- if $\beta > 1-\alpha$: $\lim_{t\to 0^+} t^\beta y'(t) = 0$,

- if $\beta = 1 - \alpha$: $\lim_{t \to 0^+} t^\beta y'(t) = \frac{\lambda y_0 + f(0)}{\Gamma(\alpha)}$ and
- if $\beta < 1 - \alpha$: $\lim_{t \to 0^+} t^\beta y'(t) = \infty$.

**Remark 3.2.2.** *From this assertion, we can theoretically deduce the value of $\alpha$, with $0 < \alpha < 1$.*

*If we define the function $h(\beta) = \lim_{t \to 0^+} t^\beta y'(t)$, from the results above it is*

$$h(\beta) = \begin{cases} 0 & \beta > 1 - \alpha \\ \frac{\lambda y_0 + f(0)}{\Gamma(\alpha)} & \beta = 1 - \alpha \end{cases}$$

*and $\lim_{\beta \to (1-\alpha)^-} h(\beta) = \infty$.*

*In consequence, the value of $\alpha$ can be determined from $\alpha = 1 - \beta$, where*

$$\beta = \min\{\gamma \in \mathbb{R} : h(\gamma) < \infty\}.$$

*Note that this formula is only theoretical since $y'$ is calculated assuming a known value of $\alpha$.*

#### *3.2.1.2 Numerical Determination of $\alpha$*

We propose a more realistic situation in which we excite the system and, afterwards, we measure the response obtained.

To simulate this situation for the linear case, we proceed as follows:

- from known values of $\alpha$ and $\lambda$, and a response $y(t)$, we calculate the excitation $f(t)$ from (3.2.1):

$$f(t) = D^\alpha y(t) - \lambda y(t).$$

- For the calculated $f(t)$ we compute the response $y(t, \beta)$ from (3.2.2), for a generic derivation order $\beta$, considering several terms of the series for the Mittag-Leffler function $E_{\beta,\beta}$.

- For $m$ instants $t_i$ we simulated noisy data,

$$\tilde{y}(t_i) = y(t_i, \alpha) + e(t)$$

  adding to the solution the noise $e(t)$ calculated by means of a normal distribution with mean in the exact value $y(t_i, \alpha)$ and 2% of it as the variance.

- We minimize

$$J(\beta) = \sum_{i=1}^{m} [y(t_i, \beta) - \tilde{y}(t_i)]^2$$

  To perform the computations we used Mathematica$^{TM}$ and its pre-built function *Minimize* to find $\arg\min_\beta J(\beta)$.

### 3.2.2 The General Case

For the general case

$$\begin{cases} D^\alpha y(t) &= g(t, y), \quad t \in [0, T] \\ y(0) &= y_0 \end{cases}$$

there is not an explicit formula for the solution $y(t)$ in terms of $g$ and $y_0$. Consequently, in order to proceed as we previously did, we look for an approximate solution $y^*(t)$ and minimize

$$J(\beta) = \sum_{i=1}^{m} [y^*(t_i, \beta) - \tilde{y}(t_i)]^2$$

We propose two different ways to calculate $y^*(t)$: the Adomian Decomposition Method and a Picard type approximation.

#### *3.2.2.1 Adomian Decomposition Method*

The Adomian Decomposition Method (ADM), developed by G. Adomian in the '80s (see [1]), has been successfully used to solve numerous problems in different areas (see [25] for an extensive bibliography and [9] for various applications). Since it is relatively easy to apply and in some cases leads to the exact solution of the problem, it is a valuable alternative tool to solve differential, integro-differential, and algebraic equations. Convergence is guaranteed in small intervals. We briefly describe it below.

For a general equation

$$F(y(t)) = h(t)$$

where $F$ is a differential, integro-differential or algebraic operator, $h$ is a known function and $y(t)$ is the solution to be determined, the method provides an analytical approximate solution.

We suppose that an analytical solution exists,

$$y(t) = \sum_{n=0}^{\infty} y_n(t),$$

and that the operator $F$ can be decomposed as

$$F(y) = L(y) + R(y) + N(y)$$

where $L$ is a linear and easily invertible operator, $R$ is linear, and $N$ is the non-linear component of $F$ that admits an expansion by means of the so-called Adomian polynomials,

$$N(y) = \sum_{n=0}^{\infty} A_n(y_0, ..., y_n).$$

Then

$$L(y) = h(t) - R(y) - N(y),$$

$$y = L^{-1}[h(t) - R(y) - N(y)]$$

and

$$\sum_{n=0}^{\infty} y_n(t) = L^{-1}[h(t)] - \sum_{n=0}^{\infty} L^{-1}[R(y_n)] - \sum_{n=0}^{\infty} L^{-1}[A_n(y_0, ..., y_n)]$$

A recursive scheme is proposed to build the solution,

$$\begin{cases} y_0(t) & = L^{-1}[h(t)] \\ y_{n+1}(t) & = -L^{-1}[R(y_n)] - L^{-1}[A_n(y_0, ..., y_n)] \text{ for } n \geq 0. \end{cases} \tag{3.2.5}$$

Initial or boundary conditions are taken into account when inverting the operator $L$ in the first equation in (3.2.5). The partial sums

$$y_{ADM(N)}(t) = \sum_{n=0}^{N} y_n(t)$$

provide an approximation to the solution $y$.

Among the advantages of this method, the author points out that neither simplifications nor linearization procedures are involved and it provides the exact solution when a closed formula exists. In the general case it gives a very good analytical approximation of the solution (or of one of them, in case that more than one solution exists). Regarding the Adomian polynomials, they must be constructed for each case. If $N$ is an analytical operator, it can be shown that a simple way to obtain these polynomials is

$$A_n = \frac{1}{n!}\frac{\partial^n}{\partial \varepsilon^n} N(y_0 + \varepsilon y_1 + \varepsilon^2 y_2 + ... + \varepsilon^n y_n)|_{\varepsilon=0}$$

Note $A_n$ depends only on $y_0, y_1, ..., y_n$.

*3.2.2.2 Picard Type Approximation*

We recall that the fractional differential equation

$$D^\alpha y(t) \quad = g(t, y), \quad t \in [0, T]$$

is equivalent to the integral equation (see [7], Lemma 6.2)

$$y(t) = y(0) + \frac{1}{\Gamma(1-\alpha)} \int_0^t g(\tau, y(\tau))(t-\tau)^{-\alpha} d\tau. \qquad (3.2.6)$$

If $g : [0, T] \to \mathbb{R}$ is piecewise continuous and Lipschitz in its second argument, i.e.,

$$|g(t, y_1) - g(t, y_2)| \leq M|y_1 - y_2|$$

for $M \in \mathbb{R}^+$, the Volterra integral equation (3.2.6) has a unique solution.

Under this hypothesis, we can apply the well-known Picard successive approximation method to obtain an approximation of it:

$$\begin{cases} y_0(t) &= y(0) \\ y_{n+1}(t) &= y(0) + \frac{1}{\Gamma(1-\alpha)} \int_0^t g(\tau, y_n(\tau))(t-\tau)^{-\alpha} d\tau \text{ for } n \geq 0. \end{cases}$$

---

## 3.3 Numerical Examples

**Example 3.3.1.** For the FVIP

$$\begin{cases} D^\alpha y(t) + y(t) &= \frac{1}{2}\sqrt{\pi} J_0(\sqrt{t}) + \sin(\sqrt{t}) + 1, \quad t \in (0,1) \\ y(0) &= 1 \end{cases}$$

with $J_0(\cdot)$ the Bessel function of the first kind, the exact solution for $\alpha = \frac{1}{2}$ is $y(t) = \sin(\sqrt{t}) + 1$.

By means of (3.2.2) we can express the solution depending on the fractional order,

$$y(t, \beta) = E_\beta(\lambda t^\beta) + \int_0^t f(t-\tau)\tau^{\beta-1} E_{\beta,\beta}(\lambda \tau^\beta) d\tau.$$

In this example, $f(t) = \frac{1}{2}\sqrt{\pi} J_0(\sqrt{t}) + \sin(\sqrt{t}) + 1$.

We obtained $y^*(t, \beta)$, an approximation of $y(t, \beta)$, by

$$y^*(t, \beta) = \tilde{E}_\beta(\lambda t^\beta) + \int_0^t \tilde{f}(t-\tau)\tau^{\beta-1} \tilde{E}_{\beta,\beta}(\lambda \tau^\beta) d\tau,$$

where $\tilde{f}(z)$ is the sum of the first five terms of the Maclaurin series of $f$ and, for $\tilde{E}_\beta(z)$ and $\tilde{E}_{\beta,\beta}(z)$, 10 terms of the respective Mittag-Leffler series were considered.

Through the known exact solution we simulated $m$ equally spaced noisy measurements $\tilde{y}_j(t_i)$ – for $t_i \in [0,1]$, $i = 1,\ldots,m$ – along 10 essays – $j = 1,\ldots,10$.

We calculated

$$\alpha_j = \arg\min_{\beta} \sum_{i=1}^{m} [y^*(t_i,\beta) - \tilde{y}_j(t_i)]^2$$

for each essay, and by averaging, we obtained

- for $m = 3$: $\alpha^* = (0.49 \pm 0.01)$,
- for $m = 5$: $\alpha^* = (0.495 \pm 0.009)$,
- for $m = 10$: $\alpha^* = (0.500 \pm 0.007)$.

The resulting approximation value $\alpha^*$ is satisfactory.

**Example 3.3.2.** For the FVIP

$$\begin{cases} D^\alpha y(t) + 2y(t) &= \frac{2048}{195} \frac{t^{\frac{13}{4}}}{\Gamma(\frac{1}{4})} + 2t^4, \quad t \in (0,1) \\ y(0) &= 0, \end{cases}$$

the exact solution, for $\alpha = \frac{3}{4}$, is $y(t) = t^4$.

Through the known exact solution, we simulated $m$ equally spaced noisy measurements $\tilde{y}_j(t_i)$ – for $t_i \in [0,1]$, with $i = 1,\ldots,m$ – along 10 essays, $j = 1,\ldots,10$. We then proceeded as described in Example 3.3.1.

For each essay and, by averaging, we obtained

- for $m = 3$: $\alpha^* = (0.74 \pm 0.03)$,
- for $m = 5$: $\alpha^* = (0.75 \pm 0.02)$,

- for $m = 10$: $\alpha^* = (0.76 \pm 0.01)$.

**Example 3.3.3.** For

$$\begin{cases} D^\alpha y(t) + y(t) &= \frac{\sqrt{\pi}}{2} + \sqrt{t} \quad t \in (0,1] \\ y(0) &= 0, \end{cases}$$

the exact solution, for $\alpha = \frac{1}{2}$, is $y(t) = \sqrt{t}$. In this case, we used ADM to approximate $y(t, \beta)$ and, afterwards, compared it with the exact solution.

In Figure 3.1, we present 10 approximations to the solution obtained using the ADM.

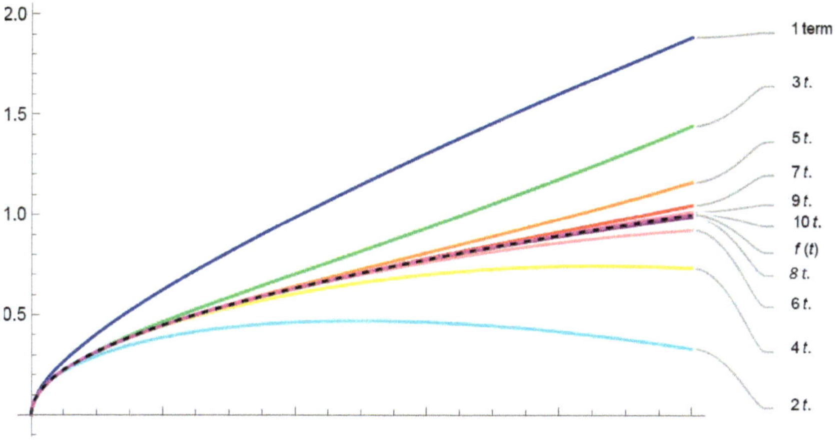


**FIGURE 3.1**. Ten approximations to the solution of Example 3 by means of ADM.

Supposing the fractional order is unknown, we obtained $y^*(t, \beta)$ using 10 terms of ADM:

$$\begin{cases} y_0(t,\beta) & = y_0 + \frac{1}{\Gamma(\beta)} \int_0^t (t-\tau)^{\beta-1}(\frac{\sqrt{\pi}}{2} + \sqrt{\tau})d\tau \\ y_{n+1}(t,\beta) & = \frac{\lambda}{\Gamma(\beta)} \int_0^t (t-\tau)^{\beta-1} y_n(\tau,\beta)d\tau \quad n \geq 0 \end{cases}$$

$$y(t,\beta) \approx y_{ADM(9)}(t,\beta) = \sum_{n=0}^{9} y_n(t,\beta).$$

Again, through the known exact solution, we simulated $m$ equally spaced noisy measurements $\tilde{y}_j(t_i)$ - for $t_i \in [0,1]$, $i = 1,\dots,m$ - along 10 essays - $j = 1,\dots,10$.

We calculated

$$\alpha_j = \arg\min_{\beta} \sum_{i=1}^{m} [y_{ADM(9)}(t_i,\beta) - \tilde{y}_j(t_i)]^2$$

for each essay, and by averaging, we obtained

- for $m = 3$: $\alpha^* = (0.49 \pm 0.03)$,
- for $m = 5$: $\alpha^* = (0.50 \pm 0.02)$,
- for $m = 10$: $\alpha^* = (0.505 \pm 0.008)$.

**Example 3.3.4.** For

$$\begin{cases} D^\alpha y(t) & = y^2 \quad t \in (0,1] \\ y(0) & = 1, \end{cases}$$

the exact solution, for $\alpha = \frac{1}{2}$, is unknown.

In Figure 3.2, we present 10 approximations to the solution obtained using the ADM with $\alpha = \frac{1}{2}$.

Summing up 10, 11 or 12 terms the resulting approximated solutions are practically indistinguishable.

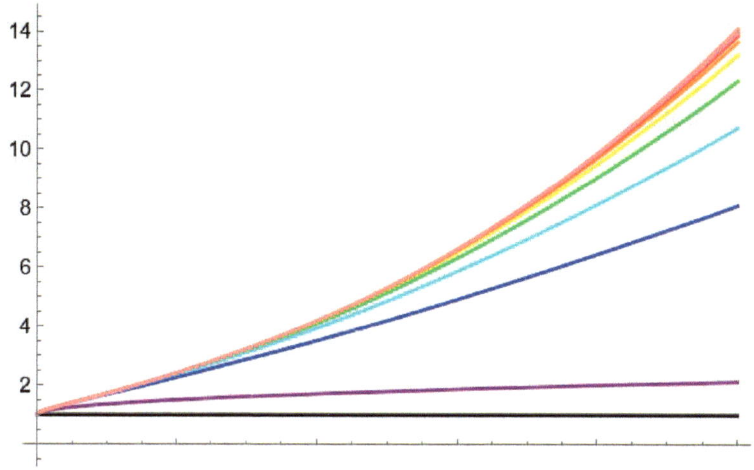


**FIGURE 3.2**. Ten approximations to the solution of Example 4 by ADM, using $\alpha = \frac{1}{2}$.

In this case, we constructed the Adomian polynomials from the partial sums of the Cauchy product for $[\sum_{i=0}^{\infty} y_i(t)][\sum_{k=0}^{\infty} y_k(t)]$:

$$A_n(y_0(t), y_1(t), ..., y_n(t)) = \sum_{i=0}^{n} y_i(t) y_{n-i}(t)$$

We obtained $y^*(t, \beta)$ by means of 10 iterations,

$$\begin{cases} y_0(t,\beta) &= y_0 \\ y_{n+1}(t,\beta) &= \frac{1}{\Gamma(1-\beta)} \sum_{i=0}^{n} \int_0^t y_i(\tau,\beta) y_{n-i}(\tau,\beta)(t-\tau)^{-\beta} d\tau \text{ for } n \geq 0, \end{cases}$$

and calculated the approximation

$$y^*(t,\beta) = y_{ADM(9)}(t,\beta) = \sum_{n=0}^{9} y_n(t,\beta).$$

We considered $\tilde{y}(t) = y_{ADM(11)}(t, \frac{1}{2}) = \sum_{n=0}^{11} y_n(t, \frac{1}{2})$ as the exact solution.

We simulated $m$ equally spaced noisy measurements $\tilde{y}_j(t_i)$—for $t_i \in [0, 1]$, $i = 1, \ldots, m$—along 10 essays, $j = 1, \ldots, 10$ and calculated

$$\alpha_j = \arg\min_{\beta} \sum_{i=1}^{m} [y_{ADM(9)}(t_i, \beta) - y_{ADM(11)}(t_i, \frac{1}{2})]^2$$

for each essay, and by averaging, we obtained

- for $m = 3$: $\alpha^* = (0.498 \pm 0.006)$,
- for $m = 5$: $\alpha^* = (0.499 \pm 0.004)$,
- for $m = 10$: $\alpha^* = (0.498 \pm 0.003)$.

**Example 3.3.5.** For

$$\begin{cases} D^{\alpha} y(t) & = ty \quad t \in (0, 1] \\ y(0) & = 1, \end{cases}$$

the exact solution, for $\alpha = \frac{2}{3}$, is unknown.

In Figure 3.3, we show 10 approximations of the solution by means of the Picard procedure, using $\alpha = \frac{2}{3}$.

Approximations of the solution beyond the seventh one are practically indistinguishable.

We obtained $y^*(t, \beta) = y_7(t, \beta)$ by means of seven iterations,

$$\begin{cases} y_0(t, \beta) & = y(0) \\ y_{n+1}(t, \beta) & = y(0) + \frac{1}{\Gamma(1-\beta)} \int_0^t \tau . y_n(\tau, \beta)(t - \tau)^{-\beta} d\tau, \quad 0 \leq n \leq 6. \end{cases}$$

We considered $\tilde{y}(t) = y_{10}(t, \frac{2}{3})$ as the exact solution and simulated $m$ equally spaced noisy measurements $\tilde{y}_j(t_i)$—for $t_i \in [0, 1]$, $i =$

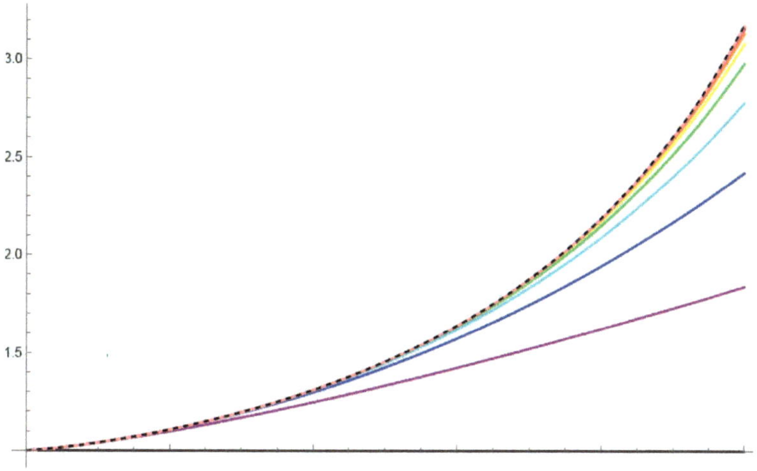


**FIGURE 3.3**. Ten approximations to the solution of Example 5 by means of Picard procedure, using $\alpha = \frac{2}{3}$.

$1, \ldots, m$—along eight essays—$j = 1, \ldots, 8$ and calculated

$$\alpha_j = \arg\min_{\beta} \sum_{i=1}^{m} [y_7(t_i, \beta) - y_{10}(t_i, \frac{2}{3})]^2$$

for each essay.

By averaging the results over the eight essays we obtained, for $m = 10$, $\alpha^* = (0.666 \pm 0.007)$.

---

## 3.4 Conclusions

When a fractional model is proposed to describe a realistic situation, the order of the fractional derivative is a parameter to be determined. In general, it should be chosen to best fit experimental data.

In this work, we present a method to identify the order of derivation, $\alpha$, of a model described by the Caputo derivative. Based on an approximate solution that depends on $\alpha$, we propose a minimization procedure to determine the value that best fits noisy data from the system response. When a formula for the exact solution of the problem is not available, we seek an approximate solution to the integral equivalent formulation using the Adomian Decomposition Method or a Picard-type approximation.

We developed various numerical examples with very good results.

# References

[1] Adomian, G., (1994). *Solving Frontier Problems of Physics: The Decomposition Method*, Kluwer Academic Publishers.

[2] Baleanu, D., Balas, V., & Praveen, A., (2022). *Fractional Order Systems and Applications in Engineering*, Academic Press.

[3] Baleanu, D., Diethelm, K., Scalas, E., & Trujillo, J., (2012). *Fractional Calculus: Models and Numerical Methods*, World Scientific.

[4] Chen, W., Zhang, X., & Korošak, D., (2010). Investigation on Fractional and Fractal Derivative Relaxation-Oscillation Models. *International Journal of Nonlinear Sciences and Numerical Simulation, 11*(1), 3–9.

[5] Cheng, J., Nakagawa, J., Yamamoto, M., & Yamazaki, T., (2009). Uniqueness in an inverse problem for a one-dimensional fractional diffusion equation. *Inverse Problems, 25*(11), 115002.

[6] Concezzi, M., & Spigler, R., (2018). Identifying the fractional orders in anomalous diffusion models from real data. *Fractal and Fractional, 2*(1), 14.

[7] Diethelm, K., (2010). *The Analysis of Fractional Differential Equations. An Application-Oriented Exposition Using Differential Operators of Caputo Type, Lecture Notes in Mathematics*, Springer-Verlag Berlin Heidelberg.

[8] Dorcak, L., Gonzalez, E., Terpak, J., Valsa, J., & Pivka, L., (2013). Identification of Fractional-order Dynamical Systems Based on Nonlinear Function Optimization. *International Journal of Pure and Applied Mathematics, 89*(2), 225–250.

[9] Duan, J., Rach, R., Baleanu, D., & Wazwaz, A., (2012). A review of the Adomian Decomposition method and its application to fractional

differential equation. *Communications in Fractional Calculus, 3*(2), 73–99.

[10] Guo, Y., Ma, B., & Wu, R., (2016). On sensitivity analysis of parameters for fractional differential equations with Caputo derivatives. *Electronic Journal of Qualitative Theory of Differential Equations, 118*, 1–17.

[11] Hatano, Y., Nakagawa, J., Wang, S., & Yamamo, M., 2013). Determination of order in fractional diffusion equation. *Journal of Math-for-Industry, 5*(2013A-7), 51–57.

[12] Inc, M., (2008). The approximate and exact solutions of the space- and time-fractional Burgers equations with initial conditions by variational iteration method. *Journal of Mathematical Analysis and Applications, 345*(1), 476–484.

[13] Jano, J., (2016). Determination of the order of fractional derivative and a kernel in an inverse problem for a generalized time fractional diffusion equation. *Electronic Journal of Differential Equations, 199*, 1–28.

[14] Javed, I., Ahmadb, A., Hussaind, M., & Iqbala, S., (n. d.). Some Solutions of Fractional Order Partial Differential Equations Using Adomian Decomposition Method, arXiv:1712.09207[math.NA].

[15] Jing, L., & Guo, B., (2013). Parameter identification in fractional differential equations. *Acta Mathematica Scientia, 33*(3), 855–864.

[16] Kilbas, A., Srivastava, H., & Trujillo, J., (2006). *Theory and Application of Fractional Differential Equations*, Elsevier.

[17] Krasnoschok, M., Pereverzyev, S., Siryk, S., & Vasylyeva, N., (2020). Determination of the fractional order in semilinear subdiffusion equations. *Fractional Calculus and Applied Analysis, 23*, 694–722.

[18] Li, Ch., & Chen, A., (2018). Numerical methods for fractional partial differential equations. *International Journal of Computer Mathematics, 95*(6–7), 1048–1099.

[19] Li, Z., Liu, Y., & Yamamoto, M., (2019). Inverse Problems of Determining Parameters of the Fractional Partial Differential Equations, in Kochubei, A., & Luchko, Y. (eds.), *Handbook*

*of Fractional Calculus with Applications, Volume 2: Fractional Differential Equations*, De Gruyter, 431–442.

[20] Meerschaert, M., & Tadjeran, Ch., (2006). Finite difference approximations for two-sided space-fractional partial differential equations. *Applied Numerical Mathematics, 56*(1), 80–90.

[21] Miller, K., & Ross, B., (1993). *An Introduction to The Fractional Calculus and Fractional Differential Equations*, Wiley.

[22] Momani, S,. & Odibat, Z., (2007). Homotopy perturbation method for nonlinear partial differential equations of fractional order. *Physics Letters A, 365*(5–6), 345–350.

[23] Oldham, K., & Spanier, J., (1974). *The Fractional Calculus*, Academic Press.

[24] Podlubny, I., (1999). *Fractional differential equations*, Academic Press.

[25] Rach, R., (2012). A bibliography of the theory and applications of the Adomian decomposition method, 1961–2011. *Kybernetes, 41*(7-8).

[26] Seminara, S., Troparevsky, M. I., & Fabio, M., (2023). Fractional order estimation for a differential equation with Caputo derivative. *Matemática Aplicada, Computacional e Industrial, 9.*

[27] Sun, H., Zhang, Y., Baleanu, D., Chen, W., & Chen, Y., (2018). A new collection of real-world applications of fractional calculus in science and engineering. *Communications in Nonlinear Science and Numerical Simulation, 64*(11), 213–231.

[28] Xu, W., Sun, H., Chen, W., & Chen H., (2018). Transport properties of concrete-like granular materials interacted by their microstructures and particle components. *International Journal of Modern Physics B, 32*(18), 1840011.

[29] Zhou, S., Cao, J., & Chen, Y., (2013). Genetic Algorithm-Based Identification of Fractional-Order Systems. *Entropy, 15*, 1624–1642.

# Chapter 4

## New Class of Integro-Differential Equations with Non-Singular Differential Operators

**Seda İğret Araz[1,2] and Abdon Atangana[2,3]**
[1]*Department Mathematics, Faculty of Education, Siirt University, Turkey*
[2]*Institute for Groundwater Studies, Faculty of Natural and Agricultural Sciences, University of the FreeState, South Africa*
[3]*Department of Medical Research, China Medical University Hospital, China Medical University, Taichung, Taiwan*

**Abstract**
Very recently, a new fractional calculus was proposed with the aim of updating and extending the limitations of the existing frameworks. More precisely, a fractional differential operator based on the exponential decay law was introduced, which led to the formulation of a new fractional integral. The associated fractional integral is a superposition of the function and its classical integral. On the other hand, a fractional derivative based on the generalized Mittag-Leffler function was introduced. This version also resulted in a new fractional integral, which is the superposition of the function and its Riemann-Liouville fractional integral. It must be acknowledged that these two approaches have opened new avenues for investigation, both in theory and in applications. In this chapter, we introduce a new class of partial differential integro-differential equations with mixed operators, particularly differential operators with non-singular kernels. We establish the conditions under which the equations admit

unique exact solutions in both cases, where the kernels are linear and non-linear. Furthermore, we present the numerical solution using a newly developed numerical scheme and include examples with simulations.

## Introduction

Integro-differential equations have been studied intensively in recent years, as they model various physical problems arising in many fields of science, engineering, and technology. In science, they appear in disciplines such as chemistry, geology, physics, hydrology, geohydrology, biology, biotechnology, and genetics (see [4]–[6], [8]–[10], and [12]). Within the framework of classical differentiation and integration, numerous studies have been conducted. These include investigations into the existence of unique solutions, analytical methods for solving such equations, and numerical methods for nonlinear cases (see [2], [3], [16], and [17]). Typically, mathematicians use a differential operator and its anti-derivative to construct integro-differential equations, whether in the ordinary or partial case. This approach has extended into the development of fractional differentiation using power-law kernels. Recently, the concept of designing integro-differential equations with mixed operators was introduced, proposing the combination of differential and integral operators of different types. For example, the differential operator could be classical, while the integral operator might be fractional. Although these ideas have not yet gained widespread attention in the literature, they hold significant potential for opening new avenues of investigation. In previous studies, these classes of equations were explored using differential operators with both singular and non-singular kernels. For instance, the Caputo derivative was combined with the Caputo-Fabrizio derivative to form a new class of integro-differential equations ([1], [7], [11], [14], and [15]). Equations were also constructed using the Caputo-Fabrizio integral alongside the Caputo derivative or the Atangana-Baleanu fractional derivative with the Riemann-Liouville integral. These studies have shown great promise, yielding many new results, with further findings still anticipated. Techniques to establish the existence and uniqueness of exact solutions for these equations have been proposed, along with numerical schemes.

In this chapter, we focus on differential operators with non-singular kernels and their associated integrals. For instance, we examine cases where the fractional derivative is of Caputo-Fabrizio type and the fractional integral is of Atangana-Baleanu type [5], or where the fractional differential operator is of Atangana-Baleanu type and the fractional integral is of Caputo-Fabrizio type [10]. Our aim is to provide a comprehensive analysis, including existence and uniqueness results, as well as the development of numerical schemes.

---

## 4.1 Preliminaries

In this section, we present the definitions and properties of fractional derivatives, such as the Atangana-Baleanu and Caputo-Fabrizio derivatives [5, 10].

**Definition 4.1.1.** *Let $u(x,t) \in C(0,l) \times W_2^1(0,T), \alpha \in [0,1]$ then, the definition of the Atangana-Baleanu fractional derivative [5] in Caputo sense is given by;*

$$ {}_{a}^{ABC}D_t^{\alpha} u(x,t) = \frac{1}{1-\alpha} \int_a^t \frac{\partial}{\partial \tau} u(x,\tau) E_{\alpha} \left[ -\frac{\alpha}{1-\alpha} (t-\tau)^{\alpha} \right] d\tau. \tag{4.1.1} $$

**Definition 4.1.2.** *Let $u(x,t) \in C(0,l) \times C(0,T), \alpha \in [0,1]$ then, the definition of the Atangana-Baleanu fractional derivative [5] in Riemann-Liouville sense becomes*

$$ {}_{a}^{ABR}D_t^{\alpha} u(x,t) = \frac{1}{1-\alpha} \frac{\partial}{\partial t} \int_a^t u(x,\tau) E_{\alpha} \left[ -\frac{\alpha}{1-\alpha} (t-\tau)^{\alpha} \right] d\tau \tag{4.1.2} $$

*where*

$$ E_{\alpha}(-t^{\alpha}) = \sum_{k=0}^{\infty} \frac{(-t)^{\alpha k}}{\Gamma(\alpha k + 1)} \tag{4.1.3} $$

*is the Mittag-Leffler function.*

**Definition 4.1.3.** *Let $u(x,t) \in C(0,l) \times W_2^1(0,T)$, then the Caputo-Fabrizio derivative [10] with fractional order $0 < \alpha \leq 1$ is given as follows.*

$$ {}_{0}^{CF}D_t^{\alpha} u(x,t) = \frac{1}{1-\alpha} \int_0^t \frac{\partial}{\partial \tau} u(x,\tau) \exp\left( \frac{-\alpha(t-\tau)}{1-\alpha} \right) d\tau. \tag{4.1.4} $$

**Definition 4.1.4.** *Laplace transform for the Caputo-Fabrizio fractional derivative is given by*

$$ \begin{aligned} \left[ {}_{0}^{CF}D_t^{\alpha+1} u(x,t) \right](s) &= \frac{1}{1-\alpha} \left[ u^{(\alpha+n)}(x,t) \right] \exp\left( -\frac{\alpha}{1-\alpha} t \right) \\ &= \frac{s^{n+1}[u(x,t)] - S}{s + \alpha(1-s)} \end{aligned} \tag{4.1.5} $$

*where*

$$S = s^n u(x,0) - s^{n-1} u'(x,0) - ... - u^{(n)}(x,0) \tag{4.1.6}$$

*for* $\alpha \in (0,1]$ *and* $n \in N$.

**Definition 4.1.5.** *Laplace transform for the Atangana-Baleanu fractional derivative is defined by*

$$\begin{aligned}\left[{}_a^{ABC}D_t^{\alpha} u(x,t)\right](s) &= \frac{1}{1-\alpha}\left[\int_a^t \frac{\partial}{\partial \tau} u(x,\tau) E_{\alpha}\left[-\frac{\alpha}{1-\alpha}(t-\tau)^{\alpha}\right] d\tau\right] \\ &= \frac{1}{1-\alpha}\frac{s^{\alpha}\left[u(x,t)\right](s) - s^{\alpha-1}u(x,0)}{s^{\alpha}+\frac{\alpha}{1-\alpha}}. \end{aligned} \tag{4.1.7}$$

**Definition 4.1.6.** *The definition of the Atangana-Baleanu fractional integral is given as*

$${}_a^{AB}I_t^{\alpha}\left[u(x,t)\right] = (1-\alpha)u(x,t) + \frac{\alpha}{\Gamma(\alpha)}\int_a^t u(x,\tau)(t-\tau)^{\alpha-1} d\tau. \tag{4.1.8}$$

**Definition 4.1.7.** *The definition of the Caputo-Fabrizio fractional integral is given as*

$${}_a^{CF}I_t^{\alpha}\left[u(x,t)\right] = (1-\alpha)u(x,t) + \alpha\int_a^t u(x,\tau)\, d\tau. \tag{4.1.9}$$

---

## 4.2 Model with Caputo-Fabrizio Derivative and Atangana-Baleanu Integral

In this section, we consider an integro-differential equation with a singular kernel. The integro-differential equation under study is given as

$$\frac{1}{1-\alpha}\int_0^t u_\tau(x,\tau)\exp\left[-\frac{\alpha}{1-\alpha}(t-\tau)\right]d\tau = f(x,t) + {}_0^{AB}J_t^{\alpha}F(x,t,u(x,t)). \tag{4.2.10}$$

where $u_\tau(x,\tau) = \frac{\partial}{\partial\tau}u(x,\tau)$. Then, with the definition of the Atangana-Baleanu integral (4.2.10) is converted to

$$\begin{aligned}\frac{1}{1-\alpha}\int_0^t u_\tau(x,\tau)\exp\left[-\frac{\alpha}{1-\alpha}(t-\tau)\right]d\tau &= f(x,t) \\ &+(1-\alpha)F(x,t,u(x,t)) \\ &+\frac{\alpha}{\Gamma(\alpha)}\int_0^t (t-\tau)^{\alpha-1} \\ &\times F(x,\tau,u(x,\tau))\,d\tau.\end{aligned} \tag{4.2.11}$$

Now, applying the Caputo-Fabrizio integral on both sides, we obtain:

$$\begin{aligned}u(x,t)-u(x,0) &= (1-\alpha)\left[\begin{array}{c} f(x,t)+(1-\alpha)F(x,t,u) \\ +\frac{\alpha}{\Gamma(\alpha)}\int_0^t (t-\tau)^{\alpha-1} \\ \times F(x,\tau,u)\,d\tau \end{array}\right] \\ &+\alpha\int_0^t\left[\begin{array}{c} f(x,\tau)\,d\tau \\ +(1-\alpha)F(x,\tau,u)\,d\tau \\ +\frac{\alpha}{AB(\alpha)}\int_0^l (l-\tau)^{\alpha-1}\,d\tau dl \end{array}\right]\end{aligned} \tag{4.2.12}$$

Rearranging, we get:

$$\begin{aligned}u(x,t)-u(x,0) &= (1-\alpha)f(x,t)+\frac{(1-\alpha)^2}{\Gamma(\alpha)}F(x,t,u) \\ &+\frac{\alpha(1-\alpha)}{\Gamma(\alpha)}\int_0^t (t-\tau)^{\alpha-1}F(x,\tau,u(x,\tau))\,d\tau\end{aligned} \tag{4.2.13}$$

$$+\alpha \int_0^t f(x,\tau)\,d\tau + \alpha(1-\alpha) F(x,t,u(x,t))$$
$$+\frac{\alpha^2}{\Gamma(\alpha+1)} \int_0^l (t-\tau)^{\alpha} F(x,\tau,u(x,\tau))\,d\tau.$$

After simplification, we have the following new class of Volterra equation

$$\begin{aligned} u(x,t) - u(x,0) &= (1-\alpha) f(x,t) + \alpha \int_0^t f(x,\tau)\,d\tau \qquad (4.2.14)\\ &+\frac{\alpha(1-\alpha)}{\Gamma(\alpha)} \int_0^t (t-\tau)^{\alpha-1} F(x,\tau,u(x,\tau))\,d\tau \\ &+\left((1-\alpha)^2 + \alpha(1-\alpha)\right) F(x,t,u(x,t)) \\ &+\frac{\alpha^2}{\Gamma(\alpha+1)} \int_0^l (t-\tau)^{\alpha} F(x,\tau,u(x,\tau))\,d\tau. \end{aligned}$$

Let $\Pi_a^b = \overline{T_a[t_0]} \times \overline{U_b[u_0^0]}$ where

$$\overline{T_a[t_0]} = [t_0 - a, t_0 + a], \quad \overline{U_b[u_0^0]} = [u(0,0) - b, u(0,0) + b]. \qquad (4.2.15)$$

This is a compact cylinder where $F$ is defined. We choose this investigation

$$M_1 = \sup_{\Pi_a^b} \|F\|, \quad M_2 = \sup_{\Pi_a^b} \|f\|. \qquad (4.2.16)$$

Let $L$ be the Lipschitz constant of $F$ with respect to the function $u$. We opt to continue our investigation within the framework of Banach's fixed point theorem using the metric on $\Pi\left(T_a(t_0), U_b\left(u_0^0\right)\right)$ induced by the uniform norm

$$\|\lambda\|_A = \sup_{t\in T_a} |\lambda(t)|. \qquad (4.2.17)$$

To proceed, we define the following operator between two function spaces of continuous functions, Picard's operator, as follows

$$\Lambda : \Pi\left(T_a\left(t_0\right), U_b\left(u_0^0\right)\right) \to \Pi\left(T_a\left(t_0\right), U_b\left(u_0^0\right)\right). \tag{4.2.18}$$

Then, we get:

$$\begin{aligned}\Lambda v\left(x,t\right) &= u\left(x,0\right) + \left(1-\alpha\right) f\left(x,t\right) + \frac{\left(1-\alpha\right)^2}{\Gamma\left(\alpha\right)} F\left(x,t,v\right) \\ &+\alpha \int_0^t f\left(x,\tau\right) d\tau + \alpha\left(1-\alpha\right) F\left(x,t,v\right) \\ &+\frac{\alpha\left(1-\alpha\right)}{\Gamma\left(\alpha\right)} \int_0^t \left(t-\tau\right)^{\alpha-1} F\left(x,\tau,v\right) d\tau \qquad (4.2.19) \\ &+\frac{\alpha^2}{\Gamma\left(\alpha+1\right)} \int_0^l \left(t-\tau\right)^{\alpha} F\left(x,\tau,v\left(x,\tau\right)\right) d\tau.\end{aligned}$$

Now, we evaluate

$$\left\|\Lambda v\left(x,t\right) - u\left(x,0\right)\right\| < b \tag{4.2.20}$$

such that we can show the first step that $\Lambda$ is the operator maps a complete non-empty metric space $Y$ into itself and additionally that $\Lambda$ is a mapping. Therefore

$$\begin{aligned}\left\|\Lambda v\left(x,t\right) - u\left(x,0\right)\right\| &\leq \left(1-\alpha\right)\left\|f\left(x,t\right)\right\| \qquad (4.2.21) \\ &+\left(\frac{\left(1-\alpha\right)^2}{\Gamma\left(\alpha\right)} + \alpha\left(1-\alpha\right)\right)\left\|Fx,t,v\right\| \\ &+\alpha \int_0^t \left\|f\left(x,\tau\right)\right\| d\tau \\ &+\frac{\alpha\left(1-\alpha\right)}{\Gamma\left(\alpha\right)} \int_0^t \left(t-\tau\right)^{\alpha-1}\left\|F\left(x,\tau,v\right)\right\| d\tau\end{aligned}$$

$$+\frac{\alpha^2}{\Gamma(\alpha+1)}\int_0^l (t-\tau)^{\alpha}\left\|F(x,\tau,v)\right\| d\tau.$$

Using the hypothesis, we have:

$$\left\|\Lambda v(x,t)-u(x,0)\right\| \leq \left[\begin{array}{c} (1-\alpha)M_1+\frac{\alpha^2}{\Gamma(\alpha+2)}M_2t^{\alpha+1} \\ +\left(\frac{(1-\alpha)^2}{\Gamma(\alpha)}+\alpha(1-\alpha)\right)M_2 \\ +\alpha M_1 t+\frac{\alpha(1-\alpha)}{\Gamma(\alpha)}M_2\frac{t^{\alpha}}{\alpha} \end{array}\right]. \quad (4.2.22)$$

Thus,

$$\left\|\Lambda v(x,t)-u(x,0)\right\| < b \Longrightarrow a < \frac{b}{\Phi} \quad (4.2.23)$$

where,

$$\begin{aligned} \Phi &= (1-\alpha)M_1+\left(\frac{(1-\alpha)^2}{\Gamma(\alpha)}+\alpha(1-\alpha)\right)M_2 \\ &\quad +\alpha M_1+\frac{\alpha(1-\alpha)}{\Gamma(\alpha)}M_2. \end{aligned} \quad (4.2.24)$$

we get:

$$\left\|\Lambda v(x,t)-\Lambda u(x,t)\right\| \leq \left[\begin{array}{c} \left(\frac{(1-\alpha)^2}{\Gamma(\alpha)}+\alpha(1-\alpha)\right) \\ \times\left\|F(x,t,v)-F(x,t,u)\right\| \\ +\frac{\alpha(1-\alpha)}{\Gamma(\alpha)}\int_0^t(t-\tau)^{\alpha-1} \\ \times\left\|F(x,\tau,v)-F(x,\tau,u)\right\| d\tau \\ +\frac{\alpha^2}{\Gamma(\alpha+1)}\int_0^l(t-\tau)^{\alpha} \\ \times\left\|F(x,\tau,v)-F(x,\tau,u)\right\| d\tau \end{array}\right]. \quad (4.2.25)$$

Employing the Lipschitz condition of $F$, we yield

$$\begin{aligned}\|\Lambda v(x,t)-\Lambda u(x,t)\| \leq &\left(\frac{(1-\alpha)^2}{\Gamma(\alpha)}+\alpha(1-\alpha)\right) \\ &\times L\|v(x,t)-u(x,t)\| \\ &+\frac{\alpha(1-\alpha)}{\Gamma(\alpha)}\frac{t^\alpha}{\alpha}L\|v(x,t)-u(x,t)\| \\ &+\frac{\alpha^2}{\Gamma(\alpha+2)}t^{\alpha+1}L\|v(x,t)-u(x,t)\|\end{aligned} \tag{4.2.26}$$

$$\begin{aligned}&\leq\left[\begin{array}{c}\frac{(1-\alpha)^2}{\Gamma(\alpha)}+\alpha(1-\alpha) \\ +\frac{(1-\alpha)}{\Gamma(\alpha)}t^\alpha+\frac{\alpha^2}{\Gamma(\alpha+2)}t^{\alpha+1}\end{array}\right]\|v(x,t)-u(x,t)\| \\ &\leq\kappa\|v(x,t)-u(x,t)\|.\end{aligned}$$

Thus $\Lambda$ is Lipschitz.

### *4.2.1 Existence and Uniqueness for F Linear*

In this section, we assume that $F$ is linear; thus,

$$F(x,t,u(x,t))=F(x,t)\,u(x,t). \tag{4.2.27}$$

To start, we present the following theorem.

**Theorem 4.2.1.** *If $\Lambda$ is a contraction, the equation under study has a unique solution. Thus if $F(x,t)\,u(x,t)$ absolutely integrable, and*

$$\begin{aligned}\|g(x,t)\| &< g_1(x,t) \\ \|F(x,t)\| &< F_1(x,t)\end{aligned} \tag{4.2.28}$$

*with* $g_1(x,t)$ *and* $F_1(x,t)$ *continuous positive functions, then there exists positive function* $u_p(x,t)$ *continuous such that:*

$$\begin{aligned} u_p(x,t) &= (1-\alpha)\, g_1(x,t) + \alpha \int_0^t g_1(x,\tau)\, d\tau \qquad (4.2.29) \\ &+ \left( \frac{(1-\alpha)^2}{\Gamma(\alpha)} + \alpha(1-\alpha) \right) F_1(x,t,u_p(x,t)) \\ &+ \frac{\alpha(1-\alpha)}{\Gamma(\alpha)} \int_0^t (t-\tau)^{\alpha-1} F(x,\tau,u_p(x,\tau))\, d\tau \\ &+ \frac{\alpha^2}{\Gamma(\alpha+1)} \int_0^l (t-\tau)^{\alpha} F(x,\tau,u_p(x,\tau))\, d\tau \end{aligned}$$

*with*

$$\|u(x,t)\| < u_p(x,t). \qquad (4.2.30)$$

*Proof.* By the formula of $u(x,t)$, we have:

$$\begin{aligned} \|u(x,t)\| &\leq (1-\alpha)\|g(x,t)\| + \alpha \int_0^t \|f(x,\tau)\|\, d\tau \qquad (4.2.31) \\ &+ \left( \frac{(1-\alpha)^2}{\Gamma(\alpha)} + \alpha(1-\alpha) \right) \|F(x,t,u)\| \\ &+ \frac{\alpha(1-\alpha)}{\Gamma(\alpha)} \int_0^t (t-\tau)^{\alpha-1} \|F(x,\tau,u)\|\, d\tau \\ &+ \frac{\alpha^2}{\Gamma(\alpha+1)} \int_0^l (t-\tau)^{\alpha} \|F(x,\tau,u)\|\, d\tau. \end{aligned}$$

Thus using the linearity of $F$ and also the hypothesis, we obtain

$$\begin{aligned} \|u(x,t)\| &\leq (1-\alpha)\|g(x,t)\| + \alpha \int_0^t \|f(x,\tau)\|\, d\tau \qquad (4.2.32) \\ &+ \left( \frac{(1-\alpha)^2}{\Gamma(\alpha)} + \alpha(1-\alpha) \right) \|F\|\, \|u\| \end{aligned}$$

$$+\frac{\alpha(1-\alpha)}{\Gamma(\alpha)}\int_0^t (t-\tau)^{\alpha-1}\|F\|\,\|u\|\,d\tau$$
$$+\frac{\alpha^2}{\Gamma(\alpha+1)}\int_0^l (t-\tau)^{\alpha}\|F\|\,\|u\|\,d\tau.$$

Also, we write

$$\begin{aligned}\|u(x,t)\| &< (1-\alpha)g_1(x,t)+\alpha\int_0^t g_1(x,\tau)\,d\tau \\ &+\left(\frac{(1-\alpha)^2}{\Gamma(\alpha)}+\alpha(1-\alpha)\right)F_1(x,t)\|u\| \\ &+\frac{\alpha(1-\alpha)}{\Gamma(\alpha)}\int_0^t (t-\tau)^{\alpha-1}F_1(x,\tau)\|u\|\,d\tau \\ &+\frac{\alpha^2}{\Gamma(\alpha+1)}\int_0^l (t-\tau)^{\alpha}F_1(x,\tau)\|u\|\,d\tau.\end{aligned} \tag{4.2.33}$$

Then, we have:

$$\begin{aligned}u_p(x,t)-\|u(x,t)\| &> \left(\frac{(1-\alpha)^2}{\Gamma(\alpha)}+\alpha(1-\alpha)\right) \\ &\times F_1(x,t)\left(u_p(x,t)-\|u(x,t)\|\right) \\ &+\frac{\alpha(1-\alpha)}{\Gamma(\alpha)}\int_0^t (t-\tau)^{\alpha-1}F_1(x,\tau) \\ &\times\left(u_p(x,t)-\|u(x,t)\|\right)d\tau \\ &+\frac{\alpha^2}{\Gamma(\alpha+1)}\int_0^l (t-\tau)^{\alpha}F_1(x,\tau) \\ &\times\left(u_p(x,t)-\|u(x,t)\|\right)d\tau.\end{aligned} \tag{4.2.34}$$

But, we know that

$$u_p(x,t)-\|u(x,t)\|>0. \tag{4.2.35}$$

Thus, with $F_1(x,t)$ being continuous, we can conclude that $\forall (x,t) \in [0,X] \times [0,T]$

$$u_p(x,t) - \|u(x,t)\| > 0. \tag{4.2.36}$$

This completes the proof.

**Theorem 4.2.2.** *If* $g(x,t)$, $\Delta g(x,t)$, $F(x,t)$ *and* $\Delta F(x,t)$ *are continuous functions and bounded such that:*

$$\begin{aligned} \|g(x,t)\| &< g_1, \ \|\Delta g(x,t)\| < \Delta g_1 \\ \|F(x,t)\| &< F_1, \ \|\Delta F(x,t)\| < \Delta F_1. \end{aligned} \tag{4.2.37}$$

*We let* $u_\Lambda(x,t)$ *to be the exact solution of*

$$\begin{aligned} u_\Lambda(x,t) &= (1-\alpha)(g(x,t) - \Delta g(x,t)) \\ &+ \left(\frac{(1-\alpha)^2}{\Gamma(\alpha)} + \alpha(1-\alpha)\right)[(F-\Delta F)(x,t)]\,u_\Lambda \\ &+ \frac{\alpha(1-\alpha)}{\Gamma(\alpha)}\int_0^t (t-\tau)^{\alpha-1}[(F-\Delta F)(x,\tau)]\,u_\Lambda d\tau \\ &+ \alpha\int_0^t [g(x,\tau) - \Delta g(x,\tau)]\,d\tau \\ &+ \frac{\alpha^2}{\Gamma(\alpha+1)}\int_0^l (t-\tau)^{\alpha}[(F-\Delta F)(x,\tau)]\,u_\Lambda d\tau, \end{aligned} \tag{4.2.38}$$

*then*

$$\|u_\Lambda(x,t) - u(x,t)\| = o(\Delta g) + o(\Delta F). \tag{4.2.39}$$

*Proof.* From the result presented in the above theorem, there exists a positive function

$$u_p(x,t) = (1-\alpha)g_1 + \left(\frac{(1-\alpha)^2}{\Gamma(\alpha)} + \alpha(1-\alpha)\right)F_1 u_p(x,t)$$

$$+\alpha g_1 \int_0^t d\tau + \frac{\alpha(1-\alpha)}{\Gamma(\alpha)} \int_0^t (t-\tau)^{\alpha-1} F_1 u_p(x,\tau)\, d\tau$$
$$+\frac{\alpha^2}{\Gamma(\alpha+1)} \int_0^l (t-\tau)^{\alpha} F_1 u_p(x,\tau)\, d\tau \tag{4.2.40}$$

for simplicity, we put

$$G(x,t) = (1-\alpha) g_1 + \alpha g_1 \int_0^t d\tau. \tag{4.2.41}$$

Thus, we have:

$$\begin{aligned} u_p(x,t) &= G(x,t) + \left( \frac{(1-\alpha)^2}{\Gamma(\alpha)} + \alpha(1-\alpha) \right) F_1 u_p(x,t) \\ &+ \frac{\alpha(1-\alpha)}{\Gamma(\alpha)} F_1 \int_0^t (t-\tau)^{\alpha-1} u_p(x,\tau)\, d\tau \qquad (4.2.42) \\ &+ \frac{\alpha^2}{\Gamma(\alpha+1)} F_1 \int_0^l (t-\tau)^{\alpha} u_p(x,\tau)\, d\tau. \end{aligned}$$

Now applying the Laplace transform on both sides, we obtain

$$\begin{aligned} \tilde{u}_p(x,D) &= G(x,D) + \alpha^2 D^{-\alpha-1} F_1 u_p(x,D) \qquad (4.2.43) \\ &+ \left( \frac{(1-\alpha)^2}{\Gamma(\alpha)} + \alpha(1-\alpha) \right) F_1 u_p(x,D) \\ &+ \alpha(1-\alpha) F_1 D^{-\alpha} u_p(x,D). \end{aligned}$$

Then

$$\tilde{u}_p(x,D) \left[ -A_\alpha - B_\alpha D^{-\alpha} + C_\alpha D^{-\alpha-1} \right] = G(x,D) \tag{4.2.44}$$

and

$$\tilde{u}_p(x, D) = \frac{G(x, D)}{C_\alpha D^{-\alpha-1} - A_\alpha - B_\alpha D^{-\alpha}}. \tag{4.2.45}$$

To continue, we consider the following equality

$$\tilde{u}_p(x, D) = \frac{G(x, D)\, D^{\alpha+1}}{C_\alpha - B_\alpha D - A_\alpha D^{\alpha+1}} \tag{4.2.46}$$

and

$$\tilde{u}_p(x, D) = \int_0^l G(x, t-\tau)\, \Lambda_\alpha(x, \tau)\, d\tau \tag{4.2.47}$$

where

$$\Lambda_\alpha(x, \tau) = L^{-1}\left(\frac{D^{\alpha+1}}{C_\alpha - B_\alpha D - A_\alpha D^{\alpha+1}}\right), \tag{4.2.48}$$

where $L^{-1}$ describes the inverse Laplace transform. Puttin all together and taking the norm, we can have the following:

$$\|u_p(x, t)\| < G E_\alpha\left(-\frac{B_\alpha}{A_\alpha} t^\alpha\right). \tag{4.2.49}$$

So from the above, we have the following:

$$\|u_\Lambda(x, t)\| \leq (G + \Delta G)\, E_\alpha\left[-\left(\frac{B_\alpha}{A_\alpha} + \Delta\frac{B_\alpha}{A_\alpha}\right) t^\alpha\right]. \tag{4.2.50}$$

But $u_\Lambda(x, t)$ is approximate solution of initial equation then,

$$\begin{aligned} u_\Lambda(x, t) &= G(x, t) + \left(\frac{(1-\alpha)^2}{\Gamma(\alpha)} + \alpha(1-\alpha)\right) F(x, t)\, u_\Lambda \\ &+ \frac{\alpha(1-\alpha)}{\Gamma(\alpha)} \int_0^t (t-\tau)^{\alpha-1} F(x, \tau)\, u_\Lambda d\tau \\ &+ \frac{\alpha^2}{\Gamma(\alpha+1)} \int_0^l (t-\tau)^\alpha F(x, \tau)\, u_\Lambda d\tau \end{aligned} \tag{4.2.51}$$

such that the remainder should be

$$
\begin{aligned}
\Pi_{\Pi}(x,t) &= G(x,t) + \left(\frac{(1-\alpha)^2}{\Gamma(\alpha)} + \alpha(1-\alpha)\right) F(x,t)\, u_{\Lambda} \\
&+ \frac{\alpha(1-\alpha)}{\Gamma(\alpha)} \int_0^t (t-\tau)^{\alpha-1} F(x,\tau)\, u_{\Lambda} d\tau \qquad (4.2.52) \\
&+ \frac{\alpha^2}{\Gamma(\alpha+1)} \int_0^l (t-\tau)^{\alpha} F(x,\tau)\, u_{\Lambda} d\tau - u_{\Lambda}.
\end{aligned}
$$

Now, replacing $u_{\Lambda}(x,t)$ by its value we obtain

$$
\begin{aligned}
\Pi_{\Pi}(x,t) &= -\Delta G(x,t) + \left(\frac{(1-\alpha)^2}{\Gamma(\alpha)} + \alpha(1-\alpha)\right) \Delta F(x,t)\, u_{\Lambda} \\
&+ \frac{\alpha(1-\alpha)}{\Gamma(\alpha)} \int_0^t (t-\tau)^{\alpha-1} \Delta F(x,\tau)\, u_{\Lambda} d\tau \\
&+ \frac{\alpha^2}{\Gamma(\alpha+1)} \int_0^l (t-\tau)^{\alpha} \Delta F(x,\tau)\, u_{\Lambda} d\tau. \qquad (4.2.53)
\end{aligned}
$$

Thus, the following is written:

$$
\begin{aligned}
u(x,t) - u_{\Lambda}(x,t) &= \Pi_{\Pi}(x,t) + \left(\frac{(1-\alpha)^2}{\Gamma(\alpha)} + \alpha(1-\alpha)\right) \\
&\times \Delta F(x,t)\left[u(x,t) - u_{\Lambda}(x,t)\right] \\
&+ \frac{\alpha(1-\alpha)}{\Gamma(\alpha)} \int_0^t (t-\tau)^{\alpha-1} \Delta F(x,\tau) \\
&\times \left[u(x,\tau) - u_{\Lambda}(x,\tau)\right] d\tau \\
&+ \frac{\alpha^2}{\Gamma(\alpha+1)} \int_0^l (t-\tau)^{\alpha} \Delta F(x,\tau) \\
&\times \left[u(x,\tau) - u_{\Lambda}(x,\tau)\right] d\tau \qquad (4.2.54)
\end{aligned}
$$

and

$$\begin{aligned}
\|u(x,t)-u_{\Lambda}(x,t)\| \leq \|\Pi_{\Pi}(x,t)\| + \left(\frac{(1-\alpha)^{2}}{\Gamma(\alpha)}+\alpha(1-\alpha)\right) \\
\times\|\Delta F(x,t)\|\,\|u(x,t)-u_{\Lambda}(x,t)\| \\
+\frac{\alpha(1-\alpha)}{\Gamma(\alpha)}\int_{0}^{t}(t-\tau)^{\alpha-1}\|\Delta F(x,\tau)\| \\
\times\|u(x,\tau)-u_{\Lambda}(x,\tau)\|\,d\tau \\
+\frac{\alpha^{2}}{\Gamma(\alpha+1)}\int_{0}^{l}(t-\tau)^{\alpha}\|\Delta F(x,\tau)\| \\
\times\|u(x,\tau)-u_{\Lambda}(x,\tau)\|\,d\tau. \qquad (4.2.55)
\end{aligned}$$

We arrange the above as follows:

$$\begin{aligned}
\|u(x,t)-u_{\Lambda}(x,t)\| &\leq \|\Pi_{\Pi}(x,t)\| \qquad (4.2.56)\\
&+\|\Delta F(x,t)\|\left\{\begin{array}{c}\frac{(1-\alpha)^{2}}{\Gamma(\alpha)}+\alpha(1-\alpha)\\ +\frac{(1-\alpha)^{2}t^{\alpha}}{\Gamma(\alpha)}+\frac{\alpha^{2}t^{\alpha+1}}{\Gamma(\alpha+2)}\end{array}\right\}\\
&\times\|u(x,\tau)-u_{\Lambda}(x,\tau)\|.
\end{aligned}$$

Thus, we obtain:

$$\|u(x,\tau)-u_{\Lambda}(x,\tau)\| = O(\Delta G)+O(\Delta F). \qquad (4.2.57)$$

This completes the proof.

Now, we consider the case where $F$ is nonlinear. Due to nonlinearity, we present the derivation of the existence and uniqueness using iterative approach, thus, we consider the following iterative formula:

$$u_{n+1}(x,t) = G(x,t)+\left(\frac{(1-\alpha)^{2}}{\Gamma(\alpha)}+\alpha(1-\alpha)\right)F(x,t,u_{n})$$

$$+\frac{\alpha(1-\alpha)}{\Gamma(\alpha)}\int_0^t (t-\tau)^{\alpha-1} F(x,\tau,u_n)\,d\tau$$
$$+\frac{\alpha^2}{\Gamma(\alpha+1)}\int_0^l (t-\tau)^{\alpha} F(x,\tau,u_n)\,d\tau \qquad (4.2.58)$$

where the initial $u_1(x,t) = G(x,t)$. Taking the difference of $u_{n+1}(x,t)$ and $u_n(x,t)$, we achieve the following:

$$\begin{aligned}\mu_{n+1}(x,t) &= u_{n+1}(x,t)-u_n(x,t)=\left(\frac{(1-\alpha)^2}{\Gamma(\alpha)}+\alpha(1-\alpha)\right)\\ &\quad\times\left[F(x,t,u_n(x,t))-F(x,t,u_{n-1}(x,t))\right] \qquad (4.2.59)\\ &\quad+\frac{\alpha(1-\alpha)}{\Gamma(\alpha)}\int_0^t (t-\tau)^{\alpha-1}\left[\begin{array}{c}F(x,\tau,u_n)\\ -F(x,\tau,u_{n-1})\end{array}\right]d\tau\\ &\quad+\frac{\alpha^2}{\Gamma(\alpha+1)}\int_0^l (t-\tau)^{\alpha}\left[\begin{array}{c}F(x,\tau,u_n)\\ -F(x,\tau,u_{n-1})\end{array}\right]d\tau.\end{aligned}$$

We let,

$$u_{n+1}(x,t)=\sum_{j=0}^{n+1}\mu_{j+1}. \qquad (4.2.60)$$

Now,

$$\begin{aligned}\|\mu_{n+1}(x,t)\| &\le \left(\frac{(1-\alpha)^2}{\Gamma(\alpha)}+\alpha(1-\alpha)\right) \qquad (4.2.61)\\ &\quad\times\|F(x,t,u_n(x,t))-F(x,t,u_{n-1}(x,t))\|\\ &\quad+\frac{\alpha(1-\alpha)}{\Gamma(\alpha)}\int_0^t (t-\tau)^{\alpha-1}\\ &\quad\times\|F(x,\tau,u_n(x,\tau))-F(x,\tau,u_{n-1}(x,\tau))\|\,d\tau\\ &\quad+\frac{\alpha^2}{\Gamma(\alpha+1)}\int_0^l (t-\tau)^{\alpha}\\ &\quad\|F(x,\tau,u_n(x,\tau))-F(x,\tau,u_{n-1}(x,\tau))\|\,d\tau.\end{aligned}$$

In the previous section, we considered $F$ to have Lipschitz conditions then the above can be:

$$\begin{aligned}
\left\|\mu_{n+1}(x,t)\right\| &\leq \left(\frac{(1-\alpha)^2}{\Gamma(\alpha)}+\alpha(1-\alpha)\right)L\left\|u_n-u_{n-1}\right\| \\
&+\frac{(1-\alpha)t^{\alpha}}{\Gamma(\alpha)}L\left\|u_n-u_{n-1}\right\| \qquad (4.2.62)\\
&+\frac{\alpha^2 t^{\alpha+1}}{\Gamma(\alpha+2)}L\left\|u_n-u_{n-1}\right\| \\
&\leq \left[\begin{array}{c}\left(\frac{(1-\alpha)^2}{\Gamma(\alpha)}+\alpha(1-\alpha)\right)\\ +\frac{(1-\alpha)t^{\alpha}}{\Gamma(\alpha)}+\frac{\alpha^2t^{\alpha+1}}{\Gamma(\alpha+2)}\end{array}\right]L\left\|\mu_n\right\|
\end{aligned}$$

$$\leq \left[\begin{array}{c}\left(\frac{(1-\alpha)^2}{\Gamma(\alpha)}+\alpha(1-\alpha)\right)\\ +\frac{(1-\alpha)t^{\alpha}}{\Gamma(\alpha)}+\frac{\alpha^2t^{\alpha+1}}{\Gamma(\alpha+2)}\end{array}\right]^2 L^2\left\|\mu_{n-1}\right\|$$

$$\vdots$$

$$\begin{aligned}
\left\|\mu_{n+1}(x,t)\right\| &\leq \left[\begin{array}{c}\left(\frac{(1-\alpha)^2}{\Gamma(\alpha)}+\alpha(1-\alpha)\right)\\ +\frac{(1-\alpha)t^{\alpha}}{\Gamma(\alpha)}+\frac{\alpha^2t^{\alpha+1}}{\Gamma(\alpha+2)}\end{array}\right]^n L^n\left\|\mu_0\right\| \\
&\leq \left[\begin{array}{c}\left(\frac{(1-\alpha)^2}{\Gamma(\alpha)}+\alpha(1-\alpha)\right)\\ +\frac{(1-\alpha)t^{\alpha}}{\Gamma(\alpha)}+\frac{\alpha^2t^{\alpha+1}}{\Gamma(\alpha+2)}\end{array}\right]^n \\
&\quad\times L^n \max_{(x,t)\in(0,.X)\times(0,T)}\left|G(x,t)\right|.
\end{aligned}$$

Since $L < 1$ then, when $n \to \infty$, $L \to 0$. Thus we get:

$$u(x,t) = \sum_{j=0}^{n} \mu_{j+1}(x,t). \tag{4.2.63}$$

It is clear that $u(x,t)$ exists and is continuous, thus we consider $B_n(x,t)$ to be

$$u(x,t) = u_n(x,t) + B_n(x,t). \tag{4.2.64}$$

Then, one can have the following:

$$\begin{aligned} u(x,t) - u_n(x,t) &= G(x,t) + \left( \frac{(1-\alpha)^2}{\Gamma(\alpha)} + \alpha(1-\alpha) \right) \\ &\quad \times \left[ F(x,t,u(x,t) - B_n(x,t)) \right] \qquad (4.2.65) \\ &\quad + \frac{\alpha(1-\alpha)}{\Gamma(\alpha)} \int_0^t (t-\tau)^{\alpha-1} \\ &\quad \times \left[ F(x,\tau,u(x,\tau) - B_n(x,\tau)) \right] d\tau \end{aligned}$$

$$+ \frac{\alpha^2}{\Gamma(\alpha+1)} \int_0^l (t-\tau)^{\alpha} \left[ F(x,\tau,u(x,\tau) - B_n(x,\tau)) \right] d\tau.$$

However,

$$\begin{aligned} & u(x,t) - G(x,t) - \left( \frac{(1-\alpha)^2}{\Gamma(\alpha)} + \alpha(1-\alpha) \right) F(x,t,u) \\ & + \frac{\alpha(1-\alpha)}{\Gamma(\alpha)} \int_0^t (t-\tau)^{\alpha-1} F(x,\tau,u)\, d\tau \qquad (4.2.66) \\ & + \frac{\alpha^2}{\Gamma(\alpha+1)} \int_0^l (t-\tau)^{\alpha} F(x,\tau,u)\, d\tau = B_n(x,t) \end{aligned}$$

$$+ \left( \frac{(1-\alpha)^2}{\Gamma(\alpha)} + \alpha(1-\alpha) \right) \left[ F(x,t,u - B_n(x,t)) \right]$$

$$+\frac{\alpha\left(1-\alpha\right)}{\Gamma\left(\alpha\right)}\int_{0}^{t}\left(t-\tau\right)^{\alpha-1}\left[F\left(x,\tau,u-B_{n}\left(x,\tau\right)\right)\right]d\tau$$
$$+\frac{\alpha^{2}}{\Gamma\left(\alpha+1\right)}\int_{0}^{l}\left(t-\tau\right)^{\alpha}\left[F\left(x,\tau,u-B_{n-1}\left(x,\tau\right)\right)\right]d\tau$$

such that:

$$\left\Vert \begin{array}{c} u\left(x,t\right)-G\left(x,t\right) \\ -\left(\frac{(1-\alpha)^{2}}{\Gamma(\alpha)}+\alpha\left(1-\alpha\right)\right)F\left(x,t,u\left(x,t\right)\right) \\ +\frac{\alpha(1-\alpha)}{\Gamma(\alpha)}\int_{0}^{t}\left(t-\tau\right)^{\alpha-1}F\left(x,\tau,u\left(x,\tau\right)\right)d\tau \\ +\frac{\alpha^{2}}{\Gamma(\alpha+1)}\int_{0}^{l}\left(t-\tau\right)^{\alpha}F\left(x,\tau,u\left(x,\tau\right)\right)d\tau \end{array}\right\Vert \leq\left\Vert B_{n}\left(x,t\right)\right\Vert$$

$$+\left(\frac{\left(1-\alpha\right)^{2}}{\Gamma\left(\alpha\right)}+\alpha\left(1-\alpha\right)\right)L\left\Vert B_{n-1}\left(x,t\right)\right\Vert$$
$$+\left[\frac{\alpha\left(1-\alpha\right)}{\Gamma\left(\alpha+1\right)}t^{\alpha}L+\frac{\alpha^{2}}{\Gamma\left(\alpha+3\right)}t^{\alpha}L\right]\left\Vert B_{n-1}\left(x,t\right)\right\Vert . \tag{4.2.67}$$

Now, $n\to\infty$ the right-hand side tends to zero also, such that:

$$\begin{aligned} u\left(x,t\right) &= G\left(x,t\right)+\left(\frac{\left(1-\alpha\right)^{2}}{\Gamma\left(\alpha\right)}+\alpha\left(1-\alpha\right)\right)F\left(x,t,u\right) \\ &\quad +\frac{\alpha\left(1-\alpha\right)}{\Gamma\left(\alpha\right)}\int_{0}^{t}\left(t-\tau\right)^{\alpha-1}F\left(x,\tau,u\right)d\tau \\ &\quad +\frac{\alpha^{2}}{\Gamma\left(\alpha+1\right)}\int_{0}^{l}\left(t-\tau\right)^{\alpha}F\left(x,\tau,u\right)d\tau. \end{aligned} \tag{4.2.68}$$

We conclude the proof by showing the uniqueness:

$$\left\Vert u\left(x,t\right)-u_{1}\left(x,t\right)\right\Vert \leq\left[\begin{array}{c} \frac{(1-\alpha)^{2}}{\Gamma(\alpha)}+\alpha\left(1-\alpha\right) \\ +\frac{(1-\alpha)}{\Gamma(\alpha)}t^{\alpha}+\frac{\alpha^{2}}{\Gamma(\alpha+2)}t^{\alpha+1} \end{array}\right] \tag{4.2.69}$$
$$\times L\left\Vert u\left(x,t\right)-u_{1}\left(x,t\right)\right\Vert$$

$$\leq \begin{bmatrix} \frac{(1-\alpha)^2}{\Gamma(\alpha)} + \alpha(1-\alpha) \\ +\frac{(1-\alpha)}{\Gamma(\alpha)}t^{\alpha} + \frac{\alpha^2}{\Gamma(\alpha+2)}t^{\alpha+1} \end{bmatrix}^2$$

$$\times L^2 \|u(x,t) - u_1(x,t)\|$$

$$\vdots$$

$$\leq \begin{bmatrix} \frac{(1-\alpha)^2}{\Gamma(\alpha)} + \alpha(1-\alpha) \\ +\frac{(1-\alpha)}{\Gamma(\alpha)}t^{\alpha} + \frac{\alpha^2}{\Gamma(\alpha+2)}t^{\alpha+1} \end{bmatrix} L^n D.$$

Thus, we have $L^n \to 0$ and the right hand side tends to zero then:

$$u(x,t) = u_1(x,t). \tag{4.2.70}$$

### *4.2.2 Numerical Solution*

In this section, we present the numerical solution of the new class of partial integro-differential equation with non-singular kernel [13]. To do this, we consider the Volterra equation at $(x_i, t_{n+1})$ thus, we have:

$$\begin{aligned} u_i^{n+1} &= u_i^0 + (1-\alpha) g_i^n + \alpha \sum_{j=0}^{n-1} g_i^j \Delta t \\ &\quad + (1-\alpha)^2 F(x_{i,}, t_n, u_i^n) \end{aligned} \tag{4.2.71}$$

$$+\alpha(1-\alpha)\sum_{j=0}^{n-1} F\left(x_{i,}, t_j, u_i^j\right) + \frac{(1-\alpha)}{\Gamma(\alpha+2)}(\Delta t)^{\alpha}$$

$$\times \sum_{j=0}^{n-1} \left[ \begin{array}{c} F\left(x_i, t_j, u_i^j\right) \\ \times \left\{ \begin{array}{c} (n-j+1)^{\alpha}(n-j+2+\alpha) \\ -(n-j)^{\alpha}(n-j+2+2\alpha) \end{array} \right\} \\ -F\left(x_i, t_{j-1}, u_i^{j-1}\right) \\ \times \left\{ \begin{array}{c} (n-j+1)^{\alpha+1} \\ -(n-j)^{\alpha}(n-j+1+\alpha) \end{array} \right\} \end{array} \right]$$

$$+\frac{\alpha^2}{\Gamma(\alpha+3)}(\Delta t)^{\alpha+1}$$

$$\times \sum_{j=0}^{n} \left[ \begin{array}{c} F\left(x_i, t_j, u_i^j\right) \\ \times \left\{ \begin{array}{c} (n-j+1)^{\alpha+1}(n-j+3+\alpha) \\ -(n-j)^{\alpha+1}(n-j+3+2\alpha) \end{array} \right\} \\ -F\left(x_i, t_{j-1}, u_i^{j-1}\right) \\ \times \left\{ \begin{array}{c} (n-j+1)^{\alpha+2} \\ -(n-j)^{\alpha+2}(n-j+2+\alpha) \end{array} \right\} \end{array} \right].$$

---

## 4.3 Model with Atangana-Baleanu Derivative and Caputo-Fabrizio Integral

In this section, we present the analysis of the new partial differential equation where the derivative is Atangana-Baleanu and the integral is that of Caputo-Fabrizio.The general equation is given as:

$$\frac{1}{1-\alpha}\int_0^t \frac{\partial u(x,\tau)}{\partial \tau} E_\alpha\left[-\frac{\alpha}{1-\alpha}(t-\tau)^\alpha\right] d\tau = f(x,t) \qquad (4.3.72)$$

$$+{}_0^{CF}I_t^{\alpha} F(x,t,u(x,t)).$$

Employing the definition of the Caputo-Fabrizio integral, we have the following:

$$\frac{1}{1-\alpha}\int_0^t \frac{\partial u(x,\tau)}{\partial \tau} E_{\alpha}\left[-\frac{\alpha}{1-\alpha}(t-\tau)^{\alpha}\right] d\tau = f(x,t) \tag{4.3.73}$$
$$+(1-\alpha)F(x,t,u(x,t)) + \alpha\int_0^t F(x,\tau,u(x,\tau))\,d\tau.$$

After applying the Atangana-Baleanu integral on both sides, the above can be arranged as:

$$u(x,t) - u(x,0) = (1-\alpha)\left[\begin{array}{c} f(x,t) + (1-\alpha)F(x,t,u(x,t)) \\ +\alpha\int_0^t F(x,\tau,u(x,\tau))\,d\tau \end{array}\right] \tag{4.3.74}$$

$$+\alpha\int_0^t \left[\begin{array}{c} f(x,\tau) + (1-\alpha)F(x,\tau,u(x,\tau)) \\ +\alpha\int_0^l F(x,\tau,u(x,\tau))\,d\tau \end{array}\right]$$
$$\times(t-\tau)^{\alpha-1}\,d\tau.$$

Rearranging, we get:

$$\begin{aligned} u(x,t) - u(x,0) &= (1-\alpha)f(x,t) + (1-\alpha)^2 F(x,t,u) \quad (4.3.75) \\ &\quad +\alpha\int_0^t f(x,\tau)(t-\tau)^{\alpha-1}\,d\tau \\ &\quad +\alpha(1-\alpha)\int_0^t F(x,\tau,u(x,\tau))\,d\tau \\ &\quad +\frac{\alpha(1-\alpha)}{\Gamma(\alpha)}\int_0^l F(x,\tau,u)(t-\tau)^{\alpha-1}\,d\tau \\ &\quad +\frac{\alpha^2}{\Gamma(\alpha+1)}\int_0^l (t-\tau)^{\alpha} F(x,\tau,u)\,d\tau. \end{aligned}$$

simplifying the above with the choice

$$G(x,t) = u(x,0) + (1-\alpha) f(x,t) + \alpha \int_0^t f(x,\tau)(t-\tau)^{\alpha-1} d\tau, \tag{4.3.76}$$

then, we have:

$$\begin{aligned} u(x,t) = G(x,t) + (1-\alpha)^2 F(x,t,u(x,t)) \\ +\alpha(1-\alpha)\int_0^t F(x,\tau,u(x,\tau))\, d\tau \\ +\frac{\alpha(1-\alpha)}{\Gamma(\alpha)}\int_0^l (t-\tau)^{\alpha-1} F(x,\tau,u(x,\tau))\, d\tau \\ +\frac{\alpha^2}{\Gamma(\alpha+1)}\int_0^l (t-\tau)^{\alpha} F(x,\tau,u(x,\tau))\, d\tau. \end{aligned} \tag{4.3.77}$$

and

$$\begin{aligned} \|u(x,t) - G(x,t)\| \le (1-\alpha)^2 \|F(x,t,u(x,t))\| \\ +\alpha(1-\alpha)\int_0^t \|F(x,\tau,u(x,\tau))\|\, d\tau \\ +\frac{\alpha(1-\alpha)}{\Gamma(\alpha)}\int_0^l (t-\tau)^{\alpha-1} \|F(x,\tau,u(x,\tau))\|\, d\tau \\ +\frac{\alpha^2}{\Gamma(\alpha+1)}\int_0^l (t-\tau)^{\alpha} \|F(x,\tau,u(x,\tau))\|\, d\tau. \end{aligned} \tag{4.3.78}$$

We need $F$ to be bounded such that:

$$\begin{aligned} \|u(x,t) - G(x,t)\| \le\ & (1-\alpha)^2 M_F + \alpha(1-\alpha) M_F t \\ & +\frac{\alpha(1-\alpha)}{\Gamma(\alpha)} M_F t^{\alpha} + \frac{\alpha^2}{\Gamma(\alpha+2)} M_F t^{\alpha+1} \end{aligned} \tag{4.3.79}$$

$$\leq \left( \begin{array}{c} (1-\alpha)^2 + \alpha(1-\alpha)t \\ +\frac{\alpha(1-\alpha)}{\Gamma(\alpha)} M_F t^\alpha + \frac{\alpha^2}{\Gamma(\alpha+2)} t^{\alpha+1} \end{array} \right) M_F < \infty.$$

We consider the following mapping:

$$\begin{aligned} Tv &= G(x,t) + (1-\alpha)^2 F(x,t,v(x,t)) \qquad (4.3.80) \\ &+\alpha(1-\alpha)\int_0^t (t-\tau)^{\alpha-1} F(x,\tau,v(x,\tau))\,d\tau \\ &+\frac{\alpha^2}{\Gamma(\alpha+1)}\int_0^l (t-\tau)^\alpha F(x,\tau,v(x,\tau))\,d\tau. \end{aligned}$$

and then,

$$\begin{aligned} \|Tv-Tu\| &= \left\| \begin{array}{c} (1-\alpha)^2 [F(x,t,v(x,t)) - F(x,t,u(x,t))] \\ +\alpha(1-\alpha)\int_0^t (t-\tau)^{\alpha-1} [F(x,\tau,v) - F(x,\tau,u)]\,d\tau \\ +\frac{\alpha^2}{\Gamma(\alpha+1)}\int_0^l (t-\tau)^\alpha [F(x,\tau,v) - F(x,\tau,u)]\,d\tau. \end{array} \right\| \\ &\leq (1-\alpha)^2 \|F(x,t,v(x,t)) - F(x,t,u(x,t))\| \qquad (4.3.81) \\ &+\alpha(1-\alpha)\int_0^t (t-\tau)^{\alpha-1} \|F(x,\tau,v) - F(x,\tau,u)\|\,d\tau \\ &+\frac{\alpha^2}{\Gamma(\alpha+1)}\int_0^l (t-\tau)^\alpha \|F(x,\tau,v) - F(x,\tau,u)\|\,d\tau. \end{aligned}$$

We need $F$ to be a contraction such that:

$$\begin{aligned} \|Tv-Tu\| &\leq (1-\alpha)^2 L\|v(x,t)-u(x,t)\| \qquad (4.3.82) \\ &+\alpha(1-\alpha)tL\|v(x,t)-u(x,t)\| \\ &+\alpha(1-\alpha)\frac{t^\alpha L}{\Gamma(\alpha)}\|v(x,t)-u(x,t)\| \\ &+\frac{\alpha^2 t^{\alpha+1} L}{\Gamma(\alpha+2)}\|v(x,t)-u(x,t)\|. \end{aligned}$$

### *4.3.1 Existence and Uniqueness*

In this subsection, we present the existence and uniqueness of the exact solution. To start this process, we first assume that the function $F$ is linear that it can be written like this:

$$F(x,t,u(x,t)) = F(x,t)\,u(x,t). \tag{4.3.83}$$

Thus, our equation becomes:

$$\begin{aligned} u(x,t) &= G(x,t) + (1-\alpha)^2 F(x,t,u(x,t)) \\ &\quad + \alpha(1-\alpha)\int_0^t F(x,\tau,u(x,\tau))\,d\tau \\ &\quad + \frac{(1-\alpha)}{AB(\alpha)\Gamma(\alpha)}\frac{\alpha}{M(\alpha)}\int_0^l (t-\tau)^{\alpha-1} F(x,\tau,u(x,\tau))\,d\tau \\ &\quad + \frac{\alpha^2}{AB(\alpha)\Gamma(\alpha+1)M(\alpha)}\int_0^l (t-\tau)^{\alpha} F(x,\tau,u(x,\tau))\,d\tau. \end{aligned} \tag{4.3.84}$$

**Theorem 4.3.1.** *If $F(x,t)\,u(x,t)$ is absolutely integrable and*

$$\begin{aligned} \|G(x,t)\| &< g(x,t) \\ \|F(x,t)\| &< F_1(x,t) \end{aligned} \tag{4.3.85}$$

*where $g(x,t)$ and $F_1(x,t)$ are positively continuous functions, then there exists a function $u_1(x,t)$ such that:*

$$\|u(x,t)\| < u_1(x,t). \tag{4.3.86}$$

*Proof.* Using equation (4.3.84), we write:

$$\|u(x,t)\| = \left\| \begin{array}{c} G(x,t) + (1-\alpha)^2 F(x,t,u(x,t)) \\ +\alpha(1-\alpha)\int_0^t F(x,\tau,u(x,\tau))\,d\tau \\ +\frac{\alpha(1-\alpha)}{\Gamma(\alpha)}\int_0^l (t-\tau)^{\alpha-1} F(x,\tau,u(x,\tau))\,d\tau \\ +\frac{\alpha^2}{\Gamma(\alpha+1)}\int_0^l (t-\tau)^{\alpha} F(x,\tau,u(x,\tau))\,d\tau \end{array} \right\|. \tag{4.3.87}$$

Since $F$ is linear, we have:

$$\begin{aligned} \|u(x,t)\| \leq\ & \|G(x,t)\| + (1-\alpha)^2 \|F(x,t)\|\,\|u(x,t)\| \\ & +\alpha(1-\alpha)\int_0^t \|F(x,\tau)\|\,\|u(x,\tau)\|\,d\tau \\ & +\frac{\alpha(1-\alpha)}{\Gamma(\alpha)}\int_0^l (t-\tau)^{\alpha-1}\|F(x,\tau)\|\,\|u(x,\tau)\|\,d\tau \\ & +\frac{\alpha^2}{\Gamma(\alpha+1)}\int_0^l (t-\tau)^{\alpha}\|F(x,\tau)\|\,\|u(x,\tau)\|\,d\tau. \end{aligned} \tag{4.3.88}$$

Thus, applying the hypothesis, we get:

$$\begin{aligned} \|u(x,t)\| \leq\ & G(x,t) + (1-\alpha)^2 F_1(x,t)\,\|u(x,t)\| \\ & +\alpha(1-\alpha)\int_0^t F_1(x,\tau)\,\|u(x,\tau)\|\,d\tau \\ & +\frac{\alpha(1-\alpha)}{\Gamma(\alpha)}\int_0^l (t-\tau)^{\alpha-1} F_1(x,\tau)\,\|u(x,\tau)\|\,d\tau \\ & +\frac{\alpha^2}{\Gamma(\alpha+1)}\int_0^l (t-\tau)^{\alpha} F_1(x,\tau)\,\|u(x,\tau)\|\,d\tau. \end{aligned} \tag{4.3.89}$$

Then, we have:

$$\begin{aligned} u_1(x,t) - \|u(x,t)\| >\ & (1-\alpha)^2 F_1(x,t)(u_1 - \|u\|) \\ & +\alpha(1-\alpha)\int_0^t F_1(x,\tau)(u_1 - \|u\|)\,d\tau \end{aligned} \tag{4.3.90}$$

$$+\frac{\alpha(1-\alpha)}{\Gamma(\alpha)}\int_0^l (t-\tau)^{\alpha-1} F_1(x,\tau)(u_1-\|u\|)\,d\tau$$
$$+\frac{\alpha^2}{\Gamma(\alpha+1)}\int_0^l (t-\tau)^{\alpha} F_1(x,\tau)(u_1-\|u\|)\,d\tau.$$

We know that

$$u_1-\|u\|>0. \tag{4.3.91}$$

From continuity of the function $F$, we can follow that $\forall(x,t)\in(0,X)\times(0,T)$

$$u_1(x,t)-\|u\|>0. \tag{4.3.92}$$

**Theorem 4.3.2.** *If* $G(x,t)$, $F(x,t)$, $\Delta G(x,t)$ *and* $\Delta F(x,t)$ *are continuous and bounded, then let* $u_p(x,t)$ *be the exact solution of equation*

$$u_p(x,t)=G(x,t)+\Delta G(x,t) \tag{4.3.93}$$
$$+(1-\alpha)^2\left[F(x,t,u_p(x,t))-\Delta F(x,t,u_p(x,t))\right]$$
$$+\alpha(1-\alpha)\int_0^t\left[F(x,\tau)-\Delta F(x,\tau)\right]u_p(x,\tau)\,d\tau$$
$$+\frac{\alpha(1-\alpha)}{\Gamma(\alpha)}\int_0^l(t-\tau)^{\alpha-1}\left[F(x,\tau)-\Delta F(x,\tau)\right]u_p(x,\tau)\,d\tau$$
$$+\frac{\alpha^2}{\Gamma(\alpha+1)}\int_0^l(t-\tau)^{\alpha}\left[F(x,\tau)-\Delta F(x,\tau)\right]u_p(x,\tau)\,d\tau$$

*and*

$$\|u_p(x,t)-u(x,t)\|=o(\Delta g)+o(\Delta F) \tag{4.3.94}$$

*with* $u_p(x,t)$ *being the exact solution of the (4.3.84).*

Now we assume that $F(x,t,u(x,t))$ is nonlinear, in this case, we consider the following iterative formula

$$\begin{aligned} u_{n+1}(x,t) &= G(x,t) + (1-\alpha)^2 F(x,t,u_n(x,t)) \\ &+\alpha(1-\alpha)\int_0^t (t-\tau)^{\alpha-1} F(x,\tau,u_n(x,\tau))\,d\tau \\ &+\frac{\alpha^2}{\Gamma(\alpha+1)}\int_0^l (t-\tau)^{\alpha} F(x,\tau,u_n(x,\tau))\,d\tau. \end{aligned} \tag{4.3.95}$$

We consider the difference between two consecutive sequences to be

$$\Omega_{n+1} = u_{n+1}(x,t) - u_n(x,t). \tag{4.3.96}$$

The iteration will be performed by taking

$$u_1(x,t) = G(x,t) \tag{4.3.97}$$

and then

$$\begin{aligned} \Omega_{n+1}(x,t) &= (1-\alpha)^2 \left[F(x,t,u_n) - F(x,t,u_{n-1})\right] \\ &+\alpha(1-\alpha)\int_0^t (t-\tau)^{\alpha-1} \left[F(x,\tau,u_n) - F(x,\tau,u_{n-1})\right] d\tau \\ &+\frac{\alpha^2}{\Gamma(\alpha+1)}\int_0^l (t-\tau)^{\alpha} \left[F(x,\tau,u_n) - F(x,\tau,u_{n-1})\right] d\tau. \end{aligned} \tag{4.3.98}$$

We put it for simplicity and the reason for the existence of the exact solution:

$$u_{n+1}(x,t) = \sum_{j=0}^{n+1} \Omega_{j+1}(x,t). \tag{4.3.99}$$

We now evaluate the following $\|\Omega_{n+1}(x,t)\|$ to see if the considered $\left(u_{n+1}(x,t)_{n\in\mathbb{N}}\right)$ could form a Cauchy series:

$$\|\Omega_{n+1}(x,t)\| \leq (1-\alpha)^2 \|F(x,t,u_n) - F(x,t,u_{n-1})\|$$
$$+\alpha(1-\alpha)\int_0^t (t-\tau)^{\alpha-1}\|F(x,\tau,u_n) - F(x,\tau,u_{n-1})\| d\tau$$
$$(4.3.100)$$
$$+\frac{\alpha^2}{\Gamma(\alpha+1)}\int_0^l (t-\tau)^{\alpha}\|F(x,\tau,u_n) - F(x,\tau,u_{n-1})\| d\tau.$$

Since $F$ is a contraction as we wished before, the above can be converted to:

$$\|\Omega_{n+1}(x,t)\| \leq (1-\alpha)^2 L\|u_n(x,t) - u_{n-1}(x,t)\| \quad (4.3.101)$$
$$+\frac{\alpha(1-\alpha)}{\Gamma(\alpha)}t^{\alpha}L\|u_n(x,\tau) - u_{n-1}(x,\tau)\|$$
$$+\frac{\alpha^2}{\Gamma(\alpha+2)}t^{\alpha+1}L\|u_n(x,\tau) - u_{n-1}(x,\tau)\|.$$

If we arrange the above, we get the following:

$$\|\Omega_{n+1}(x,t)\| \leq L\|u_n(x,t) - u_{n-1}(x,t)\| \times\left\{\begin{array}{c}(1-\alpha)^2 + \frac{\alpha(1-\alpha)}{\Gamma(\alpha)}t^{\alpha} \\ +\frac{\alpha^2}{\Gamma(\alpha+2)}t^{\alpha+1}\end{array}\right\}. \quad (4.3.102)$$

The above produces

$$\|\Omega_{n+1}(x,t)\| \leq L^2\|u_{n-1}(x,t) - u_{n-2}(x,t)\| \times\left\{\begin{array}{c}(1-\alpha)^2 + \frac{\alpha(1-\alpha)}{\Gamma(\alpha)}t^{\alpha} \\ +\frac{\alpha^2}{\Gamma(\alpha+2)}t^{\alpha+1}\end{array}\right\}^2 \quad (4.3.103)$$

$$\leq L^3 \|u_{n-2}(x,t) - u_{n-3}(x,t)\| \times \left\{ \begin{array}{c} (1-\alpha)^2 + \frac{\alpha(1-\alpha)}{\Gamma(\alpha)} t^{\alpha} \\ + \frac{\alpha^2}{\Gamma(\alpha+2)} t^{\alpha+1} \end{array} \right\}^3$$

$$\vdots$$

$$\leq L^{n+1} \|\Omega_1\| \left\{ \begin{array}{c} (1-\alpha)^2 + \frac{\alpha(1-\alpha)}{\Gamma(\alpha)} t^{\alpha} \\ + \frac{\alpha^2}{\Gamma(\alpha+2)} t^{\alpha+1} \end{array} \right\}^{n+1}$$

and then

$$\|\Omega_{n+1}(x,t)\| \leq L^{n+1} \delta_{\alpha}(t) \max_{(x,t)\in[0,X]\times[0,T]} |G(x,t)| . \tag{4.3.104}$$

When $n \to \infty$, the right hand side tends to zero such that:

$$\|\Omega_{n+1}(x,t)\| \to 0 \tag{4.3.105}$$

which makes $(\Omega_{n+1}(x,t))_{n\in\mathbb{N}}$ to be a Cauchy series. This helps us to conclude that by putting

$$u(x,t) = \sum_{j=0}^{n} \Omega_{j+1}(x,t) . \tag{4.3.106}$$

$u(x,t)$ exists and is continuous. Therefore, considering $K_n(x,t)$ to be the error, between exact and approximate solution, we can get

$$u(x,t) = u_n(x,t) \pm K_n(x,t) . \tag{4.3.107}$$

Then

$$u(x,t) - u_n(x,t) = G(x,t) + (1-\alpha)^2 F(x,t,u(x,t) - K_n(x,t)) \tag{4.3.108}$$

$$+\alpha(1-\alpha)\int_0^t (t-\tau)^{\alpha-1} F(x,\tau,u(x,\tau)-K_n(x,\tau))\,d\tau$$
$$+\frac{\alpha(1-\alpha)}{\Gamma(\alpha)}\int_0^l (t-\tau)^{\alpha-1} F(x,\tau,u(x,\tau)-K_n(x,\tau))\,d\tau$$
$$+\frac{\alpha^2}{\Gamma(\alpha+1)}\int_0^l (t-\tau)^{\alpha} F(x,\tau,u(x,\tau)-K_n(x,\tau))\,d\tau.$$

From the above, we can conclude that

$$\left\| \begin{array}{c} u(x,t)-G(x,t)-(1-\alpha)^2 F(x,t,u(x,t)) \\ -\alpha(1-\alpha)\int_0^t (t-\tau)^{\alpha-1} F(x,\tau,u(x,\tau))\,d\tau \\ -\frac{\alpha(1-\alpha)}{\Gamma(\alpha)}\int_0^l (t-\tau)^{\alpha-1} F(x,\tau,u(x,\tau))\,d\tau \\ -\frac{\alpha^2}{\Gamma(\alpha+1)}\int_0^l (t-\tau)^{\alpha} F(x,\tau,u(x,\tau))\,d\tau \end{array} \right\| \le \|K_n(x,t)\| \tag{4.3.109}$$

$$+(1-\alpha)^2 L\|K_{n-1}(x,t)\| + \alpha(1-\alpha)tL\|K_{n-1}(x,t)\|$$
$$+\frac{(1-\alpha)t^\alpha}{\Gamma(\alpha)}L\|K_{n-1}(x,t)\| + \frac{\alpha^2 t^{\alpha+1}}{\Gamma(\alpha+2)}L\|K_{n-1}(x,t)\|.$$

When $n \to \infty$, the right-hand side tends to zero, then

$$\begin{array}{c} u(x,t) = G(x,t)+(1-\alpha)^2 F(x,t,u(x,t)) \\ +\alpha(1-\alpha)\int_0^t (t-\tau)^{\alpha-1} F(x,\tau,u(x,\tau))\,d\tau \\ +\frac{\alpha(1-\alpha)}{\Gamma(\alpha)}\int_0^l (t-\tau)^{\alpha-1} F(x,\tau,u(x,\tau))\,d\tau \\ +\frac{\alpha^2}{\Gamma(\alpha+1)}\int_0^l (t-\tau)^{\alpha} F(x,\tau,u(x,\tau))\,d\tau. \end{array} \tag{4.3.110}$$

The uniqueness is achieved as

$$\|u_1(x,t)-u_2(x,t)\| \le L^{n+1}\left\{ \begin{array}{c} (1-\alpha)^2 + \frac{\alpha(1-\alpha)}{\Gamma(\alpha)}t^\alpha \\ +\frac{\alpha^2}{\Gamma(\alpha+2)}t^{\alpha+1} \end{array} \right\}^{n+1} D. \tag{4.3.111}$$

Also when $n \to \infty$, the right hand side tends to zero, thus

$$u_1(x,t) = u_2(x,t). \tag{4.3.112}$$

### 4.3.2 Numerical Solution

We present in this section the numerical solution of the new class of differential equation [13]. Thus at $(x_i, t_{n+1})$, we have:

$$\begin{aligned} u_i^{n+1} &= u_i^0 + G(x_i, t_n) + (1-\alpha)^2 F(x_i, t, u_i^n) \\ &+\alpha(1-\alpha)\int_0^{t_{n+1}} F(x_i, \tau, u(x_i, \tau))\, d\tau \\ &+\tfrac{\alpha(1-\alpha)}{\Gamma(\alpha)}\int_0^{t_{n+1}} (t_{n+1}-\tau)^{\alpha-1} F(x_i, \tau, u(x, \tau))\, d\tau \\ &+\tfrac{\alpha^2}{\Gamma(\alpha+1)}\int_0^{t_{n+1}} (t_{n+1}-\tau)^{\alpha} F(x_i, \tau, u(x_i, \tau))\, d\tau. \end{aligned} \tag{4.3.113}$$

Integrating each integral, we obtain:

$$\begin{aligned} u_i^{n+1} &= u_i^0 + G(x_i, t_n) + (1-\alpha)^2 F(x_{i,}, t_n, u_i^n) \\ &\quad +\alpha(1-\alpha)\sum_{j=0}^{n}\left[\begin{array}{c} \frac{3}{2}\Delta t F\left(x_i, t_j, u_i^j\right) \\ -\frac{\Delta t}{2} F\left(x_i, t_{j-1}, u_i^{j-1}\right) \end{array}\right] \end{aligned} \tag{4.3.114}$$

$$+\frac{\alpha(1-\alpha)(\Delta t)^\alpha}{\Gamma(\alpha+2)}\sum_{j=0}^{n}\left[\begin{array}{c} F\left(x_i, t_j, u_i^j\right) \\ \times\left\{\begin{array}{c} (n-j+1)^\alpha(n-j+2+\alpha) \\ -(n-j)^\alpha(n-j+2+2\alpha) \end{array}\right\} \\ -F\left(x_i, t_{j-1}, u_i^{j-1}\right) \\ \times\left\{\begin{array}{c} (n-j+1)^{\alpha+1} \\ -(n-j)^\alpha(n-j+1+\alpha) \end{array}\right\} \end{array}\right]$$

$$+\frac{\alpha^2(\Delta t)^{\alpha+1}}{\Gamma(\alpha+2)}\sum_{j=0}^{n}\left[\begin{array}{c} F\left(x_i,t_j,u_i^j\right) \\ \times\left\{\begin{array}{c}(n-j+1)^\alpha(n-j+2+\alpha) \\ -(n-j)^\alpha(n-j+2+2\alpha)\end{array}\right\} \\ -F\left(x_i,t_{j-1},u_i^{j-1}\right) \\ \times\left\{\begin{array}{c}(n-j+1)^{\alpha+1} \\ -(n-j)^\alpha(n-j+1+\alpha)\end{array}\right\}\end{array}\right]$$

Now we present the numerical scheme for the following integro-differential equation

$$u(x,t)=G(x,t)+(1-\alpha)^2u(x,t)+\alpha(1-\alpha)\int_0^t u(x,\tau)\,d\tau \tag{4.3.115}$$

$$+\frac{\alpha(1-\alpha)}{\Gamma(\alpha)}\int_0^l (t-\tau)^{\alpha-1}u(x,\tau)\,d\tau$$
$$+\frac{\alpha^2}{\Gamma(\alpha+1)}\int_0^l (t-\tau)^{\alpha}u(x,\tau)\,d\tau$$

where specially we choose $F(x,t,u(x,t)) = u(x,t)$. Thus at $(x_i,t_{n+1})$, we have:

$$\begin{array}{c} u_i^{n+1}=u_i^0+G(x_i,t_n)+(1-\alpha)^2u_i^n \\ +\alpha(1-\alpha)\int_0^{t_{n+1}}u(x_i,\tau)\,d\tau \\ +\frac{\alpha(1-\alpha)}{\Gamma(\alpha)}\int_0^{t_{n+1}}(t_{n+1}-\tau)^{\alpha-1}u(x_i,\tau)\,d\tau \\ +\frac{\alpha^2}{\Gamma(\alpha+1)}\int_0^{t_{n+1}}(t_{n+1}-\tau)^{\alpha}u(x_i,\tau)\,d\tau. \end{array} \tag{4.3.116}$$

Integrating each integral, we obtain:

$$\begin{aligned} u_i^{n+1} &= u_i^0+G(x_i,t_n)+(1-\alpha)^2u_i^n \\ &\quad+\alpha(1-\alpha)\sum_{j=0}^{n}\left[\frac{3}{2}\Delta t u_i^j-\frac{\Delta t}{2}u_i^{j-1}\right] \end{aligned} \tag{4.3.117}$$

$$+\frac{\alpha\left(1-\alpha\right)\left(\Delta t\right)^{\alpha}}{\Gamma\left(\alpha+2\right)}\sum_{j=0}^{n}\left[\begin{array}{c} u_i^j\left\{\begin{array}{c}\left(n-j+1\right)^{\alpha}\left(n-j+2+\alpha\right)\\ -\left(n-j\right)^{\alpha}\left(n-j+2+2\alpha\right)\end{array}\right\}\\ -u_i^{j-1}\left\{\begin{array}{c}\left(n-j+1\right)^{\alpha+1}\\ -\left(n-j\right)^{\alpha}\left(n-j+1+\alpha\right)\end{array}\right\}\end{array}\right]$$

$$+\frac{\alpha^2\left(\Delta t\right)^{\alpha+1}}{\Gamma\left(\alpha+2\right)}\sum_{j=0}^{n}\left[\begin{array}{c} u_i^j\left\{\begin{array}{c}\left(n-j+1\right)^{\alpha}\left(n-j+2+\alpha\right)\\ -\left(n-j\right)^{\alpha}\left(n-j+2+2\alpha\right)\end{array}\right\}\\ -u_i^{j-1}\left\{\begin{array}{c}\left(n-j+1\right)^{\alpha+1}\\ -\left(n-j\right)^{\alpha}\left(n-j+1+\alpha\right)\end{array}\right\}\end{array}\right].$$

---

## 4.4 Numerical Illustration

In this section, we present an illustrative example of the new mathematical equation, as well as the application of the proposed numerical scheme to solve the example. The mathematical equation considered here involves a differential operator of the Atangana-Baleanu type and an integral of the Caputo-Fabrizio type. The initial condition and boundary conditions are given as:

$$u\left(x,0\right)=u\left(0,t\right)=0. \tag{4.4.118}$$

We present the numerical simulations for different values of the fractional order.

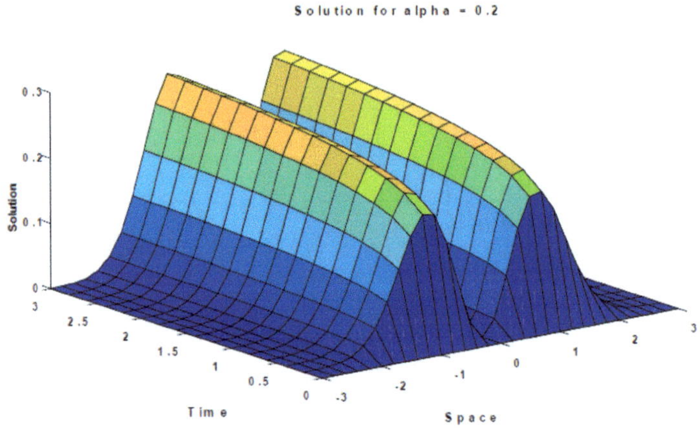


**FIGURE 4.1**. Numerical solution of integro-differential equation for $\alpha = 0.2$.

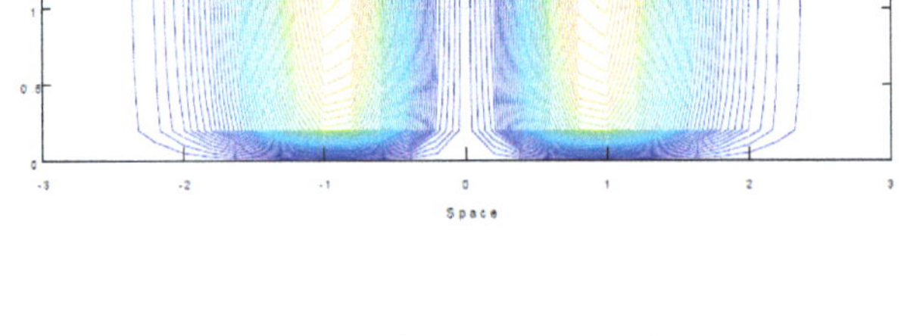


**FIGURE 4.2**. Contour plot for $\alpha = 0.2$.

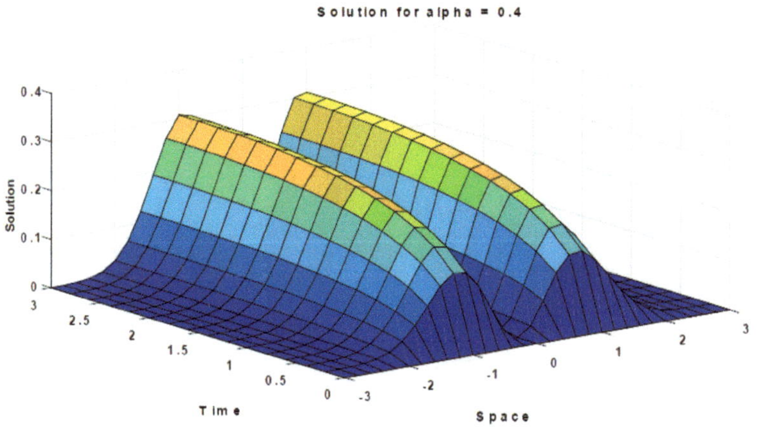


**FIGURE 4.3**. Numerical solution of integro-differential equation for $\alpha = 0.4$.

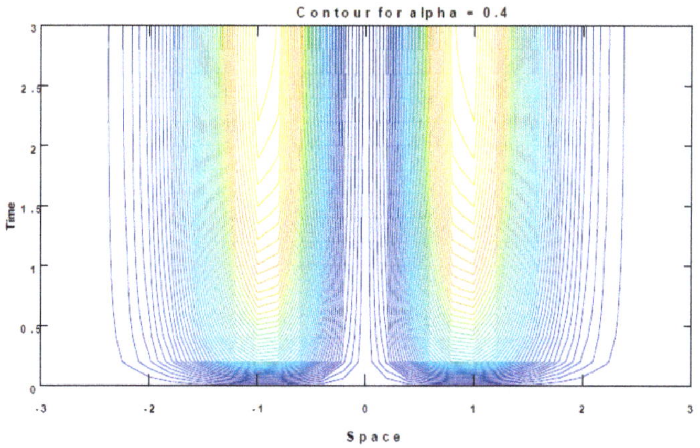


**FIGURE 4.4**. Contour plot for $\alpha = 0.4$.

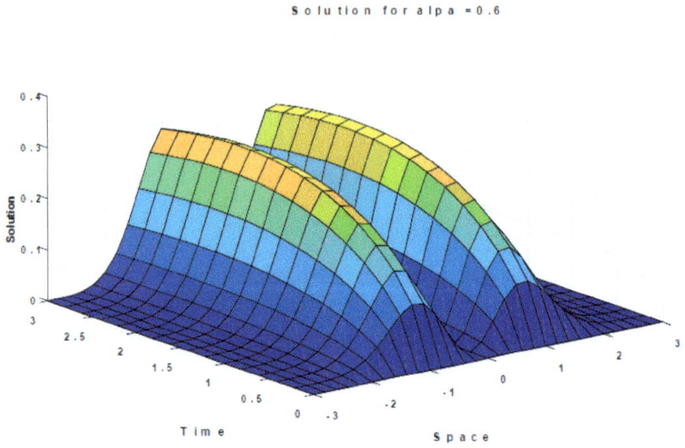


**FIGURE 4.5**. Numerical solution of integro-differential equation for $\alpha = 0.6$.

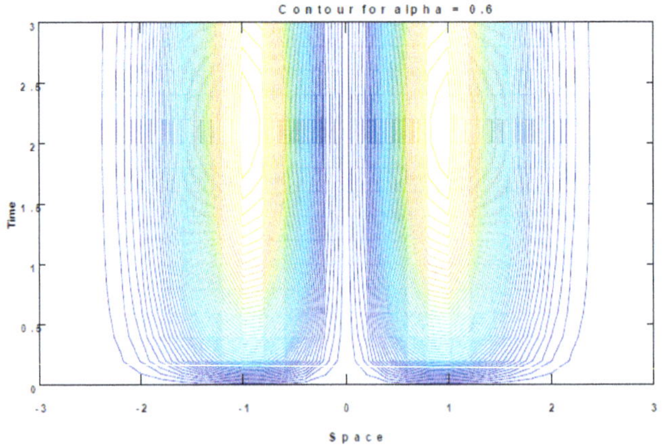


**FIGURE 4.6**. Contour plot for $\alpha = 0.6$.

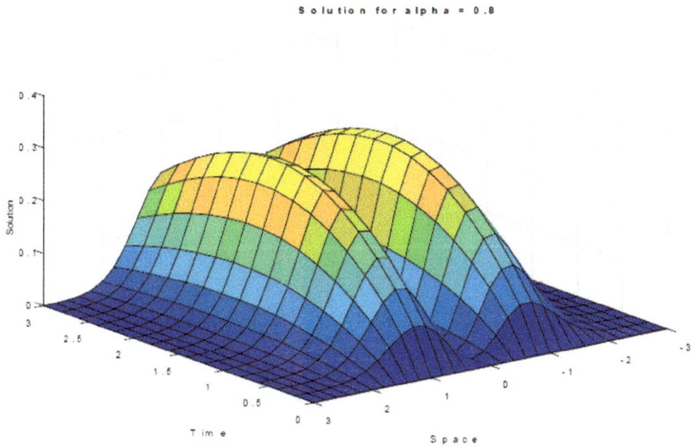


**FIGURE 4.7**. Numerical solution of integro-differential equation for $\alpha = 0.8$.

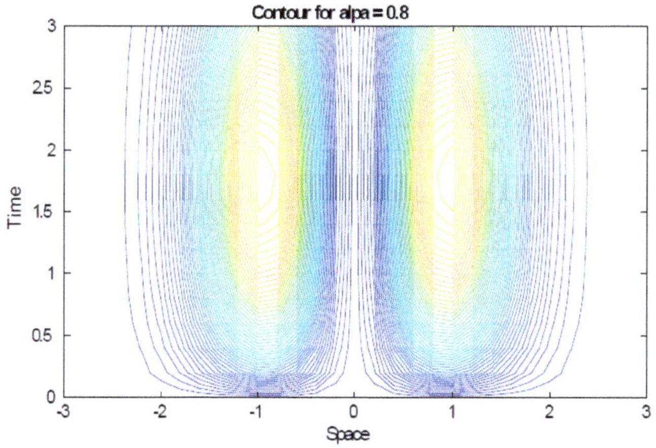


**FIGURE 4.8**. Contour plot for $\alpha = 0.8$.

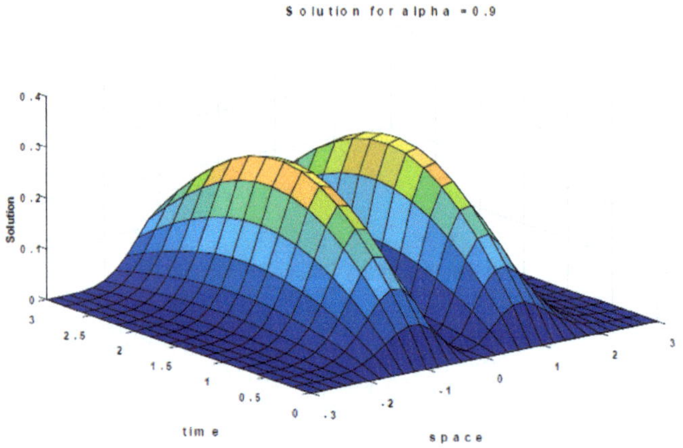


**FIGURE 4.9**. Numerical solution of integro-differential equation for $\alpha = 0.9$.

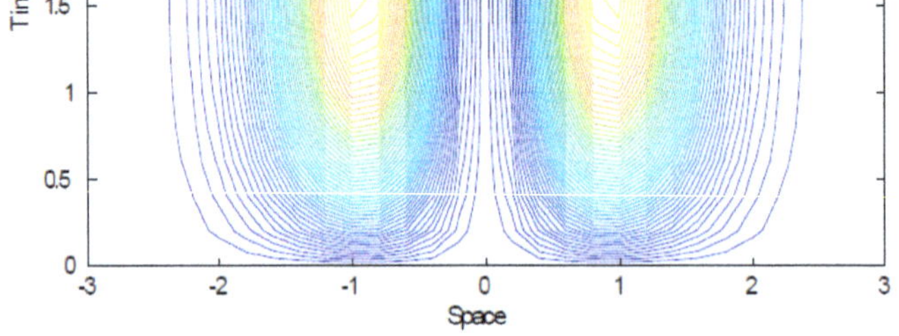


**FIGURE 4.10**. Contour plot for $\alpha = 0.9$.

## 4.5 Conclusion

In this chapter, we devoted our attention to new partial integro-differential equations with fractional mixed differential and integral operators. We considered two cases generally: in the first case, the fractional differential operator is that of Atangana-Baleanu, and the fractional integral is that of Caputo-Fabrizio. In the second case, we permuted, where the fractional differential operator is now that of Caputo-Fabrizio and the fractional integral is that of Atangana-Baleanu. For each case, we presented a detailed analysis. We derived the conditions under which the mathematical equation admits a unique solution. We also presented a new numerical scheme to solve such partial differential-integral equations with fractional mixed operators based on nonsingular kernels. Finally, we presented illustrative examples with numerical simulations.

# References


[1] Aghajani, A., Banas, J., & Jalilian, Y., (2011). Existence of solutions for a class of nonlinear Volterra singular integral equations. *Comput. Math. Appl. 62*, 1215–1227.

[2] Akbulut Arık, İ., & Tunç, C., (2022). On the periodic solutions of neutral integro-differential equations via fixed point method. *Journal of Mathematical Analysis, 13*(5), 35–48.

[3] Akbulut, İ., & Tunç, C., (2019). On the stability of solutions of neutral-differential equations of first order. *International Journal of Mathematics and Computer Science, 14*(4), 849–866.

[4] Anguraj, A., Karthikeyan, P., Rivero, M., & Trujillo, J. J., (2014). On new existence results for fractional integro-differential equations with impulsive and integral conditions. *Computers & Mathematics with Applications, 66*(12), 2587–2594.

[5] Atangana, A., & Dumitru, B., (2016). New fractional derivatives with non-local and non-singular kernel: Theory and application to heat transfer model. *Therm. Sci., 20*(2), 763–769.

[6] Baleanu, D., Mousalou, A., & Rezapour, S., (2017). A new method for investigating approximate solutions of some fractional integro-differential equations involving the Caputo-Fabrizio derivative. *Advances in Difference Equations*, 51.

[7] Balachandran, K., & Kiruthika, S., (2011). Existence results for fractional integro-differential equations with nonlocal condition via resolvent operators. *Comput. Math. Appl., 62*, 1350–1358.

[8] Becker, L. C., Burton, T., & Purnaras, I. K., (2015). Fractional Differential Equations, Transformations, and Fixed Points,

Dynamics of Continuous. *Discrete and Impulsive Systems Series A: Mathematical Analysis, 22*(5), 333–361.

[9] Burton, T., (2017). An existence theorem for a fractional differential equation using progressive contractions. *Journal of Fractional Calculus and Applications, 8*(1), 168–172.

[10] Caputo, M. , & Fabrizio, F., (2015). A new definition of fractional derivative without singular kernel. *Progr Fract Differ Appl, 1* (2), 73–85.

[11] Cui, Y., (2016). Uniqueness of solution for boundary value problems for fractional differential equations. *Appl. Math. Lett., 51*, 48–54.

[12] Mahdy, A. M. S., (2018). Numerical studies for solving fractional integro-differential equations. *Journal of Ocean Engineering and Science, 3*(2), 127–132.

[13] Mekkaoui, T., & Atangana, A., (2017). New numerical approximation of fractional derivative with non- local and non-singular kernel: Application to chaotic models, *European Physical Journal Plus, 132*, 444.

[14] Nemati, S., & Lima, P. M., (2018). Numerical solution of nonlinear fractional integro-differential equations with weakly singular kernels via a modification of hat functions. *Applied Mathematics and Computation, 327*, 79–92.

[15] Owolabi, K. M., & Atangana, A., (2017). Numerical approximation of nonlinear fractional parabolic differential equations with Caputo–Fabrizio derivative in Riemann–Liouville sense. *Chaos, Solitons and Fractals, 99*, 171–179.

[16] Shabibi, M., Postolache, M., Rezapour, S., & Vaezpour, S. M., (2016). Investigation of a multi-singular pointwise defined fractional integro-differential equation. *J. Math. Anal. 7*(5), 61–77.

[17] Tunç, C., & Akbulut, İ., (2018). Stability of a linear integro-differential equation of first order with variable delays. *Bulletin of Mathematical Analysis and Applications, 10*(2), 19–30.

# Chapter 5

## Exact Solution of Fractional Equations

**Mousa Ilie**
*Department of Mathematics, Rasht Branch, Islamic Azad University, Rasht, Iran*

**Abstract**
This chapter provides a comprehensive method to determine the general solution for solving second-order M-fractional differential equations (MFDEs). This method utilizes the general solution of homogeneous M-fractional differential equations and a specific solution of second-order linear MFDEs. The essence of solutions for second-order linear MFDEs has been proven in several theorems. The D'Alembert approach and the auxiliary equation method for solving homogeneous M-fractional differential equations are generalized. Additionally, a general solution is designed for M-fractional Euler's equidimensional equation. The Lagrange method and the indeterminate coefficient approach are proposed to solve MFDEs. Several examples are provided to further explain each part.

## Introduction

Recently, fractional calculus has been applied to obtain more precise results in studies involving differential equations. Although it gained prominence in 1974, following the first global conference on fractional calculus, it has since been recognized as a more accurate tool for modeling natural phenomena. Fractional differential equations have been studied across various fields, such as engineering and physics, particularly in control engineering, signal processing, the dynamics of viscoelastic materials, biosciences, electromagnetism, electrochemistry, fluid mechanics, diffusion processes, statistical and continuum mechanics, and the propagation of spherical flames [5, 36, 37]. Contributions from renowned mathematicians such as Liouville, Lagrange, Euler, Abel, Riemann, and Caputo have been instrumental in its development.

Many fractional integrals and fractional derivatives have been defined, each with its own limitations and strengths, providing an invaluable framework for both theoretical and applied fractional calculus [1, 3, 31–33, 35]. Sousa et al. (2017) introduced an M-fractional derivative with a Mittag-Leffler function involving one parameter, which satisfies the characteristics of non-integer order calculus [4, 41]. Sousa and Oliveira later proposed a truncated M-fractional derivative (TMFD), a type unifying four existing fractional derivative types, which satisfies the classical features of non-integer order calculus [40].

The chapter is organized as follows: Section 5.1 presents elementary definitions of fractional derivatives and integrals, such as the TMFD and the local M-fractional integral (LMFI). Section 5.2 describes the basic theoretical framework of the method. Section 5.3 explains methods such as using a well-known solution to detect another one, the D'Alembert method, the homogeneous equation characterized by constant coefficients, the auxiliary equation method, and Euler's equidimensional equation. Section 5.4 discusses differences in parameters, the Lagrange method, and the method of undetermined coefficients. Section 5.5 concludes the chapter.

---

## 5.1 Overview of M-Fractional Definitions

This section recalls definitions and some results about the M-fractional derivative, which in this study is applied.

**Definition 5.1.1.** *When $\beta > 0$, and $z \in \mathbb{C}$, the truncated Mittag-Leffler function of a parameter can be determined via [4, 40]*

$$ {}_i\mathbb{E}_\beta(z) = \sum_{k=0}^{i} \frac{z^k}{\Gamma(k\beta + 1)}. \tag{5.1.1}$$

**Definition 5.1.2.** *Considering $f : [0, \infty) \longrightarrow \mathbb{R}$, the truncated M-fractional derivative (TMFD) $f$, of order $\alpha$, can be determined with*

$$ {}_i\mathcal{D}_M^{\alpha,\beta} f(x) = \lim_{\varepsilon \to 0} \frac{f(x\, {}_i\mathbb{E}_\beta(\varepsilon x^{-\alpha})) - f(x)}{\varepsilon} \tag{5.1.2}$$

*for all $x > 0, \alpha \in (0, 1)$, and ${}_i\mathbb{E}_\beta(.)$, $\beta > 0$, is the Mittag-Leffler function with a parameter determined using Eq. (5.1.1) [40].*

*When $f$, is $\alpha$-differentiable in some $(0, a), a > 0$, and $\lim_{x \to 0^+} {}_i\mathcal{D}_M^{\alpha,\beta} f(x)$, we can determine [40]*

$$ {}_i\mathcal{D}_M^{\alpha,\beta} f(0) = \lim_{x \to 0^+} {}_i\mathcal{D}_M^{\alpha,\beta} f(x).$$

*In the presence of the M-fractional derivative of $f$, of order $\alpha$, then $f$, is $\alpha$-differentiable [40].*

TMFD can satisfy all these properties [40], When $\alpha \in (0,1)$, and $f$, and $g$, are functions $\alpha$-differentiable at $x > 0$,

**A.** (Linearity Rule) For $a, b \in \mathbb{R}$,

$$ {}_i\mathcal{D}_M^{\alpha,\beta}(af + bg) = a({}_i\mathcal{D}_M^{\alpha,\beta} f) + b({}_i\mathcal{D}_M^{\alpha,\beta} g), $$

**B.** For all $p \in \mathbb{R}$, ${}_i\mathcal{D}_M^{\alpha,\beta} x^p = \dfrac{p}{\Gamma(\beta+1)} x^{p-\alpha}$,

**C.** Regarding constant functions $f(x) = \lambda$, ${}_i\mathcal{D}_M^{\alpha,\beta} \lambda = 0$,

**D.** (Product Rule) ${}_i\mathcal{D}_M^{\alpha,\beta}(f.g) = g.\big({}_i\mathcal{D}_M^{\alpha,\beta} f\big) + f.\big({}_i\mathcal{D}_M^{\alpha,\beta} g\big)$,

**E.** (Quotient Rule) ${}_i\mathcal{D}_M^{\alpha,\beta}\Big(\dfrac{f}{g}\Big) = \dfrac{g.({}_i\mathcal{D}_M^{\alpha,\beta} f) - f.({}_i\mathcal{D}_M^{\alpha,\beta} g)}{g^2}$,

**F.** (Chain Rule) when function $f$, ordinary differentiable at $g(x)$, followed by ${}_i\mathcal{D}_M^{\alpha,\beta}(fog) = f'(g(x).({}_i\mathcal{D}_M^{\alpha,\beta} g)$,

**G.** ${}_i\mathcal{D}_M^{\alpha,\beta} f(x) = \dfrac{x^{1-\alpha}}{\Gamma(\beta+1)} \dfrac{df}{dx}$.

**Definition 5.1.3.** *When $\beta > 0$, $\alpha \in (n, n+1]$, for some $n \in \mathbb{N}$, and $f$, $n$-times differentiable (in the sense classical) for $x > 0$. The local M-derivative of order $n$, of function $f$, can be determined via*

$$ {}_i\mathcal{D}_M^{\alpha,\beta;n} f(x) = \lim_{\varepsilon \to 0} \frac{f^{(n)}(x\,{}_i\mathbb{E}_\beta(\varepsilon x^{-\alpha})) - f^{(n)}(x)}{\varepsilon}, \tag{5.1.3} $$

*when the limit is present [40].*

**Definition 5.1.4.** *When $f : [a, \infty) \to \mathbb{R}$, $a \geq 0$, the local M-fractional integral (LMFI) of $f$, order $\alpha$, can be determined by*

$$ {}_M\mathcal{T}_a^{\alpha,\beta} f(x) = \Gamma(\beta+1) \int_a^x \frac{f(t)}{t^{1-\alpha}} dt \tag{5.1.4} $$

*This is a common Riemann improper integral, and* $\alpha \in (0,1)$, *and* $\beta > 0$ *[40]. For simplicity, we have,* ${}_{M}\mathcal{T}_{0}^{\alpha,\beta} f(x) = {}_{M}\mathcal{T}^{\alpha,\beta} f(x)$. *These are some of the best results [40, 41]. When* $\alpha \in (0,1)$, *and* $f$, *is* $\alpha$*-differentiable at* $x > 0$,

**H.** *(Inverse theorem)* ${}_{i}\mathcal{D}_{M}^{\alpha,\beta}\big({}_{M}\mathcal{T}_{a}^{\alpha,\beta} f(x)\big) = f(x)$,

**I.** *(calculus fundamental theorem)* ${}_{M}\mathcal{T}_{a}^{\alpha,\beta}\big({}_{i}\mathcal{D}_{M}^{\alpha,\beta} f(x)\big) = f(x) - f(0)$,

**J.** *If* $\alpha \in (n, n+1]$, *for some* $n \in \mathbb{N}$, *and function* $f$, $n+1$, *times differentiable*

$$
{}_{M}\mathcal{T}_{a}^{\alpha,\beta}\big({}_{i}\mathcal{D}_{M}^{\alpha,\beta} f(x)\big) = f(x) - \sum_{k=0}^{n} \frac{f^{(k)}(a)}{k!}(x-a)^{k}.
$$

---

## 5.2 Fundamental of the Method

In this section, the viewpoint of how to solve MFDEs in the format of several required definitions and several theorems has been stated and proved.

**Definition 5.2.1.** *The general second-order linear MFDEs is*

$$
{}_{i}\mathcal{D}_{M}^{\alpha,\beta}\big({}_{i}\mathcal{D}_{M}^{\alpha,\beta} u(x)\big) + A(x)\,{}_{i}\mathcal{D}_{M}^{\alpha,\beta} u(x) + B(x)u(x) = D(x), \; x > 0. \tag{5.2.5}
$$

$A(x)$, $B(x)$, *and* $D(x)$, *represent* $\alpha$*-differentiable functions and* $u(x)$, *indicates the unidentified function.*

**Definition 5.2.2.** *When* $D(x) = 0$, *MFDEs (14.5), decreases to the homogeneous MFDEs*

$$
{}_{i}\mathcal{D}_{M}^{\alpha,\beta}\big({}_{i}\mathcal{D}_{M}^{\alpha,\beta} u(x)\big) + A(x)\,{}_{i}\mathcal{D}_{M}^{\alpha,\beta} u(x) + B(x)u(x) = 0, \; x > 0. \tag{5.2.6}
$$

**Definition 5.2.3.** *When $\alpha \in (0,1)$, $\beta > 0$, and $f$, and $g$, are functions $\alpha$-differentiable at $x > 0$, M-fractional Wronskian of two functions $f(x)$, and $g(x)$, can be determined thru [18, 39],*

$$W_{\alpha,\beta}(f(x), g(x)) = \begin{vmatrix} f(x) & g(x) \\ {}_i\mathcal{D}_M^{\alpha,\beta} f(x) & {}_i\mathcal{D}_M^{\alpha,\beta} g(x) \end{vmatrix}$$
$$= f(x).\,{}_i\mathcal{D}_M^{\alpha,\beta}(g(x)) - g(x).\,{}_i\mathcal{D}_M^{\alpha,\beta}(f(x)).$$

**Theorem 5.2.1.** *If $u_h(x, C_1, C_2)$, indicates the homogeneous MFDEs general solution (5.2.6) and $u_p(x)$, indicates a particular solution MFDEs (5.2.5), $u_h(x, C_1, C_2) + u_p(x)$, represents the MFDEs general solution (5.2.5).*

*Proof.* Consider $u(x)$, as the MFDEs solution (5.2.5), and $u_p(x)$, represents a specific solution of MFDE (5.2.5), and an easy calculation indicates that $u(x) - u_p(x)$, can be a solution of the homogeneous MFDEs (5.2.6),

$$\begin{aligned}
&{}_i\mathcal{D}_M^{\alpha,\beta}\Big({}_i\mathcal{D}_M^{\alpha,\beta}(u(x) - u_p(x))\Big) + A(x)\,{}_i\mathcal{D}_M^{\alpha,\beta}(u(x) - u_p(x)) + B(x)(u(x) - u_p(x)) \\
&= \Big[{}_i\mathcal{D}_M^{\alpha,\beta}\Big({}_i\mathcal{D}_M^{\alpha,\beta}u(x)\Big) + A(x)\,{}_i\mathcal{D}_M^{\alpha,\beta}u(x) + B(x)u(x)\Big] \\
&- \Big[{}_i\mathcal{D}_M^{\alpha,\beta}\Big({}_i\mathcal{D}_M^{\alpha,\beta}u_p(x)\Big) + A(x)\,{}_i\mathcal{D}_M^{\alpha,\beta}u_p(x) + B(x)u_p(x)\Big] \\
&= D(x) - D(x) = 0.
\end{aligned}$$

When $u_h(x, C_1, C_2)$, indicates a general solution for Eq. (5.2.6), it is an appropriate selection for the constants $C_1$, and $C_2$ [18, 39]

$$u(x) - u_p(x) = u_h(x, C_1, C_2) \text{ or } u(x) = u_h(x, C_1, C_2) + u_p(x).$$

**Theorem 5.2.2.** *If $u_1(x)$, and $u_2(x)$, indicate two solutions of the homogeneous MFDEs (5.2.6), $C_1u_1(x) + C_2u_2(x)$, can be a solution for the constants $C_1$, and $C_2$.*

*Proof.* This statement is obtained from

$$\begin{aligned}
&{}_i\mathcal{D}_M^{\alpha,\beta}\Big({}_i\mathcal{D}_M^{\alpha,\beta}(C_1u_1(x) + C_2u_2(x))\Big) + A(x)\,{}_i\mathcal{D}_M^{\alpha,\beta}(C_1u_1(x) + C_2u_2(x))\\
&+ B(x)(C_1u_1(x) + C_2u_2(x))\\
&= C_1\Big({}_i\mathcal{D}_M^{\alpha,\beta}\Big({}_i\mathcal{D}_M^{\alpha,\beta}u_1(x)\Big) + A(x)\,{}_i\mathcal{D}_M^{\alpha,\beta}u_1(x) + B(x)u_1(x)\Big)\\
&+ C_2\Big({}_i\mathcal{D}_M^{\alpha,\beta}\Big({}_i\mathcal{D}_M^{\alpha,\beta}u_2(x)\Big) + A(x)\,{}_i\mathcal{D}_M^{\alpha,\beta}u_2(x) + B(x)u_2(x)\Big)\\
&= C_1 \cdot 0 + C_2 \cdot 0 = 0,
\end{aligned}$$

when $u_1(x)$, and $u_2(x)$, are solutions of the homogeneous MFDEs (5.2.6) [18, 39].

**Theorem 5.2.3.** *When $u_1(x)$, and $u_2(x)$, indicate two solutions of the homogeneous MFDEs (5.2.6), on an interval $(a, b)$, their M-fractional Wronskian $W = W_{\alpha,\beta}(u_1(x), u_2(x))$, can be zero or never zero on $(a, b)$.*

*Proof.* Considering

$${}_i\mathcal{D}_M^{\alpha,\beta}W = u_1\,{}_i\mathcal{D}_M^{\alpha,\beta}\big({}_i\mathcal{D}_M^{\alpha,\beta}u_2\big) - u_2\,{}_i\mathcal{D}_M^{\alpha,\beta}\big({}_i\mathcal{D}_M^{\alpha,\beta}u_2\big).$$

Next, since $u_1(x)$, and $u_2(x)$, are both solutions of the homogeneous MFDEs (5.2.6), we have

$$\begin{aligned}
{}_i\mathcal{D}_M^{\alpha,\beta}\Big({}_i\mathcal{D}_M^{\alpha,\beta}u_1(x)\Big) + A(x)\,{}_i\mathcal{D}_M^{\alpha,\beta}u_1(x) + B(x)u_1(x) = 0\\
{}_i\mathcal{D}_M^{\alpha,\beta}\Big({}_i\mathcal{D}_M^{\alpha,\beta}u_2(x)\Big) + A(x)\,{}_i\mathcal{D}_M^{\alpha,\beta}u_2(x) + B(x)u_2(x) = 0
\end{aligned}$$

The first equation multiplying by $u_2$, is subtracted from the second one by $u_1$,

$$\left(u_1\, {}_i\mathcal{D}_M^{\alpha,\beta}\left({}_i\mathcal{D}_M^{\alpha,\beta}u_2\right) - u_2\, {}_i\mathcal{D}_M^{\alpha,\beta}\left({}_i\mathcal{D}_M^{\alpha,\beta}u_1\right)\right) + A(x)\left(u_1\, {}_i\mathcal{D}_M^{\alpha,\beta}u_2 - u_2\, {}_i\mathcal{D}_M^{\alpha,\beta}u_1\right) = 0.$$

Or ${}_i\mathcal{D}_M^{\alpha,\beta}W + A(x)W = 0$. This is the first-order MFDEs with a general solution $W = W_{\alpha,\beta}(x_0)e^{-{}_M\mathcal{T}_a^{\alpha,\beta}A(x)}$, and because the exponential factor is not zero, the proof can be completed [18, 27, 39].

**Theorem 5.2.4.** *When $u_1(x)$, and $u_2(x)$, indicate solutions of the homogeneous MFDEs (5.2.6), on an interval $(a, b)$, they are associated with this interval when their M-fractional Wronskian $W = W_{\alpha,\beta}(u_1(x), u_2(x))$, is zero.*

*Proof.* Considering $u_1(x)$, and $u_2(x)$, to be linearly dependent, then

$$W_{\alpha,\beta}(u_1(x), u_2(x)) = 0.$$

When the functions are zero on $(a, b)$, the conclusion can be obvious, and with no loss of generality, it can be concluded that none is identically zero. Also, their linear dependence indicates that each is a constant multiple of the other. Thus, $u_2 = Cu_1$, for some constant, so ${}_i\mathcal{D}_M^{\alpha,\beta}u_2 = C\, {}_i\mathcal{D}_M^{\alpha,\beta}u_1$. After eliminating $C$, $u_1\, {}_i\mathcal{D}_M^{\alpha,\beta}u_2 - u_2\, {}_i\mathcal{D}_M^{\alpha,\beta}u_1 = 0$, is obtained indicating half of the theorem. $W_{\alpha,\beta}(u_1(x), u_2(x)) = 0$, proves linearly dependent. When $u_1(x)$, is zero on $(a, b)$, the functions show linear dependence. When $u_1(x)$, does not vanish identically on $(a, b)$, $u_1(x)$, does not vanish on some subinterval $(c, d)$, of $(a, b)$. Because the M-fractional Wronskian is zero on $(a, b)$, it is dividable by $u_1^2$, to obtain $\frac{u_1\, {}_i\mathcal{D}_M^{\alpha,\beta}u_2 - u_2\, {}_i\mathcal{D}_M^{\alpha,\beta}u_1}{u_1^2} = 0$, on $(c, d)$. Here, is another way of writing ${}_i\mathcal{D}_M^{\alpha,\beta}\left(\frac{u_2}{u_1}\right) = 0$, and after local M-fractional integration, $\frac{u_2}{u_1} = C$, or $u_2 = Cu_1$, are obtained for some constant $C$, and all $x$, in

$(c, d)$. Because $u_2$, and $Cu_1$, are similar in value in $(c, d)$, they possess similar M-fractional derivatives, thus, $u_2 = Cu_1$, all $x$, in $(a, b)$, which concludes the argument [18, 39].

**Theorem 5.2.5.** *$u_1(x)$, and $u_2(x)$, show a linear independence of the homogeneous MFDEs (5.2.6), on the interval $(a, b)$. Thus, $C_1u_1(x) + C_2u_2(x)$, can be the general solution of the homogeneous MFDEs (5.2.6), on the mentioned interval.*

*Proof.* Consider $u(x)$, which is a solution of (5.2.6), on $(a, b)$. Constant $C_1$, and $C_2$, is obtainable so that $u(x) = C_1u_1(x)+C_2u_2(x)$, for all $x$, in $[a, b]$. Since $C_1u_1(x)+C_2u_2(x)$, and $u(x)$, can be solutions of (5.2.6), on $(a, b)$, that's enough to show that for some point $x_0$, in $(a, b)$, can be found $C_1$, and $C_2$, with $C_1u_1(x_0) + C_2u_2(x_0) = u(x_0)$, and $C_1 \, {}_i\mathcal{D}_M^{\alpha,\beta}u_1(x_0) + C_2 \, {}_i\mathcal{D}_M^{\alpha,\beta}u_2(x_0) = {}_i\mathcal{D}_M^{\alpha,\beta}u(x_0)$. To achieve this system being solvable for $C_1$, and $C_2$, the following determinant should not be zero,

$$\begin{vmatrix} u_1(x_0) & u_2(x_0) \\ {}_i\mathcal{D}_M^{\alpha,\beta}u_1(x_0) & {}_i\mathcal{D}_M^{\alpha,\beta}u_2(x_0) \end{vmatrix} = u_1(x_0) \cdot {}_i\mathcal{D}_M^{\alpha,\beta}u_2(x_0) - u_2(x_0) \cdot {}_i\mathcal{D}_M^{\alpha,\beta}u_1(x_0).$$

This leads us to investigate the function M-fractional Wronskian of $u_1(x), u_2(x)$, at $x_0$. According to theorems (5.2.3), and (5.2.4), it is clear that $W_{\alpha,\beta}(u_1(x_0), u_2(x_0))$, are not zero [18, 39].

---

## 5.3 Determination of a General Solution of a Homogeneous M-Fractional Differential Equation

Here, we obtain a general solution of the homogeneous MFDEs and solve Euler's equation.

### 5.3.1 The D'Alembert Approach

Considering $u_1(x)$, as a well-known nonzero solution of the homogeneous MFDEs (5.2.6), $u_2(x) = v(x)u_1(x)$, can be a solution of (5.2.6), $v(x)$, is an unknown function [18, 39]. So

$$
\begin{aligned}
{}_i\mathcal{D}_M^{\alpha,\beta} u_2(x) &= v(x) \cdot {}_i\mathcal{D}_M^{\alpha,\beta} u_1(x) + u_1(x)\, {}_i\mathcal{D}_M^{\alpha,\beta}(v(x)),\\
{}_i\mathcal{D}_M^{\alpha,\beta}\Big({}_i\mathcal{D}_M^{\alpha,\beta} u_2(x)\Big) &= v(x) \cdot {}_i\mathcal{D}_M^{\alpha,\beta}\Big({}_i\mathcal{D}_M^{\alpha,\beta} u_1(x)\Big)\\
&\quad + 2\, {}_i\mathcal{D}_M^{\alpha,\beta} u_1(x) \cdot {}_i\mathcal{D}_M^{\alpha,\beta} v(x)\\
&\quad + u_1(x) \cdot {}_i\mathcal{D}_M^{\alpha,\beta}\Big({}_i\mathcal{D}_M^{\alpha,\beta} v(x)\Big),
\end{aligned}
$$

through substituting the mentioned findings into Eq. (5.2.6),

$$
\begin{aligned}
&v(x)\Big({}_i\mathcal{D}_M^{\alpha,\beta}\big({}_i\mathcal{D}_M^{\alpha,\beta} u_1(x)\big) + A(x)\, {}_i\mathcal{D}_M^{\alpha,\beta} u_1(x) + B(x)u_1(x)\Big)\\
&+ u_1(x)\, {}_i\mathcal{D}_M^{\alpha,\beta}\Big({}_i\mathcal{D}_M^{\alpha,\beta} v(x)\Big)\\
&+ {}_i\mathcal{D}_M^{\alpha,\beta} v(x)\Big(A(x)u_1(x) + 2\, {}_i\mathcal{D}_M^{\alpha,\beta} u_1(x)\Big) = 0.
\end{aligned}
$$

Since $u_1(x)$, indicates a solution of the homogeneous MFDEs (5.2.6), it reduces

$$
u_1(x)\, {}_i\mathcal{D}_M^{\alpha,\beta}\Big({}_i\mathcal{D}_M^{\alpha,\beta} v(x)\Big) +_i \mathcal{D}_M^{\alpha,\beta} v(x)\Big(A(x)u_1(x) + 2\, {}_i\mathcal{D}_M^{\alpha,\beta} u_1(x)\Big) = 0
$$

or

$$
\frac{{}_i\mathcal{D}_M^{\alpha,\beta}\Big({}_i\mathcal{D}_M^{\alpha,\beta} v(x)\Big)}{{}_i\mathcal{D}_M^{\alpha,\beta} v(x)} = -2\frac{{}_i\mathcal{D}_M^{\alpha,\beta} u_1(x)}{u_1(x)} - A(x).
$$

A local M-fractional integration now gives

$$\ln({}_i\mathcal{D}_M^{\alpha,\beta}v(x)) = -2\ln(u_1(x)) - {}_M\mathcal{I}_a^{\alpha,\beta}A(x),$$

so

$${}_i\mathcal{D}_M^{\alpha,\beta}v(x) = \frac{1}{u_1^2(x)}e^{-{}_M\mathcal{I}_a^{\alpha,\beta}A(x)}$$

and

$$v(x) = {}_M\mathcal{I}_a^{\alpha,\beta}\frac{1}{u_1^2(x)}e^{-{}_M\mathcal{I}_a^{\alpha,\beta}A(x)}. \tag{5.3.7}$$

Then, the general solution of homogeneous MFDEs of (5.2.6), is [18, 26, 39]

$$u_h(x) = C_1u_1(x) + C_2\,{}_M\mathcal{I}_a^{\alpha,\beta}\frac{1}{u_1^2(x)}e^{-{}_M\mathcal{I}_a^{\alpha,\beta}A(x)}u_1(x).$$

**Example 5.3.1.** *When* $u_1(x) = 3\Gamma(\beta+1)\sqrt[3]{x}$, *is a solution of the below homogeneous M-fractional equation*

$${}_i\mathcal{D}_M^{\frac{1}{3},\beta}\Big({}_i\mathcal{D}_M^{\frac{1}{3},\beta}u(x)\Big) - \frac{2}{3\Gamma(\beta+1)\sqrt[3]{x}}\,{}_i\mathcal{D}_M^{\frac{1}{3},\beta}u(x) + \frac{2}{9(\Gamma(\beta+1))^2\sqrt[3]{x^2}}u(x) = 0.$$

*According to (5.3.7), we have*

$$v(x) = 3\Gamma(\beta+1)\sqrt[3]{x}.$$

*Therefore*

$$u_2(x) = 9(\Gamma(\beta+1))^2\sqrt[3]{x^2}.$$

*The following general solution is obtained,*

$$u_h(x) = (C_1 + C_2 3\Gamma(\beta+1)\sqrt[3]{x}) 3\Gamma(\beta+1)\sqrt[3]{x}.$$

**Example 5.3.2.** $u_1(x) = \left(2\Gamma(\beta+1)\sqrt{x}\right)^{\frac{1}{2}} \sin(2\Gamma(\beta+1)\sqrt{x})$, *is a solution of M-fractional equation*

$$\begin{aligned} {}_i\mathcal{D}_M^{\frac{1}{2},\beta}\left({}_i\mathcal{D}_M^{\frac{1}{2},\beta} u(x)\right) + \left(\frac{1}{2\Gamma(\beta+1)\sqrt{x}}\right) {}_i\mathcal{D}_M^{\frac{1}{2},\beta} u(x) \\ + \left(1 - \frac{1}{4(\Gamma(\beta+1))^2 x}\right) u(x) = 0. \end{aligned}$$

*Based on the D'Alembert method* $v(x)$, *and the second solution* $u_2(x)$, *can be achieved as follows,*

$$\begin{aligned} v(x) &= -\cot(2\Gamma(\beta+1)\sqrt{x}), \\ u_2(x) &= -(2\Gamma(\beta+1)\sqrt{x})^{-\frac{1}{2}} \cos(2\Gamma(\beta+1)\sqrt{x}). \end{aligned}$$

*Thus, the general solution of the mentioned equation is*

$$u_h(x) = (2\Gamma(\beta+1)\sqrt{x})^{-\frac{1}{2}} \left(C_1 \sin(2\Gamma(\beta+1)\sqrt{x}) + C_2 \cos(2\Gamma(\beta+1)\sqrt{x})\right).$$

### *5.3.2 The Auxiliary Equation Method*

A complete discussion of the homogeneous MFDEs (5.2.6), can be presented now for the special case where $a$, and $b$, are constants [18, 39].

$${}_i\mathcal{D}_M^{\alpha,\beta}\left({}_i\mathcal{D}_M^{\alpha,\beta} u(x)\right) + a\, {}_i\mathcal{D}_M^{\alpha,\beta} u(x) + b u(x) = 0. \tag{5.3.8}$$

M-fractional derivatives of the exponential function $e^{r\left(\frac{\Gamma(\beta+1)}{\alpha} x^{\alpha}\right)}$, are all constant multiples of the function itself. It leads us to consider

$$u(x) = e^{r\left(\frac{\Gamma(\beta+1)}{\alpha}x^{\alpha}\right)}, \tag{5.3.9}$$

as a possible solution for M-fractional differential (5.3.8), we have

$${}_i\mathcal{D}_M^{\alpha,\beta} u(x) = re^{r\left(\frac{\Gamma(\beta+1)}{\alpha}x^{\alpha}\right)}, \tag{5.3.10}$$

and

$${}_i\mathcal{D}_M^{\alpha,\beta}\left({}_i\mathcal{D}_M^{\alpha,\beta} u(x)\right) = r^2 e^{r\left(\frac{\Gamma(\beta+1)}{\alpha}x^{\alpha}\right)}. \tag{5.3.11}$$

Substituting (5.3.9), (5.3.10), and (5.3.11), into (5.3.8), yields to

$$(r^2 + ar + b)e^{r\left(\frac{\Gamma(\beta+1)}{\alpha}x^{\alpha}\right)} = 0, \tag{5.3.12}$$

and because $e^{r\left(\frac{\Gamma(\beta+1)}{\alpha}x^{\alpha}\right)}$, is not zero (5.3.12), holds if $r$, satisfies the below auxiliary equation [18, 39]

$$r^2 + ar + b = 0. \tag{5.3.13}$$

The roots $r_1$, and $r_2$, of algebraic (5.3.13), are distinct actual numbers if $a^2 - 4b > 0$.

Thus, the two solutions $u_1(x) = e^{r_1\left(\frac{\Gamma(\beta+1)}{\alpha}x^{\alpha}\right)}$, and $u_2(x) = e^{r_2\left(\frac{\Gamma(\beta+1)}{\alpha}x^{\alpha}\right)}$, are obtained.

Since the ratio $\frac{e^{r_1\left(\frac{\Gamma(\beta+1)}{\alpha}x^{\alpha}\right)}}{e^{r_2\left(\frac{\Gamma(\beta+1)}{\alpha}x^{\alpha}\right)}} = e^{(r_1-r_2)\left(\frac{\Gamma(\beta+1)}{\alpha}x^{\alpha}\right)}$, is not constant, these solutions show a linear independence and

$$u_h(x) = C_1 e^{r_1\left(\frac{\Gamma(\beta+1)}{\alpha}x^{\alpha}\right)} + C_2 e^{r_2\left(\frac{\Gamma(\beta+1)}{\alpha}x^{\alpha}\right)},$$

can be the general solution of the homogeneous MFDEs (5.3.8).

When $r_1 = r_2$, only one solution $u_1(x) = e^{r_1\left(\frac{\Gamma(\beta+1)}{\alpha}x^\alpha\right)}$, is obtained. Thus, a second linearly independent solution is found by the D'Alembert approach

$$u_2(x) = \Big(\frac{\Gamma(\beta+1)}{\alpha}x^\alpha\Big)e^{r_1\left(\frac{\Gamma(\beta+1)}{\alpha}x^\alpha\right)},$$

and the general solution of the homogeneous MFDEs (5.3.8), is

$$u_h(x) = \Big(C_1 + C_2\Big(\frac{\Gamma(\beta+1)}{\alpha}x^\alpha\Big)\Big)e^{r_1\left(\frac{\Gamma(\beta+1)}{\alpha}x^\alpha\right)}.$$

When the roots $r_1$, and $r_2$, are distinct complex numbers, they are written as $p \pm iq$, and using the Euler's formula $e^{i\theta} = \cos\theta + i\sin\theta$, our two solutions of MFDEs (5.3.8), are as follows

$$e^{r_1\left(\frac{1}{\alpha}x^\alpha\right)} = e^{p\left(\frac{\Gamma(\beta+1)}{\alpha}x^\alpha\right)}\Big(\cos q\Big(\frac{\Gamma(\beta+1)}{\alpha}x^\alpha\Big) + i\sin q\Big(\frac{\Gamma(\beta+1)}{\alpha}x^\alpha\Big)\Big),$$

$$e^{r_2\left(\frac{1}{\alpha}x^\alpha\right)} = e^{p\left(\frac{\Gamma(\beta+1)}{\alpha}x^\alpha\right)}\Big(\cos q\Big(\frac{\Gamma(\beta+1)}{\alpha}x^\alpha\Big) - i\sin q\Big(\frac{\Gamma(\beta+1)}{\alpha}x^\alpha\Big)\Big).$$

So, the solution of the homogeneous MFDEs (5.3.8), will be obtained as the following

$$u_h(x) = e^{p\left(\frac{\Gamma(\beta+1)}{\alpha}x^\alpha\right)}\Big(C_1\cos q\Big(\frac{\Gamma(\beta+1)}{\alpha}x^\alpha\Big) + C_2\sin q\Big(\frac{\Gamma(\beta+1)}{\alpha}x^\alpha\Big)\Big).$$

**Example 5.3.3.** *Consider the following the homogeneous MFDE with constant coefficients*

$${}_i\mathcal{D}_M^{\alpha,\beta}\big({}_i\mathcal{D}_M^{\alpha,\beta}u(x)\big) - 3\,{}_i\mathcal{D}_M^{\alpha,\beta}u(x) + 2u(x) = 0.$$

*Below is the general solution of the above equation:*

$$u_h(x) = C_1 e^{\left(\frac{\Gamma(\beta+1)}{\alpha}x^\alpha\right)} + C_2 e^{2\left(\frac{\Gamma(\beta+1)}{\alpha}x^\alpha\right)}.$$

**Example 5.3.4.** *The general solution of homogeneous MFDE with constant coefficients of*

$${}_i\mathcal{D}_M^{\alpha,\beta}\left({}_i\mathcal{D}_M^{\alpha,\beta}u(x)\right) + 4\,{}_i\mathcal{D}_M^{\alpha,\beta}u(x) + 4u(x) = 0,$$

*are obtained as*

$$u_h(x) = \left(C_1 + C_2\left(\frac{\Gamma(\beta+1)}{\alpha}x^\alpha\right)\right)e^{-2\left(\frac{\Gamma(\beta+1)}{\alpha}x^\alpha\right)}.$$

**Example 5.3.5.** *Let's consider following homogeneous MFDE with constant coefficients*

$${}_i\mathcal{D}_M^{\alpha,\beta}\left({}_i\mathcal{D}_M^{\alpha,\beta}u(x)\right) - 2\,{}_i\mathcal{D}_M^{\alpha,\beta}u(x) + 3u(x) = 0.$$

*Signature approach offers*

$$u_h(x) = e^{\left(\frac{\Gamma(\beta+1)}{\alpha}x^\alpha\right)}\left(C_1\cos\sqrt{2}\left(\frac{\Gamma(\beta+1)}{\alpha}x^\alpha\right) + C_2\sin\sqrt{2}\left(\frac{\Gamma(\beta+1)}{\alpha}x^\alpha\right)\right).$$

### *5.3.3 The Euler's Equidimensional MFDE*

The homogeneous M-fractional differential equation,

$$\left(\frac{\Gamma(\beta+1)}{\alpha}x^\alpha\right)^2 {}_i\mathcal{D}_M^{\alpha,\beta}\left({}_i\mathcal{D}_M^{\alpha,\beta}\right) + a\left(\frac{\Gamma(\beta+1)}{\alpha}x^\alpha\right){}_i\mathcal{D}_M^{\alpha,\beta}u(x) + bu(x) = 0,\ \ x > 0, \qquad (5.3.14)$$

where $a$, and $b$, are constant, is called the Euler's MFDE [18, 39]. Using the change independent variable [18]

$$z = \ln\Big(\frac{\Gamma(\beta+1)}{\alpha}x^{\alpha}\Big), \tag{5.3.15}$$

we have

$$\begin{aligned} {}_i\mathcal{D}_M^{\alpha,\beta}u(x) &= \Big(\frac{\Gamma(\beta+1)}{\alpha}x^{\alpha}\Big)^{-1}\frac{du}{dz}, \\ {}_i\mathcal{D}_M^{\alpha,\beta}\Big({}_i\mathcal{D}_M^{\alpha,\beta}u(x)\Big) &= -\Big(\frac{\Gamma(\beta+1)}{\alpha}x^{\alpha}\Big)^{-2}\frac{du}{dz} \\ &\quad + \Big(\frac{\Gamma(\beta+1)}{\alpha}x^{\alpha}\Big)^{-2}\frac{d^2u}{dz^2}, \end{aligned} \tag{5.3.16}$$

substituting Eqs. (5.3.15), and (5.3.16), into (5.3.14), leads to

$$\frac{d^2u}{dz^2} + (a-1)\frac{du}{dz} + bu = 0. \tag{5.3.17}$$

Equation (5.3.17), is a normal differential equation that has a constant coefficient, and based on this approach, the auxiliary equation is

$$r^2 + (a-1)r + b = 0 \tag{5.3.18}$$

$r_1$, and $r_2$, are roots of Eq. (5.3.18), [19]. If these two are distinct actual numbers, the below solution of (5.3.14), is obtainable,

$$u_h(x) = C_1\Big(\frac{\Gamma(\beta+1)}{\alpha}x^{\alpha}\Big)^{r_1} + C_2\Big(\frac{\Gamma(\beta+1)}{\alpha}x^{\alpha}\Big)^{r_2}.$$

If $r_1 = r_2$, we drive

$$u_h(x) = \Big(C_1 + C_2\ln\Big(\frac{\Gamma(\beta+1)}{\alpha}x^{\alpha}\Big)\Big)\Big(\frac{\Gamma(\beta+1)}{\alpha}x^{\alpha}\Big)^{r_1}.$$

And if $r_1$, and $r_2$, are distinct complex numbers as $p \pm q$, then the general solution of (5.3.14), is derived

$$u_h(x) = \Big(\frac{\Gamma(\beta+1)}{\alpha}x^\alpha\Big)^p \Big[C_1 \cos q \ln\Big(\frac{\Gamma(\beta+1)}{\alpha}x^\alpha\Big) + C_2 \sin q \ln\Big(\frac{\Gamma(\beta+1)}{\alpha}x^\alpha\Big)\Big].$$

**Example 5.3.6.** *Considering the below homogeneous MFDE*

$$\Big(\frac{\Gamma(\beta+1)}{\alpha}x^\alpha\Big)\,{}_i\mathcal{D}_M^{\alpha,\beta}\Big({}_i\mathcal{D}_M^{\alpha,\beta}u(x)\Big) - 2\Big(\frac{\Gamma(\beta+1)}{\alpha}x^\alpha\Big)\,{}_i\mathcal{D}_M^{\alpha,\beta}u(x) + 2u(x) = 0.$$

*The auxiliary equation is*

$$r^2 - 3r + 2 = 0,$$

*the roots are $r_1 = 1$, and $r_2 = 2$.*
*Thus, the general solution can be*

$$u_h(x) = C_1\frac{\Gamma(\beta+1)}{\alpha}x^\alpha + C_2\Big(\frac{\Gamma(\beta+1)}{\alpha}x^\alpha\Big)^2.$$

**Example 5.3.7.** *Consider following Euler's MFDE*

$$\Big(\frac{\Gamma(\beta+1)}{\alpha}x^\alpha\Big)^2\,{}_i\mathcal{D}_M^{\alpha,\beta}\Big({}_i\mathcal{D}_M^{\alpha,\beta}u(x)\Big) - 3\Big(\frac{\Gamma(\beta+1)}{\alpha}x^\alpha\Big)\,{}_i\mathcal{D}_M^{\alpha,\beta}u(x) + 4u(x) = 0.$$

*The auxiliary root is*

$$r_1 = r_2 = 2.$$

*Therefore, the general solution can be*

$$u_h(x) = \Big(C_1 + C_2 \ln\Big(\frac{\Gamma(\beta+1)}{\alpha}x^\alpha\Big)\Big)\Big(\frac{\Gamma(\beta+1)}{\alpha}x^\alpha\Big)^2.$$

**Example 5.3.8.** *Let's consider Euler's MFDE as follows*

$$\Big(\frac{\Gamma(\beta+1)}{\alpha}x^\alpha\Big)^2\,{}_i\mathcal{D}_M^{\alpha,\beta}\Big({}_i\mathcal{D}_M^{\alpha,\beta}u(x)\Big) + 3\Big(\frac{\Gamma(\beta+1)}{\alpha}x^\alpha\Big)\,{}_i\mathcal{D}_M^{\alpha,\beta}u(x) + 2u(x) = 0.$$

*The auxiliary equation root is*

$$r_1 = -1 + i, \quad or \quad r_2 = -1 - i.$$

*Consequently, we have,*

$$u_h(x) = \left(\frac{\Gamma(\beta+1)}{\alpha}x^{\alpha}\right)^{-1}\left[C_1 \cos\ln\left(\frac{\Gamma(\beta+1)}{\alpha}x^{\alpha}\right) + C_2 \sin\ln\left(\frac{\Gamma(\beta+1)}{\alpha}x^{\alpha}\right)\right].$$

---

## 5.4 Determination of a Specific Solution of MFDE

Here, we introduce differences in the parameters and undetermined coefficient methods to determine a specific solution for nonhomogeneous MFDEs.

### *5.4.1 Variation of Parameters or Lagrange Approach*

Consider $u_1(x)$, and $u_2(x)$, as two linearly independent solutions, of homogeneous M-fractional differential equation of the second-order fractional MFDEs (5.2.5), then, the particular solution $u_p(x)$, can be

$$u_p(x) = \vartheta_1(x)u_1(x) + \vartheta_2(x)u_2(x), \tag{5.4.19}$$

$\vartheta_1(x)$, and $\vartheta_2(x)$, are unfamiliar functions [18, 39]. Calculating the M-fractional derivative of (5.4.19), the following equation is obtained

$$\begin{aligned} {}_i\mathcal{D}_M^{\alpha,\beta} u_p(x) = & \left(u_1\, {}_i\mathcal{D}_M^{\alpha,\beta}\vartheta_1(x) + u_2\, {}_i\mathcal{D}_M^{\alpha,\beta}\vartheta_2(x)\right) \\ & + \left(\vartheta_1\, {}_i\mathcal{D}_M^{\alpha,\beta}u_1(x) + \vartheta_2\, {}_i\mathcal{D}_M^{\alpha,\beta}u_2(x)\right). \end{aligned} \tag{5.4.20}$$

For no use of second-order M-fractional derivative, it can be supposed that

$$u_1\,{}_i\mathcal{D}_M^{\alpha,\beta}\vartheta_1(x) + u_2\,{}_i\mathcal{D}_M^{\alpha,\beta}\vartheta_2(x). \tag{5.4.21}$$

So,

$${}_i\mathcal{D}_M^{\alpha,\beta}u_p(x) = \vartheta_1\,{}_i\mathcal{D}_M^{\alpha,\beta}u_1(x) + \vartheta_2\,{}_i\mathcal{D}_M^{\alpha,\beta}u_2(x), \tag{5.4.22}$$

therefore,

$$\begin{aligned}{}_i\mathcal{D}_M^{\alpha,\beta}\left({}_i\mathcal{D}_M^{\alpha,\beta}u_p(x)\right) &= \vartheta_1\,{}_i\mathcal{D}_M^{\alpha,\beta}\left({}_i\mathcal{D}_M^{\alpha,\beta}u_1(x)\right)\\ &\quad+\left({}_i\mathcal{D}_M^{\alpha,\beta}\vartheta_1(x)\right)\left({}_i\mathcal{D}_M^{\alpha,\beta}u_1(x)\right)\\ &\quad+\left({}_i\mathcal{D}_M^{\alpha,\beta}\vartheta_2(x)\right)\left({}_i\mathcal{D}_M^{\alpha,\beta}u_2(x)\right)\\ &\quad+\vartheta_2\,{}_i\mathcal{D}_M^{\alpha,\beta}\left({}_i\mathcal{D}_M^{\alpha,\beta}u_2(x)\right).\end{aligned} \tag{5.4.23}$$

By substituting (5.4.19), (5.4.22), and (5.4.23), into MFDEs (5.2.5), and some manipulation, we get:

$$\begin{aligned}&\vartheta_1\left({}_i\mathcal{D}_M^{\alpha,\beta}({}_i\mathcal{D}_M^{\alpha,\beta}u_1(x)) + A(x)\,{}_i\mathcal{D}_M^{\alpha,\beta}u_1(x) + B(x)u_1(x)\right)\\ &+\vartheta_2\left({}_i\mathcal{D}_M^{\alpha,\beta}({}_i\mathcal{D}_M^{\alpha,\beta}u_2(x)) + A(x)\,{}_i\mathcal{D}_M^{\alpha,\beta}u_2(x) + B(x)u_2(x)\right)\\ &+\left({}_i\mathcal{D}_M^{\alpha,\beta}\vartheta_1(x)\right)\left({}_i\mathcal{D}_M^{\alpha,\beta}u_1(x)\right)+\left({}_i\mathcal{D}_M^{\alpha,\beta}\vartheta_2(x)\right)\left({}_i\mathcal{D}_M^{\alpha,\beta}u_2(x)\right)\\ &= D(x).\end{aligned} \tag{5.4.24}$$

Since $u_1(x)$, and $u_2(x)$, are solutions of homogeneous MFDEs (5.2.6), expressions in parentheses are zero, and (5.4.24), reduced to:

$$\left({}_i\mathcal{D}_M^{\alpha,\beta}\vartheta_1(x)\right)\left({}_i\mathcal{D}_M^{\alpha,\beta}u_1(x)\right)+\left({}_i\mathcal{D}_M^{\alpha,\beta}\vartheta_2(x)\right)\left({}_i\mathcal{D}_M^{\alpha,\beta}u_2(x)\right) = D(x). \tag{5.4.25}$$

Considering (5.4.21), and (5.4.25), the following results are obtained

$$\vartheta_1(x) = {}_M\mathcal{T}_a^{\alpha,\beta}\Big(\frac{-u_2(x)D(x)}{W_{\alpha,\beta}(u_1(x),u_2(x))}\Big),$$

and

$$\vartheta_2(x) = {}_M\mathcal{T}_a^{\alpha,\beta}\Big(\frac{u_1(x)D(x)}{W_{\alpha,\beta}(u_1(x),u_2(x))}\Big),$$

so,

$$\begin{aligned} u_p(x) = u_1(x)\cdot {}_M\mathcal{T}_a^{\alpha,\beta}\Big(\frac{-u_2(x)D(x)}{W_{\alpha,\beta}(u_1(x),u_2(x))}\Big) \\ + u_2(x)\cdot {}_M\mathcal{T}_a^{\alpha,\beta}\Big(\frac{u_1(x)D(x)}{W_{\alpha,\beta}(u_1(x),u_2(x))}\Big). \end{aligned}$$

**Example 5.4.1.** *Consider the below linear MFDE*

$$\begin{aligned} 9\sqrt[3]{x^2}\,{}_i\mathcal{D}_M^{\frac{1}{3},\beta}\Big({}_i\mathcal{D}_M^{\frac{1}{3},\beta}u(x)\Big) - \frac{6}{\Gamma(\beta+1)}\sqrt[3]{x}\,{}_i\mathcal{D}_M^{\frac{1}{3},\beta}u(x) \\ + \frac{2}{(\Gamma(\beta+1))^2}u(x) = 9\sqrt[3]{x^2}. \end{aligned} \tag{5.4.26}$$

*By using example 5.3.1., the homogeneous solutions of MFDE (5.4.26), are as follows,*

$$u_1(x) = 3\Gamma(\beta+1)\sqrt[3]{x},\ u_2(x) = 9(\Gamma(\beta+1))^2\sqrt[3]{x^2},$$

*and* $u_p(x) = 9(\Gamma(\beta+1))^2\sqrt[3]{x^2}(\ln(3\Gamma(\beta+1)\sqrt[3]{x}) - 1)$, *a particular solution of M-fractional (5.4.26). So the M-fractional differential equation (5.4.26), has a general solution, like*

$$u(x) = 3\Gamma(\beta+1)\sqrt[3]{x}(c_1 + c_2 3\Gamma(\beta+1)\sqrt[3]{x})$$

$$+\,9(\Gamma(\beta+1))^2\sqrt[3]{x^2}(\ln(3\Gamma(\beta+1)\sqrt[3]{x})-1).$$

**Example 5.4.2.** *We want to obtain the general solution of second-order linear MFDE*

$$x\,{}_i\mathcal{D}_M^{\frac{1}{2},\beta}\Big({}_i\mathcal{D}_M^{\frac{1}{2},\beta}u(x)\Big)+\frac{1}{2\Gamma(\beta+1)}\sqrt{x}\,{}_i\mathcal{D}_M^{\frac{1}{2},\beta}u(x)$$
$$-\frac{1}{(\Gamma(\beta+1))^2}u(x)=4(\Gamma(\beta+1))^2x^2.$$

*We know that the homogeneous solutions are* $u_1(x)=4(\Gamma(\beta+1))^2x$, *and* $u_2(x)=(4(\Gamma(\beta+1))^2x)^{-1}$, *and* $u_p(x)=\frac{1}{12}(\Gamma(\beta+1))^2x^2$, *is a specific solution of it. Thus, the general solution of an equation is*

$$u(x)=4(\Gamma(\beta+1))^2\Big(C_1x+C_2x^{-1}+\frac{1}{12}x^2\Big).$$

### *5.4.2 Undetermined Coefficients*

Undetermined coefficients are used to find $u_p(x)$, when MFDEs (5.2.5), is

$${}_i\mathcal{D}_M^{\alpha,\beta}\big({}_i\mathcal{D}_M^{\alpha,\beta}u(x)\big)+a\,{}_i\mathcal{D}_M^{\alpha,\beta}u(x)+bu(x)=D(x), \tag{5.4.27}$$

$a$, and $b$, are constant and

$$D(x)=\Big(p_0+p_1\Big(\frac{\Gamma(\beta+1)}{\alpha}x^\alpha\Big)+p_2\Big(\frac{\Gamma(\beta+1)}{\alpha}x^\alpha\Big)^2+\cdots+p_n\Big(\frac{\Gamma(\beta+1)}{\alpha}x^\alpha\Big)^n\Big)$$
$$e^{\beta\left(\frac{\Gamma(\beta+1)}{\alpha}x^\alpha\right)}\sin\gamma\Big(\frac{\Gamma(\beta+1)}{\alpha}x^\alpha\Big),$$

or

$$D(x)=\Big(p_0+p_1\Big(\frac{\Gamma(\beta+1)}{\alpha}x^\alpha\Big)+p_2\Big(\frac{\Gamma(\beta+1)}{\alpha}x^\alpha\Big)^2+\cdots+p_n\Big(\frac{\Gamma(\beta+1)}{\alpha}x^\alpha\Big)^n\Big)$$

$$e^{\beta\left(\frac{\Gamma(\beta+1)}{\alpha}x^{\alpha}\right)}\cos\gamma\left(\frac{\Gamma(\beta+1)}{\alpha}x^{\alpha}\right).$$

A particular solution is used as follows:

$$\begin{aligned}u_p(x) = &\Bigg[\left(P_0 + P_1\left(\frac{\Gamma(\beta+1)}{\alpha}x^{\alpha}\right) + P_2\left(\frac{\Gamma(\beta+1)}{\alpha}x^{\alpha}\right)^2 + \cdots + P_n\left(\frac{\Gamma(\beta+1)}{\alpha}x^{\alpha}\right)^n\right)\\ &e^{\beta\left(\frac{\Gamma(\beta+1)}{\alpha}x^{\alpha}\right)}\sin\gamma\left(\frac{\Gamma(\beta+1)}{\alpha}x^{\alpha}\right) + \left(Q_0 + Q_1\left(\frac{\Gamma(\beta+1)}{\alpha}x^{\alpha}\right)\right.\\ &\left. + Q_2\left(\frac{\Gamma(\beta+1)}{\alpha}x^{\alpha}\right)^2 + \cdots + Q_n\left(\frac{\Gamma(\beta+1)}{\alpha}x^{\alpha}\right)^n\right)\\ &e^{\beta\left(\frac{\Gamma(\beta+1)}{\alpha}x^{\alpha}\right)}\cos\gamma\left(\frac{\Gamma(\beta+1)}{\alpha}x^{\alpha}\right)\Bigg]\left(\frac{\Gamma(\beta+1)}{\alpha}x^{\alpha}\right)^n,\end{aligned} \tag{5.4.28}$$

that $P_0, P_1, \cdots, P_n, Q_1, Q_2, \cdots, Q_n$, unknown coefficients, and $n$, is the lowest non-negative integer number, that can remove homogeneous solutions, in selecting $u_p(x)$. Substituting (5.4.28), into (5.4.27), unfamiliar coefficients are obtainable.

**Example 5.4.3.** *Consider the below nonhomogeneous MFDE*

$${}_i\mathcal{D}_M^{\frac{2}{3},\beta}\left({}_i\mathcal{D}_M^{\frac{2}{3},\beta}u(x)\right) - 2\,{}_i\mathcal{D}_M^{\frac{2}{3},\beta}u(x) = 10 - 6\sqrt[3]{x^2}. \tag{5.4.29}$$

*Homogeneous solutions of (5.4.29), can be*

$$u_1(x) = 1, \quad \text{or} \quad u_2(x) = e^{3\Gamma(\beta+1)\sqrt[3]{x^2}},$$

*a specific solution of it is*

$$u_p(x) = \frac{3}{2}\left(\left(\frac{1}{\Gamma(\beta+1)} - 5\right) + \frac{3}{2}\sqrt[3]{x^2}\right)\Gamma(\beta+1)\sqrt[3]{x^2}.$$

*Thus, the general solution of (5.4.29), is*

$$u(x) = C_1+C_2e^{3\Gamma(\beta+1)\sqrt[3]{x^2}}+\frac{3}{2}\left(\left(\frac{1}{\Gamma(\beta+1)}-5\right)+\frac{3}{2}\sqrt[3]{x^2}\right)\Gamma(\beta+1)\sqrt[3]{x^2}.$$

**Example 5.4.4.** *The general solution of linear second-order MFDE*

$${}_i\mathcal{D}_M^{\frac{1}{2},\beta}\left({}_i\mathcal{D}_M^{\frac{1}{2},\beta}u(x)\right) - 2\,{}_i\mathcal{D}_M^{\frac{1}{2},\beta}u(x) + u(x) = 2e^{\Gamma(\beta+1)\sqrt{x}},$$

*can be presented in the following form*

$$u(x) = e^{2\Gamma(\beta+1)\sqrt{x}}(C_1 + C_2\Gamma(\beta+1)\sqrt{x}) + 8e^{\Gamma(\beta+1)\sqrt{x}}.$$

**Example 5.4.5.** *Consider nonhomogeneous second-order MFDE:*

$${}_i\mathcal{D}_M^{\frac{1}{2},\beta}\left({}_i\mathcal{D}_M^{\frac{1}{2},\beta}u(x)\right) + 4u(x) = 3\cos 2\Gamma(\beta+1)\sqrt{x} + 8\sqrt{x} + 4.$$

*This equation has a general solution, such as,*

$$u(x) = C_1\cos\left(4\Gamma(\beta+1)\sqrt{x}\right) + C_2\sin\left(4\Gamma(\beta+1)\sqrt{x}\right) + \cos\left(2\Gamma(\beta+1)\sqrt{x}\right) + 2\sqrt{x} + 1.$$

---

## 5.5 Conclusion

In this chapter, some methods, like using a known solution to detect another one or the D'Alembert approach, homogeneous equation with constant coefficients or auxiliary method, and Euler's equidimensional equation were explained to determine a general solution of homogeneous MFDEs. The difference in parameters, or the Lagrange method and undetermined coefficient approach to specifying a certain solution of second-order linear MFDEs, were provided. The approach results in an accurate solution, without the need for using numerical approaches. To more study on fractional calculations, refer to the articles mentioned in the references of this section as suggested by the author [2, 6–30, 34, 38].

# References


[1] Capelas de Oliveira, E., & Tenreiro Machado, J. A, (2014). A review of definitions for fractional derivatives and integrals, *Mathematical Problems in Engineering*, 2014 (238459).

[2] Eslami, M., Hosseini, K., Matinfar, M., Mirzazadeh, M., Ilie, M., & Gómez-Aguilar, J. F., (2021). A nonlinear Schrödinger equation describing the polarization mode and its chirped optical solitons, *Optical and Quantum Electronics, 53*, 1–9.

[3] Figueiredo Camargo, R., & Capelas de Oliveira, E., *Fractional Calculus (In Portuguese)*, Editora Livraria da Física, So Paulo, 2015.

[4] Goreno, R., Kilbas, A. A., Mainardi, F., & Rogosin, S. V., Mittag-Leffler, *Functions Related Topics and Applications*, Springer, Berlin, 2014.

[5] Herrmann, R., & *Fractional Calculus: An Introduction for Physicists*, World Scientific Publishing Company, Singapore, 2011.

[6] Hosseini, K., Ilie, M., Mirzazadeh, M., & Baleanu, D., (2020). A detailed study on a new (2+1)-dimensional mKdV equation involving the Caputo–Fabrizio time-fractional derivative, *Advances in Difference Equations, 331*, https://doi.org/10.1186/s13662-020-02789-5.

[7] Hosseini, K., Ilie, M., Mirzazadeh, M., & Baleanu, D., (2021). An analytic study on the approximate solution of a nonlinear time-fractional Cauchy reaction–diffusion equation with the Mittag–Leffler law, *Mathematical Methods in the Applied Sciences, 44*(8), 6247–6258.

[8] Hosseini, K., Ilie, M. Mirzazadeh, M., Baleanu, D., Park, C., & Salahshour, S., (2022). The Caputo–Fabrizio time-fractional

Sharma–Tasso–Olver–Burgers equation and its valid approximations, *Communications in Theoretical Physics, 74*(7), 075003.

[9] Hosseini, K., Ilie, M., Mirzazadeh, M., Yusuf, A., Sulaiman, T. A., Baleanu, D., & Salahshour, S., (2021). An effective computational method to deal with a time-fractional nonlinear water wave equation in the Caputo sense, *Mathematics and Computers in Simulation, 187*, 248–260.

[10] Hosseini, K., Inc, M., Shafiee, M., Ilie, M., Shafaroody, A., Yusuf, & A. Bayram, M., (2020). Invariant subspaces, exact solutions and stability analysis of nonlinear water wave equations, *Journal of Ocean Engineering and Science, 5*(1), 35–40.

[11] Hosseini, K., Mirzazadeh, M., Ilie, M., & Gómez-Aguilar, JF., (2020). Biswas–Arshed equation with the beta time derivative: optical solitons and other solutions, *Optik, 217*, 164801.

[12] Hosseini, K., Mirzazadeh, M., Ilie, M., & Radmehr, S., (2020). Dynamics of optical solitons in the perturbed Gerdjikov–Ivanov equation,*Optik, 206,* 164350.

[13] Ilie, M., Biazar, J., & Ayati, Z., (2018). Analytical solutions for conformable fractional Bratu-type equations, *International Journal of Applied Mathematical Research, 7*(1), 15–19.

[14] Ilie, M., Biazar, J., & Ayati, Z., (2019). Analytical solutions for second-order fractional differential equations via OHAM, *Journal of Fractional Calculus and Applications, 10*(1), 105–119.

[15] Ilie, M., Biazar, J., & Ayati, Z., (2018). Analytical study of exact traveling wave solutions for time-fractional nonlinear Schrödinger equations, *Optical and Quantum Electronics, 50*(12), https://doi.org/10.1007/s11082-018-1682-y.

[16] Ilie, M., Biazar, J., & Ayati, Z., (2017). Application of the Lie Symmetry Analysis for second-order fractional differential equations, *Iranian Journal of Optimization, 9*(2), 79–83.

[17] Ilie, M., Biazar, J., & Ayati, Z., (2017). General solution of Bernoulli and Riccati fractional differential equations based on conformable

fractional derivative, *International Journal of Applied Mathematical Research, 6*(2), 49–51.

[18] Ilie, M., Biazar, J., & Ayati, Z., (2018). General solution of second order fractional differential equations, *International Journal of Applied Mathematical Research, 7*(2), 56–61.

[19] Ilie, M., Biazar, J., & Ayati, Z., (2018). Lie Symmetry Analysis for the solution of first-order linear and nonlinear fractional differential equations, *International Journal of Applied Mathematical Research, 7*(2), 37–41.

[20] Ilie, M., Biazar, J., & Ayati, Z., (2019). Mellin transform and conformable fractional operator: applications, *SeMA Journal, 76*(2), 203–215.

[21] Ilie, M., Biazar, J., & Ayati, Z., (2020). Neumann method for solving conformable fractional Volterra integral equations, *Computational Methods for Differential Equations, 8*(1), 54–68.

[22] Ilie, M., Biazar, J., & Ayati, Z., (2018). Optimal homotopy asymptotic method for conformable fractional Volterra integral equations of the second kind, *49$^{th}$ Annual Iranian Mathematics Conference, August 23–26, ISC* 97180–51902.

[23] Ilie, M., Biazar, J., & Ayati, Z., (2019). Optimal Homotopy Asymptotic Method for first-order conformable fractional differential equations, *Journal of Fractional Calculus and Applications, 10*(1), 33–45.

[24] Ilie, M., Biazar, J., & Ayati, Z., (2018). Resonant solitons to the nonlinear Schrödinger equation with different forms of nonlinearities, *Optik, 164,* 201–209.

[25] Ilie, M., Biazar, J., & Ayati, Z., (2018). The first integral method for solving some conformable fractional differential equations, *Optical and Quantum Electronics, 50*(2), https://doi.org/10.1007/s11082–017–1307-x.

[26] Ilie, M., & Khoshkenar, A., (2023). A novel study on nonlinear fractional differential equations: general solution, *Iranian Journal of Optimization.*

[27] Ilie, M., & Khoshkenar, A., (2021). General solution of the Bernoulli and Riccati fractional differential equations via truncated M-fractional derivative, *International Journal of Computer Mathematics, 1*(1), 67–77, http://dx.doi.org/ 10.22124/cse.2021.19568.1014.

[28] Ilie, M., & Khoshkenar, A., (2021). Lie Symmetry Analysis method to first-order M-fractional differential equations, *Computational Sciences and Engineering, 1*(1), 67–77.

[29] Ilie, M., & Khoshkenar, A., (2021). Resonant solitons solutions to the time M-fractional Schrödinger equation, *Iranian Journal of Optimization, 13*(3), 197–210.

[30] Ilie, M., Navidi, M., & Khoshkenar, A., (2018). Analytical solutions for conformable fractional Volterra integral equations of the second kind, *$49^{th}$ Annual Iranian Mathematics Conference, August 23–26, ISC* 97180–51902.

[31] Katugampola, U. N., (2014). A new fractional derivative with classical properties,*arXiv:1410.6535v2.*

[32] Katugampola, U. N., (2016). New fractional integral unifying six existing fractional integrals, *arxiv.org/abs/1612.08596.*

[33] Khalil, R., Al Horani, M., Yousef, A., & Sababheh, M., (2014). A new definition of fractional derivative, *Journal of Computational and Applied Mathematics, 264*, 65–70.

[34] Khoshkenar, A., Ilie, M., Hosseini, K., Baleanu, D., Salahshour, S., & Lee, J. R., (2022). Further studies on ordinary differential equations involving the M-fractional derivative, *AIMS Mathematics, 7*(6), 10977–10993.

[35] Kilbas, A. A., Srivastava, H. M., & Trujillo, J. J. (2006). *Theory and Applications of the Fractional Differential Equations*, Elsevier, Amsterdam, Vol. 204.

[36] Kilbas, A. A., Srivastava, H. M., & Trujillo, J. J., (2006). *Theory and Applications of Fractional Differential Equations*, North-Holland Mathematics Studies, Elsevier, Amsterdam, Vol. 207.

[37] Podlubny, I., (1999). *Fractional Differential Equations*, Mathematics in Science and Engineering, Academic Press, San Diego, Vol. 198.

[38] Salahshour, S., Ahmadian, A., Abbasbandy, S., & Baleanu, D., (2018). M-fractional derivative under interval uncertainty: Theory, properties and applications, *Chaos, Solitons and Fractals, 117*, 84–93.

[39] Simmons, G. F., (1974). *Differential equations with applications and historical notes*, McGraw-Hill, Inc.: New York.

[40] Vanterler da Sousa, J., & Capelas de Oliveira, E., (2018). A New Truncated M-Fractional Derivative Type Unifying Some Fractional Derivative Types with Classical Properties, *International Journal of Analysis and Applications, 16*(1), 83–96.

[41] Vanterler da Sousa, J., & Capelas de Oliveira, E., (2017). M-fractional derivative with classical properties, *arXiv:1704.08187*.

# Chapter 6

## *Numerical Computation of Fractional Derivatives of Caputo Type*

**Bengt Fornberg,**[1] **Cécile Piret,**[2] **and Austin Higgins**[2]
[1]*Department of Applied Mathematics, University of Colorado, Boulder, USA*
[2]*Department of Mathematical Sciences, Michigan Technological University, Houghton, USA*

**Abstract**
This chapter describes how fractional-order derivatives of the Caputo type can be computed with high accuracy using only function values on grids with node spacing $h$. With equispaced grid points between the derivative's base and evaluation points, convergence rates can reach approximately $O(h^{10})$. If the function to be differentiated is analytic and its values are also available at grid points in the complex plane, rates exceeding $O(h^{20})$ can be readily achieved. The approach utilized here is based on the classical Gregory quadrature method and its recent generalization to complex plane contour integration.

## Introduction

While integer-order derivatives are uniquely defined, numerous definitions have been proposed for derivatives of non-integer (including complex) orders, each with different features and areas of applicability.[1] Non-integer order derivatives are typically defined as inverses of certain fractional-order integrals and, as such, are not local operators.[2] However, they are characterized by two key points—a *base point* and an *evaluation point*. The value is obtained as an integral between these two points.

Of the two most commonly used versions—Riemann-Liouville and Caputo derivatives—we focus here on the latter. These are often preferred in the context of solving fractional-order differential equations due to their simplicity when formulating initial and boundary conditions.

In cases where function values are available at Gaussian quadrature-like locations, which allow for accurate expansion in Jacobi polynomials, certain identities exist for fractional derivative (and integral) calculations.[3] These identities are utilized, for example, in Chebfun. However, since such node locations are impractical in many contexts,[4] it is essential to develop numerical methods that use grid data with spacing $h$. Previously available methods for this case have generally suffered from very low-order convergence rates, such as the first-order rate $O(h^1)$ in the case of the original Grünwald-Letnikov formula for approximating Riemann-Liouville derivatives.

---

[1] Many seemingly different fractional derivative versions are shown in [27] to belong to a two-parameter family.

[2] Local fractional-order derivatives (with somewhat different properties) have also been considered [15, 19, 20].

[3] See, for example, equation (18.17.9) in [23] and [21], Section 2.1.7. For the Chebfun system, see [5].

[4] e.g., when solving fractional differential equations. Applications include anomalous diffusion, epidemiology [2], finance [26], and geology [4].

We summarize here recent developments in two situations:

i. Data is equispaced between base and evaluation points, and

ii. For analytic functions, equispaced grid data is available in the complex plane.

Convergence rates in these two cases reach around $O(h^{10})$ and $O(h^{22})$, respectively, making it fast and easy to obtain results close to machine precision, e.g., $10^{-15}$ in double precision arithmetic. This chapter is organized into two parts, the first of which is structured as follows: We note first in Section 6.1 that the fractional derivative takes the form of an integral between a base point and an evaluation point. As an introduction to the concept of end corrections for quadrature, we describe in Section 6.2.1 the classical Gregory method and in Section 6.2.2 how this can be generalized to integrals that have a singular end point. The second part of this chapter focuses on the case where the function to be numerically differentiated is analytic and numerically available on an $h$-spaced grid in the complex plane. As a lead-in to evaluating fractional order derivatives, we describe in Section 6.3.1 a recent approach in this case for approximating regular (integer order) derivatives, and in Section 6.3.2 how this, in connection with an also recently introduced complex plane contour integration method, provides an effective method for the fractional derivative case. A number of illustrations of computed fractional derivatives of analytic functions are given. Section 6.4 contains some concluding remarks.

## 6.1 Riemann-Liouville and Caputo Derivatives

The two most commonly used fractional derivative definitions are Riemann-Liouville:

$$ {}_{a}^{RL}D_t^{\alpha} f(t) = \frac{1}{\Gamma(n-\alpha)} \frac{d^n}{dt^n} \int_a^t \frac{f(\tau)}{(t-\tau)^{\alpha+1-n}} d\tau \,, \qquad n-1<\alpha<n \tag{6.1.1} $$

and Caputo [3]:

$$ {}_{a}^{C}D_t^{\alpha} f(t) = \frac{1}{\Gamma(n-\alpha)} \int_a^t \frac{\frac{d^n}{d\tau^n} f(\tau)}{(t-\tau)^{\alpha+1-n}} d\tau \,, \qquad n-1<\alpha<n. \tag{6.1.2} $$

With no loss of generality, we let the *base point* $a$ be located at $a=0$ and (initially) assume that the *evaluation point* $t$ satisfies $t>0$. The two definitions are closely related:

$$ {}_{0}^{RL}D_t^{\alpha} f(t) = {}_{0}^{C}D_t^{\alpha} f(t) + \sum_{k=0}^{n-1} \frac{t^{k-\alpha}}{\Gamma(k+1-\alpha)} f^{(k)}(0), \tag{6.1.3} $$

but have nevertheless quite different characteristics, as illustrated in Figure 6.1 when differentiating $f(t) = e^t$ for $0 \le \alpha \le 1$, $0 \le t \le 1$.[5] As noted above, our focus on Caputo rather than on Riemann-Liouville derivatives is largely motivated by the simplicity these offer when formulating initial conditions for fractional order differential equations, as described, for example, in [25], Chapter 4 and [17], Chapters 2 and 4. We simplify the notation by writing

[5]The closed form values in this parameter range are ${}_{0}^{RL}D_t^{\alpha} e^t = e^t\left(1-\frac{\Gamma(-\alpha,t)}{\Gamma(-\alpha)}\right)$ and ${}_{0}^{C}D_t^{\alpha} e^t = e^t\left(1-\frac{\Gamma(1-\alpha,t)}{\Gamma(1-\alpha)}\right)$.

${}_0^C D_t^\alpha f(t)$ as $D^\alpha f(t)$ in what follows. Since, for $m$ integer, $D^{\alpha+m} f(t) = D^\alpha D^m f(t)$,[6] we furthermore focus on $n = 1$, i.e., $0 < \alpha < 1$.[7]

---

## 6.2 Gregory's Method: Classical and Generalized Versions

### *6.2.1 Classical Gregory Method for Regular Quadrature*

James Gregory[8] discovered already in 1670 an effective trapezoidal rule (TR) correction procedure, using a collection of mathematical tools that typically are attributed to later calculus pioneers.[9]

For simplicity, we describe it first when applied to the difference $\int_0^\infty f(x)dx - \sum_{k=0}^\infty f(k)$. With the notation $\Delta f(k) = f(k+1) - f(k)$ for the first one-sided difference, it follows that

[6] In the Riemann-Liouville case, the operators $D^\alpha$ and $D^m$ need for this result to be in reversed order. In both cases, the operators commute if $f^{(k)}(0) = 0$ for $k = 0, 1, 2, \ldots, m$.

[7] Another convenience with the Caputo version is that (6.1.2) implies that $D^\alpha\{\text{constant}\} = 0$. On the other hand (as seen in Figure 6.1), while $\lim_{\alpha\to 1} D^\alpha f(t) = f'(t)$, $\lim_{\alpha\to 0} D^\alpha f(t) = f(t) - f(0)$. In the Riemann-Liouville case, both limits are as expected.

[8] 1638-1675, Scottish mathematician and astronomer.

[9] The details of Gregory's derivation are not known, but it was likely similar to the one used here. A letter he wrote in 1670 [14] contains both (6.2.4) and (6.2.8), and he used in other contexts Taylor expansions (about 45 years before these were described by Taylor). Gregory's letter precedes by about 15 years the descriptions of calculus by Leibniz and Newton.

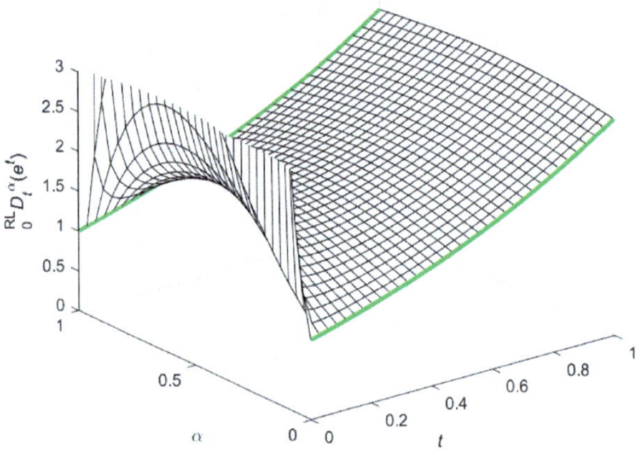


(a) Riemann-Liouville derivative

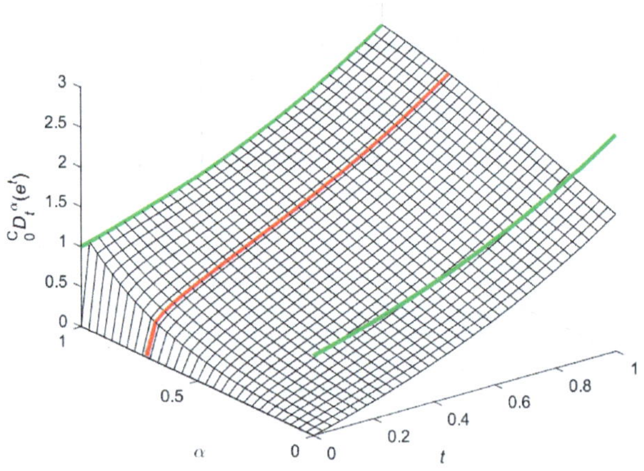


(b) Caputo derivative

**FIGURE 6.1**. Riemann-Liouville and Caputo derivatives of order $\alpha \in [0, 1]$ for $f(t) = e^t,\ t \geq 0$. The green curves show the standard derivatives for the integer order cases $\alpha = 0$ and $\alpha = 1$. The red curve in part (b) at $\alpha = 5/7$ will be seen again in Figure 6.8.

$$\begin{aligned} \Delta^0 f(0) &= f(0) \\ \Delta^1 f(0) &= f(1) - f(0) \\ \Delta^2 f(0) &= f(2) - 2f(1) + f(0) \\ \Delta^3 f(0) &= f(3) - 3f(2) + 3f(1) - f(0) \\ \vdots \quad & \quad \vdots \end{aligned} \tag{6.2.4}$$

with coefficients from Pascal's triangle. Gregory's idea was to look for an improved TR formula of the form

$$\int_0^\infty f(x)dx \sim \left(\sum_{k=0}^{\infty} f(k)\right) + \left[b_0\Delta^0 + b_1\Delta^1 + b_2\Delta^2 + \ldots\right] f(0)\,, \tag{6.2.5}$$

and then to use only a finite number of these correction terms. Substituting $f(x) = e^{\xi x}$ into (6.2.5) gives[10]

$$-\frac{1}{\xi} = \frac{1}{1-e^{\xi}} + \left[b_0 - b_1(1-e^{\xi}) + b_2(1-e^{\xi})^2 - b_3(1-e^{\xi})^3 + - \ldots\right].$$

With the further substitution

$$w = (1-e^{\xi})\,, \tag{6.2.6}$$

i.e., $\xi = \log(1-w)$, this becomes

$$\frac{1}{\log(1-w)} + \frac{1}{w} = -b_0 + b_1 w - b_2 w^2 + b_3 w^3 - + \ldots \tag{6.2.7}$$

The coefficients $b_k$ can now be calculated recursively based on the Taylor expansion of $\log(1-w)$:

$$b_0 = -\tfrac{1}{2},\ b_1 = \tfrac{1}{12},\ b_2 = -\tfrac{1}{24},\ b_3 = \tfrac{19}{720},$$

[10] with Re $\xi < 0$.

$$b_4 = -\tfrac{3}{160},\ b_5 = \tfrac{863}{60480},\ b_6 = -\tfrac{275}{24192},\dots \tag{6.2.8}$$

Using only $b_0 = -\frac{1}{2}$ turns (6.2.5) into the TR. For each further term, the accuracy order increases by one [22]. When applying the Gregory method to function values $f(k)$, it is more convenient to work with *weights* $w_k$ that multiply these function values, rather than with the $b_k$ expansion coefficients. Table 6.1 gives the Gregory *corrections* $d_k = w_k - 1$, up through $p = 10$ (with all $d_k$ equal to zero to the right of the shown table entries). In the case of a finite interval, we add this $d_k$-sequence from each end (in reversed order at the right end) to obtain actual weights to use. These two sequences can overlap; each one just has to fit into the interval. If the grid spacing is $h$ rather than one, all weights need to be multiplied by $h$. With the MATLAB function EC_weights given (in a more general context) in Section 6.2.4, the corrections shown in Table 6.1 corresponding to order $p$ can be computed numerically by:

```
w = -EC_weights(-1,1,0:p-2);    w(1) = w(1)-1;
```

We next extend the Gregory scheme in two directions, to allow for

i. fractional order singularity of the integral at an endpoint, and

ii. arbitrary location of end correction stencil nodes near or around the end points (including surrounding the end points in the complex plane).

**TABLE 6.1**. Corrections $d_k$ to the Weights All Being One, According to Gregory's Formulas, Up Through Order $p = 10$

| $p =$ | Gregory corrections $d_k$ to the weights all being one | | | | | | | | | |
|---|---|---|---|---|---|---|---|---|---|---|
| 2 | $-\frac{1}{2}$ | | | | | | | | | |
| 3 | $-\frac{7}{12}$ | $\frac{1}{12}$ | | | | | | | | |
| 4 | $-\frac{5}{8}$ | $\frac{1}{6}$ | $-\frac{1}{24}$ | | | | | | | |
| 5 | $-\frac{469}{720}$ | $\frac{59}{240}$ | $-\frac{29}{240}$ | $\frac{19}{720}$ | | | | | | |
| 6 | $-\frac{193}{288}$ | $\frac{77}{240}$ | $-\frac{7}{30}$ | $\frac{73}{720}$ | $-\frac{3}{160}$ | | | | | |
| 7 | $-\frac{41393}{60480}$ | $\frac{23719}{60480}$ | $-\frac{11371}{30240}$ | $\frac{7381}{30240}$ | $-\frac{5449}{60480}$ | $\frac{863}{60480}$ | | | | |
| 8 | $-\frac{12023}{17280}$ | $\frac{6961}{15120}$ | $-\frac{66109}{120960}$ | $\frac{33}{70}$ | $-\frac{31523}{120960}$ | $\frac{1247}{15120}$ | $-\frac{275}{24192}$ | | | |
| 9 | $-\frac{2558783}{3628800}$ | $\frac{1908311}{3628800}$ | $-\frac{299587}{403200}$ | $\frac{115963}{145152}$ | $-\frac{426809}{725760}$ | $\frac{112477}{403200}$ | $-\frac{278921}{3628800}$ | $\frac{33953}{3628800}$ | | |
| 10 | $-\frac{63887}{89600}$ | $\frac{427487}{725760}$ | $-\frac{3498217}{3628800}$ | $\frac{500327}{403200}$ | $-\frac{6467}{5670}$ | $\frac{2616161}{3628800}$ | $-\frac{24019}{80640}$ | $\frac{263077}{3628800}$ | $-\frac{8183}{1036800}$ | |
| $\vdots$ | $\vdots$ | $\vdots$ | $\vdots$ | $\vdots$ | $\vdots$ | $\vdots$ | $\vdots$ | $\vdots$ | $\vdots$ | $\ddots$ |

The middle entry on the $p = 10$ line shows the first instance of a correction $< -1$, leading to a negative quadrature weight $w_k = 1 + d_k$.

### *6.2.2 Gregory's Method Generalized to the Case of An End Point Singularity*

Evaluation of Caputo derivatives requires us to approximate

$$D^{\alpha} f(t) = \frac{1}{\Gamma(1-\alpha)} \int_0^t \frac{f'(\tau)}{(t-\tau)^{\alpha}} d\tau \; , \tag{6.2.9}$$

where we assume that $f(\tau)$ (rather than $f'(\tau)$) is numerically available, at $h$-spaced $\tau$-values.

### *6.2.3 Method of Exponential Test Functions*

Integration by parts for the integral in (6.2.9) leads us to consider[11]

$$\alpha \int \frac{f(\tau)}{(t-\tau)^{\alpha+1}} d\tau \; , \tag{6.2.10}$$

[11]Temporarily omitting the factor $\frac{1}{\Gamma(1-\alpha)}$ in (6.2.9).

and the TR-type sum

$$\alpha h \sum_{k=1}^{[\frac{t}{h}]-1} \frac{f(kh)}{(t-kh)^{\alpha+1}} \tag{6.2.11}$$

corresponding to this integral over the interior of the interval $[0, t]$ (compared to regular TR, omitting a term at each end).

**Lower End:** There is no singularity present in (6.2.10) around $\tau = 0$, and the TR end corrections can be applied by the regular Gregory approach. In order to generalize to the singular (upper end), we derive the Gregory weights in a slightly different way than used above in Section 6.2.1. End correcting at $\tau = 0$ leads us to consider $\int_0^\infty g(\tau)d\tau$ with $g(\tau) = e^{\xi\tau}$ (again with Re $\xi < 0$), in which case

$$\begin{aligned}
\int_0^\infty e^{\xi\tau} d\tau - h\sum_{k=1}^{\infty} e^{\xi kh} \\
&= \frac{h}{2}\left(1 + \coth\frac{h\xi}{2}\right) - \frac{1}{\xi} \\
&= -h\left(\frac{\zeta(0)}{0!} + \frac{\zeta(-1)}{1!}(h\xi)^1 + \frac{\zeta(-2)}{2!}(h\xi)^2\right. \\
&\quad \left. + \frac{\zeta(-3)}{3!}(h\xi)^3 + \dots\right).
\end{aligned}$$

In this expansion, the terms with $\xi^2, \xi^4, \xi^6, \dots$ vanish (due to the trivial zeros of the zeta function). For arbitrary locations of nodes in an end correction stencil, the weights are obtained from the coefficients in this Taylor expansion as described in Section 6.2.4.

**Upper End:** This end correction can be obtained by comparing (6.2.11) against $\int_0^t \frac{f'(\tau)}{(t-\tau)^\alpha} d\tau$ (without integrating by parts). To simplify the discussion, we perform the change variables $\sigma = t - \tau$, and set $g(\sigma) = f(t-\tau)$ to move the evaluation point to the origin and analyze the corresponding correction. This leads us to consider the end correction problem (around $\sigma = 0$) of the form

$$\int_0^\infty \frac{g'(\sigma)}{\sigma^\alpha} d\sigma - \alpha h \sum_{k=1}^\infty \frac{g(kh)}{(kh)^{\alpha+1}} \,. \tag{6.2.12}$$

With $g(\sigma) = e^{\xi\,\sigma}$, we obtain

$$\int_0^\infty \frac{\xi\, e^{\xi\,\sigma}}{\sigma^\alpha} d\sigma - \alpha h \sum_{k=1}^\infty \frac{e^{\xi kh}}{(kh)^{\alpha+1}} = (-\xi)^{\alpha-1} \xi \Gamma(1-\alpha) - \alpha h^{-\alpha} \mathrm{Li}_{1+\alpha}(e^{h\xi})$$

$$= -\alpha h^{-\alpha} \Bigg( \frac{\zeta(\alpha+1)}{0!} + \frac{\zeta(\alpha)(h\xi)}{1!} + \frac{\zeta(\alpha-1)(h\xi)^2}{2!} + \frac{\zeta(\alpha-2)(h\xi)^3}{3!} + \dots \Bigg). \tag{6.2.13}$$

Here, Li denotes the polylogarithm function. The two terms in the first right-hand side of (6.2.13) contain each a fractional power of $\xi$ that cancels out in their difference. Just as for the non-singular lower end (6.2.12), the result is again a Taylor series in $\xi$ of similar form[12].

**Combined Procedure:** The TR-type part of the calculation (6.2.11) is corrected separately at the lower and upper ends. At both ends, we have obtained error expressions in the form of Taylor expansions, which we will convert to correction weights. At the lower end, these weights are applied to values of $\frac{f(\tau)}{(t-\tau)^{\alpha+1}}$ (with an additional term having arisen from the integration by parts).

[12] The expansion (6.2.13) follows from equation (25.12.12) in [23].

At the upper end, the correction weights are instead applied to $f(\tau)$-values.[13] The resulting approximation can be written as

$$\int_0^t \frac{f'(\tau)}{(t-\tau)^\alpha} d\tau$$

$$= \underbrace{-\alpha h \sum_{k=1}^{[t/h]-1} \frac{f(kh)}{(t-kh)^{\alpha+1}}}_{\text{TR-type sum}} + \underbrace{\left( -\frac{f(0)}{t^\alpha} + \alpha\, h \sum_{k=1}^{n} w_k^{(bp)} \frac{f(z_k^{(bp)})}{(t-z_k^{(bp)})^{\alpha+1}} \right)}_{\text{Lower end correction}}$$
$$+ \underbrace{\left( \alpha h^{-\alpha} \sum_{k=1}^{n} w_k^{(ep)} f(t - z_k^{(ep)}) \right)}_{\text{Upper end correction}} + O(h^n). \tag{6.2.14}$$

Here, $n$ denotes the total number of stencil nodes at each end (and is not related to derivative order, as in (6.1.1), (6.1.2)), $z_k^{(bp)}$ and $z_k^{(ep)}$ are the locations of the correction stencil nodes relative to the base and evaluation points[14], and $w_k^{(bp)}$ and $w_k^{(ep)}$ the respective weights, obtained as described in the next Section 6.2.4. At both ends, the node locations for the end corrections can be chosen arbitrarily (with two options considered below; these node locations need not be related to the step size $h$ in the TR approximation).

---

[13] Here with no integration by parts term, as the integral in (6.2.12) is in its original form.

[14] We denote these $z_k^{(bp)}$, $z_k^{(ep)}$ rather than $\tau_k^{(bp)}$, $\tau_k^{(ep)}$ as they may be placed in the complex plane surrounding the base- and evaluation points.

### *6.2.4 Converting Taylor Expansions in $\xi$ to Weights in a Correction Stencil*

With correction stencil nodes located at $z_k$, $k = 1, \ldots, n$, equating the leading part of a Taylor expansion $\sum_{k=0}^{n-1} \alpha_k \xi^k$ with that of a correction stencil $\sum_{k=1}^{n} w_k e^{\xi z_k}$ leads to a linear system of Vandermonde type that is readily solved numerically. Both for (6.2.12) and (6.2.13), the weights at node locations $z_k$ can be obtained by calling the MATLAB function shown below (for $\alpha = -1$ and with $0 < \alpha < 1$, respectively):

```
function w = EC_weights(al,h,z)
%    Calculate weights in a stencil matching the Taylor expansion
%    ζ(α+1)/0!+ζ(α)(hξ)/1!+ζ(α-1)(hξ)²/2!+ζ(α-2)(hξ)³/3!+ζ(α-3)(hξ)⁴/4!+...
% Input parameters
%      al  The parameter denoted α above
%       h  Step size in TR-type approximation (with first term omitted)
%       z  Stencil node locations (array or vector)
% Output parameter
%       w  Weights at locations given by z

dm = size(z); z = z(:); sq = (0:length(z)-1);           % Make z a column
                                                         % vector
A = rot90(vander(z)); rhs = zeta(al+1-sq).*h.^sq;        % Form linear system
                                                         % and RHS
w = A\rhs'; w = reshape(w,dm);                           % Solve linear system,
end                                                      % and arrange weights
                                                         % as for z
```

### *6.2.5 Numerical Illustrations of Two Gregory-Based Calculations*

Figure 6.2 (a) shows a test function that is infinitely differentiable over $[0, 1]$, and part (b) its exact Caputo derivative of order $\alpha = 1/2$

(featuring a square root-type singularity at the origin: $D^{1/2}f(t) = -\frac{400}{\sqrt{\pi}}t^{1/2}\left(1 - \frac{400}{9}t + \frac{6400}{9}t^2 + \ldots\right)$). Figure 6.3(a) shows the error in computing $\int_0^1 f(t)dt$ as a function of the number of Gregory correction coefficients used at each end ($1 - 13$ from top and down, corresponding to accuracy orders $2 - 14$, with the first case being the trapezoidal rule TR). Part (b) shows for the same number of correction terms at each end the error in computing $D^{1/2}f(1)$ when using Gregory-type approximations to the integral (6.2.9), implemented as described in Sections 6.2.3 and 6.2.4. Unsurprisingly, as derivative calculations, in general, are much more numerically unstable than integrations[15], the rounding error level for small $h$ is notably higher, and the slope of each curve is reduced by a factor of $h^{-3/2}$. Nevertheless, rates of convergence better than $O(h^{10})$ can be observed.

### 6.2.6 Case of Short Interval

The procedure described just above requires base and evaluation points to be well separated. When this is not the case, differentiation matrices (DMs) offer an alternative. Such matrices are often convenient for expressing the effect of linear operators on a vector of data, and are widely used in the contexts of finite difference, pseudospectral, and radial basis function methods [11]. In the case

[15]While all Gregory weights are positive for accuracy orders up through $p = 9$, this is not the case for Caputo derivative approximations, since $D^{\alpha}(1) = 0$, implying that all weights closely add up to zero.

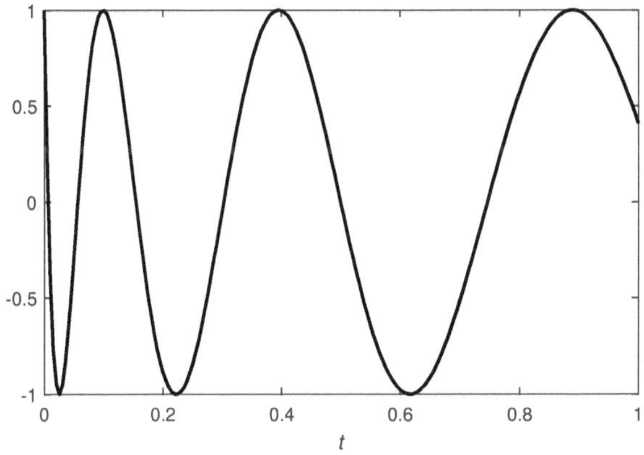


(a) $f(t)=\cos(20\sqrt{t})$

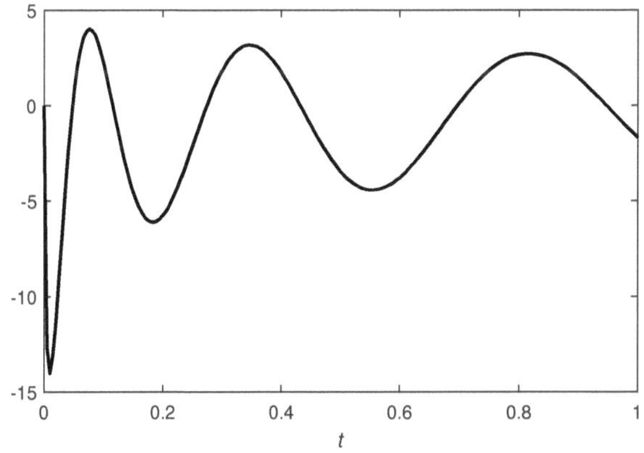


(b) $D^{1/2}f(t)=-10\sqrt{\pi}\mathrm{H}_0(20\sqrt{t})$

**FIGURE 6.2.** (a) The test function $f(t)$ used in Section 6.2.5, (b) Its Caputo fractional derivative $D^{1/2}f(t)$. Here $\mathrm{H}_0$ denotes the Struve function of order zero, alternatively expressed as a hypergeometric function: $\mathrm{H}_0(t) = \frac{2t}{\pi}\, {}_1F_2\left(1; \frac{3}{2}, \frac{3}{2}; -\frac{t^2}{4}\right)$.

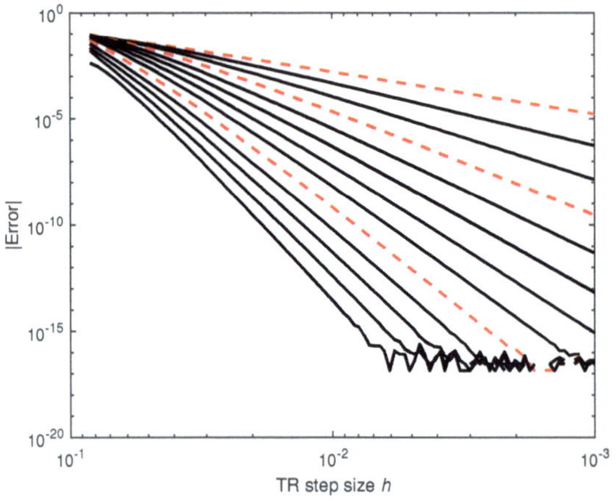


(a) Gregory integration

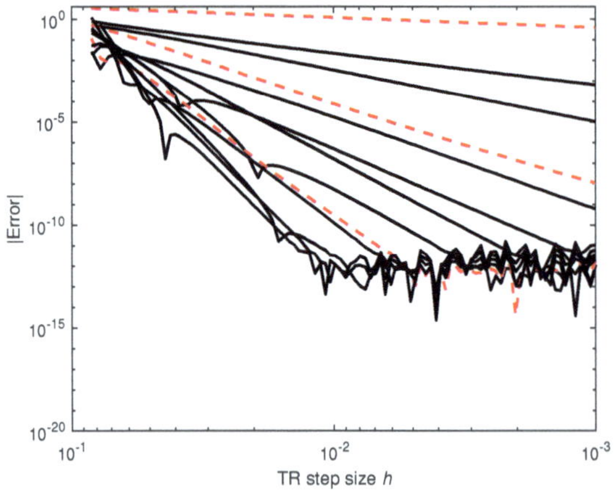


(b) Gregory-based Caputo differentiation

**FIGURE 6.3**. Errors as function of the number of Gregory-type correction coefficients applied at each interval end ($1-13$ for the curves, from top to bottom) and as function of step size $h$, when applied to the test function shown in Figure 6.2 (a) for computing (a) $\int_0^1 f(t)dt$, and (b) $D^{1/2}f(1)$. In part (a), the curves marked dashed red instead of solid blue correspond to convergence rates 2, 5, and 10, while in part (b), they show the 1/2, 3 1/2, and 8 1/2 rate cases (1 1/2 order lower for a matching number of correction coefficients).

of nodes at $t_k \geq 0$, $k = 1, 2, \ldots, n$, the entries of the DM

$$D^{\alpha} \begin{bmatrix} y_1 \\ y_2 \\ \vdots \\ y_n \end{bmatrix} = \begin{bmatrix} & & \\ & \text{DM} & \\ & & \\ & & \end{bmatrix} \begin{bmatrix} y_1 \\ y_2 \\ \vdots \\ y_n \end{bmatrix}$$

can be obtained by considering the expansions in powers of $t$ of the $n$ successive cardinal polynomials over the nodes. The fractional derivative of each such expansion is then obtained by the relation $D^{\alpha} t^m = \frac{\Gamma(m+1)}{\Gamma(m+1-\alpha)} t^{m-\alpha}$. Evaluating these at the successive nodes gives the entries down the matching DM column.[16] Approximating a Caputo derivative with base point zero, the DM can be obtained by the MATLAB code shown on the next page.

### *6.2.7 The Grünwald-Letnikov Formula*

The Riemann-Liouville derivative ${}_0^{RL}D^{\alpha} f(t)$ can be written as ${}_0^{RL}D^{\alpha} f(t) = \lim_{h \to 0} \frac{{}_0^{GL}\Delta_h^{\alpha} f(t)}{h^{\alpha}}$, where

$${}_0^{GL}\Delta_h^{\alpha} f(t) = \sum_{j=0}^{\left[\frac{t}{h}\right]} (-1)^j \binom{\alpha}{j} f(t - jh) \,. \tag{6.2.15}$$

This formula has a long history (introduced in 1868). Numerical usage of it is described, for example, in [25], Chapter 7. Following

[16]The function Diff_Matrix has been written for simplicity rather than for optimal numerical stability. For high accuracy, it can either be run in extended precision or be rearranged to use polynomial recursion instead of the polyfit function.

```
function DM = Diff_Matrix(t,a)
% Input parameters
%       t   Vector with the node locations (row or column, nodes >=0, in any
%            order)
%       a   Order of the Caputo derivative, 0 < a < 1
% Output parameters
%      DM   Differentiation matrix. Base point 0, evaluation points t
n = length(t); DM = zeros(n,n);                  % Create matrix for DM
ind = n:-1:2;  cf = gamma(ind)./gamma(ind-a);    % Coefficients for CaputoD
                                                 % of a polynomial
for k = 1:n                                      % Loop over the columns of
                                                 % the DM
    yv = zeros(1,n);           yv(k) = 1;        % Create cardinal data
    p = polyfit(t,yv,n-1);                       % Create cardinal
                                                 % polynomial
    p(1:n-1) = p(1:n-1).*cf;   p(n) = 0;         % Differentiate it
    for j = 1:n                                  % Evaluate its CaputoD at
                                                 % successive nodes
        if t(j) == 0; continue; end;             % CaputoD = 0 at base
                                                 % point t = 0
        DM(j,k) = polyval(p,t(j))*t(j)^(-a);     % Store in DM
    end
end
```

the expansion approach above (again using $f(z) = e^{\xi z}$) gives

$$GL = -\frac{1}{h^\alpha}\sum_{k=0}^{\infty}(-1)^k\binom{\alpha}{k}e^{\xi kh} = \left(\frac{e^{h\xi}-1}{h}\right)^\alpha$$
$$= \xi^\alpha\left(-1+\frac{1}{2}\alpha(h\xi)^1-\frac{1}{24}\alpha(1+3\alpha)(h\xi)^2\right.$$
$$\left.+\frac{1}{48}\alpha^2(1+\alpha)(h\xi)^3+\ldots\right). \tag{6.2.16}$$

The leading term $-\xi^\alpha$ agrees with the exact result, but already the next term reflects a first-order error.[17] Since all terms in the expansion (6.2.16) contain fractional powers of $\xi$, there is in this case no opportunity for FD-type end corrections.

---

## 6.3 Fractional Derivative Approximations of an Analytic Function Given on a Grid in the Complex Plane

### *6.3.1 Complex Plane FD Approximations*

This section recalls briefly some observations from [9], in parts earlier described in [6, 7]. Complex plane FD stencils with $N = (2n+1) \times (2n+1)$ nodes ($n = 1, 2, \ldots$) on a grid with spacing $h$ will, for the $p^{\text{th}}$ derivative ($p = 1, 2, 3, \ldots$), be accurate at least of order $O(h^{N-p})$.[18] As examples of such stencils, Table 6.2 illustrates the weights in the cases of $n = 1$ and $n = 2$ for the first two derivatives. The order of accuracy for the shown $3 \times 3$ stencils are both $O(h^8)$, and for the $5 \times 5$ stencils both $O(h^{24})$. One major difference from traditional (real axis) FD approximations is that complex plane FD weights decrease in magnitude very much faster with the distance from the stencil's center. With stencil nodes at $z_k = \mu + i\,\nu$, $-n \leq \mu, \nu \leq +n$, the decay rate contains again an algebraic factor, but now further multiplied by the extremely rapidly decreasing factor $e^{-\frac{\pi}{2}(\mu^2+\nu^2)}$. This makes the approximations remain

---

[17] A second-order accurate variation of the Grünwald–Letnikov formula is described in [18].

[18] Here and in the following, $n$ denotes stencil size and not derivative ranges (as in (6.1.1), (6.1.2)).

**TABLE 6.2.** The Weights in $3\times 3$ and $5\times 5$ Size Stencils for the First Two Derivatives

| $n = 1$; Stencil size $3 \times 3$ | $n = 2$; Stencil size $5 \times 5$ |
|---|---|
| $f'(0) \approx \frac{1}{h}\begin{bmatrix} \frac{-1-i}{40} & \frac{-i}{5} & \frac{1-i}{40} \\ \frac{-1}{5} & 0 & \frac{1}{5} \\ \frac{-1+i}{40} & \frac{i}{5} & \frac{1+i}{40} \end{bmatrix} f;$ | $\frac{1}{h}\begin{bmatrix} \frac{1+i}{477360} & \frac{4(-1-i)}{29835} & \frac{i}{1326} & \frac{4(1-i)}{29835} & \frac{-1+i}{477360} \\ \frac{4(-1-i)}{29835} & \frac{8(-1-i)}{351} & \frac{-8i}{39} & \frac{8(1-i)}{351} & \frac{4(1-i)}{29835} \\ \frac{1}{1326} & \frac{-8}{39} & 0 & \frac{8}{39} & \frac{-1}{1326} \\ \frac{4(-1+i)}{29835} & \frac{8(-1+i)}{351} & \frac{8i}{39} & \frac{8(1+i)}{351} & \frac{4(1+i)}{29835} \\ \frac{1-i}{477360} & \frac{4(-1+i)}{29835} & \frac{-i}{1326} & \frac{4(1+i)}{29835} & \frac{-1-i}{477360} \end{bmatrix} f$ |
| $f''(0) \approx \frac{1}{h^2}\begin{bmatrix} \frac{i}{20} & \frac{-2}{5} & \frac{-i}{20} \\ \frac{2}{5} & 0 & \frac{2}{5} \\ \frac{-i}{20} & \frac{-2}{5} & \frac{i}{20} \end{bmatrix} f;$ | $\frac{1}{h^2}\begin{bmatrix} \frac{-i}{477360} & \frac{8(-1+3i)}{149175} & \frac{1}{1326} & \frac{8(-1-3i)}{149175} & \frac{i}{477360} \\ \frac{8(1+3i)}{149175} & \frac{16i}{351} & \frac{-16}{39} & \frac{-16i}{351} & \frac{8(1-3i)}{149175} \\ \frac{-1}{1326} & \frac{16}{39} & 0 & \frac{16}{39} & \frac{-1}{1326} \\ \frac{8(1-3i)}{149175} & \frac{-16i}{351} & \frac{-16}{39} & \frac{16i}{351} & \frac{8(1+3i)}{149175} \\ \frac{i}{477360} & \frac{8(-1-3i)}{149175} & \frac{1}{1326} & \frac{8(-1+3i)}{149175} & \frac{-i}{477360} \end{bmatrix} f$ |

highly localized even when their stencil sizes and accuracy orders are increased.[19] With this background, it becomes natural to end correct TR in the complex plane with square stencils surrounding each end of an integration interval.

### *6.3.2 Fractional Derivative Approximations of an Analytic Functions*

We first note that, with $f(\tau)$ an analytic function, the fractional derivative

$$D^{\alpha} f(z) = \frac{1}{\Gamma(1-\alpha)} \int_0^z \frac{f'(\tau)}{(z-\tau)^{\alpha}} d\tau, \quad 0 < \alpha < 1$$

[19]The infinite order accuracy (pseudospectral) $n \to \infty$ limit is studied in [10].

becomes again an analytic function.[20] The derivative will typically have a branch point at its base point. The procedure summarized below is described in more detail in [13].

### 6.3.3 Evaluation Point $z$ Well Separated From Base Point: $z$ Complex

We assume again that $f(z)$ is non-singular at the base point $z = 0$. The main idea is to follow horizontal and vertical grid lines from the base point to the evaluation point, as used for contour integrals in [6, 7] (i.e., with correction stencils also at each path corner). A key consideration is that no correction stencil can be centered close to the evaluation point (since the integrand has a singularity there, these correction stencils would then become inaccurate). Depending on the direction by which an interval endpoint is reached, the order of weights in the stencil will change as schematically illustrated in Figure 6.4 and a different factor of $\left(\frac{h}{\triangle x + i\,\triangle y}\right)^{\alpha}$ will also have to be applied to the stencil (e.g., when the evaluation point is reached from right to left, $\triangle x = -h$ and $\triangle y = 0$ giving a factor of $(-1)^{\alpha}$.)

### 6.3.4 Evaluation Point Close to the Base Point

The end correction approach described above assumes that $f(z)$ is smooth (i.e., can be well represented by a polynomial approximation) near the endpoint. When the evaluation point $z$ is only a few grid points away from the base point, the integrand $\frac{f(\tau)}{(z-\tau)^{\alpha+1}}$ violates this, necessitating a different strategy, as is illustrated in Figure 6.5.

---

[20] Denoting $D^{\alpha} f(z) = g(z)$, it follows from integration by parts that $g'(z) = \frac{1}{\Gamma(1-\alpha)}\left(\frac{f'(0)}{z^{\alpha}} + \int_0^z \frac{f''(\tau)}{(z-\tau)^{\alpha}} d\tau\right)$. With both $g(z)$ and $g'(z)$ well defined, analyticity of $D^{\alpha} f(z) = g(z)$ follows.

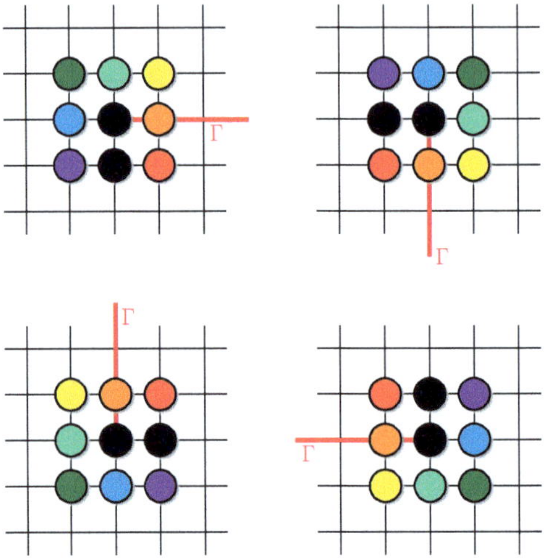


**FIGURE 6.4**. Illustration of how an end correction rotates with the direction of its path of integration Γ (typically of size $5 \times 5$ rather than $3 \times 3$ illustrated here).

We find the midpoint between base and evaluation points and approximate the Taylor expansion of $f(z)$ centered at this midpoint. The resulting fractional derivative integral can then be evaluated analytically, term by term. Two practical issues this gives rise to are described next.

#### *6.3.4.1 Numerically Stable Way to Approximate the Taylor Coefficients*

The linear system that provides Taylor coefficients from grid data is of Vandermonde type, and the task is generally very ill-conditioned. However, if the data points are equispaced around a circle centered at the expansion point, the matrix becomes a scaled version of the DFT (Discrete Fourier Transform) matrix, which is unitary, and thus perfectly conditioned. This motivates the choice of grid points marked blue in Figure 6.5, as an approximation to this circle case. For our standard choice of $n = 2$ (i.e., $5 \times 5$ size correction stencils), the 'rule of thumb' we have followed is to apply the midpoint procedure within a radius of $10h$ from the origin.[21]

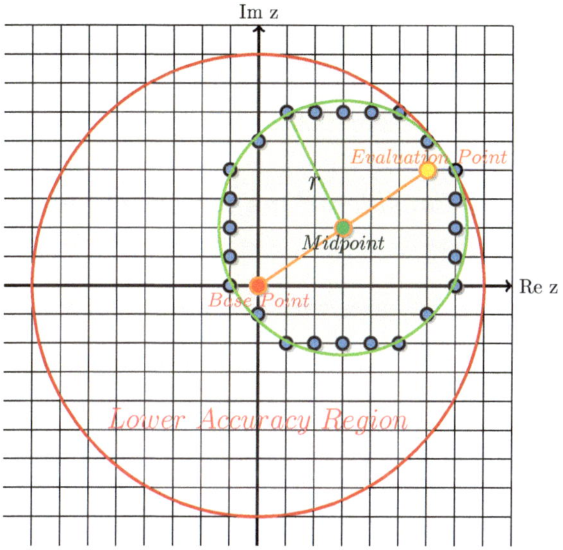


**FIGURE 6.5**. Computational strategy for when the evaluation point is too close to the base point for the approach in Section 6.3.3 to provide machine accuracy. This region is marked by a red circle here and later in the *Relative error* parts of Figures 6.8–6.12.

*6.3.4.2 Analytic Form of The Resulting Integrals*

The key formula here is

$$\int_{-b}^{b} \frac{z^k}{(b-z)^\alpha} dz = -(2b)^{1-\alpha}(-b)^k c_k, \tag{6.3.17}$$

[21]Euler-Maclaurin-based estimates suggest error levels of around $O\left(\frac{h}{|z|}\right)^{(2n+1)^2}$, indicating $r \approx 60h$ (impractically large) for $n = 1$ and $r \approx 3h$ for $n = 3$.

where

$$c_0 = \frac{1}{\alpha - 1}, \qquad c_k = \frac{k\, c_{k-1} + 1}{\alpha - (k+1)}, \quad k = 1, 2, 3, \ldots.$$

This formula can be obtained, for example, by repeated integration of parts. Applying it to the Taylor terms obtained as described above produces the value of the fractional derivative at the evaluation point.

### 6.3.5 *Additional Considerations for Functions $f(z)$ With Singularities*

A number of different cases with singularities may arise:

i. $f(z)$ has a singularity at the base point $z = 0$,

ii. $f(z)$ has poles in the complex plane,

iii. $f(z)$ has branch points in the complex plane.

Section 6 in [13] gives technical details for effective implementations also in these cases.

### 6.3.6 *Some Illustrations of Fractional Derivatives of Analytic Functions*

Figure 6.6 illustrates a case of fractional derivatives computed using the approach outlined above. The method presented here can also compute values on different sheets of multi-valued functions, as seen in Figure 6.7, using to the formula $D^{\alpha} f(z) = \frac{1}{e^{2\pi k \alpha i}} \left( \frac{1}{\Gamma(1-\alpha)} \int_0^z \frac{f'(\tau)}{(z-\tau)^{\alpha}} d\tau \right)$, where $k \in \mathbb{Z}$ and $f(z)$ is assumed to be analytic at $z = 0$.

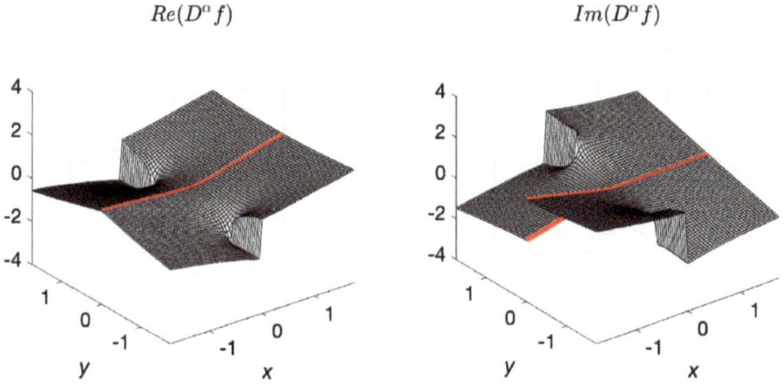


**FIGURE 6.6**. Plots of the real and imaginary parts of $D^\alpha\sqrt{1+z^2}$ with $\alpha = 0.4$, displayed and computed with $h = 0.05$. The exact value is $D^{0.4}\sqrt{1+z^2} = \frac{z^{1.6}\,{}_3F_2(1/2,1,3/2;1.3,1.8;-z^2)}{0.96\,\Gamma(0.6)}$.

In Figures 6.8–6.12, we also display the magnitude with phase angle, and the relative error of the approximation[22]. Regarding the errors, the numbers by the color bar correspond to $\log_{10}$ of the error, i.e., –16 matches roughly the machine precision. The plots were produced using $n = 2$ (i.e., size $5 \times 5$ correction stencils) and only function values at the nodes of the displayed computational domain (padded with $n = 2$ layers of nodes). The end correction scheme is used in the entire domain except within the red circles shown in the error plots, where instead, the Taylor expansion approaches described in Section 6.3.4 are used. Additional figures can be found in [13, 16].

[22]In each of these figures, the first three of the four subplots follow the display style used for a large number of analytic functions that are illustrated in [12].

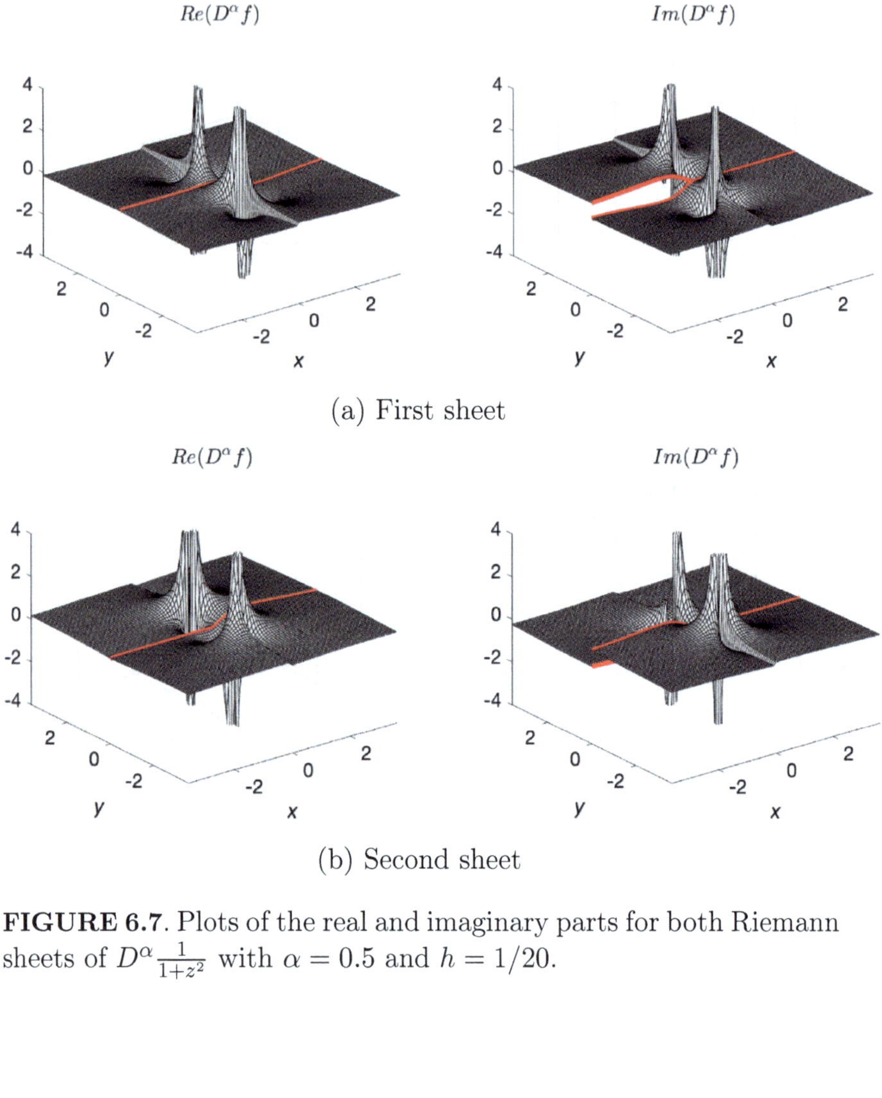


**FIGURE 6.7**. Plots of the real and imaginary parts for both Riemann sheets of $D^{\alpha}\frac{1}{1+z^2}$ with $\alpha = 0.5$ and $h = 1/20$.

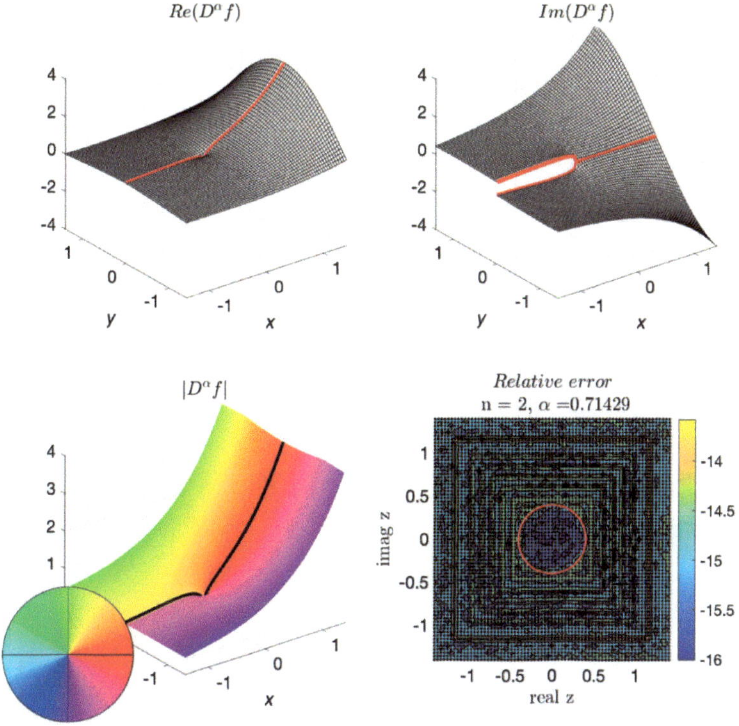


**FIGURE 6.8**. Plots of the real, imaginary, magnitude, phase portrait, and relative error of $D^{\alpha}e^{z}$ with $\alpha = 5/7$ and $h = 0.04$. The exact value is $D^{\alpha}e^{z} = e^{z}\left(1 - \frac{\Gamma(1-\alpha,z)}{\Gamma(1-\alpha)}\right)$. For the red circle in the last subplot of Figures 6.8-6.12, see the explanation in the caption of Figure 6.5. The red curve to the right of the origin in the upper left subplot (for $Re(D^{\alpha}f)$) is the same as was shown in Figure 6.1 (b). This figure is reprinted with permission from [13].

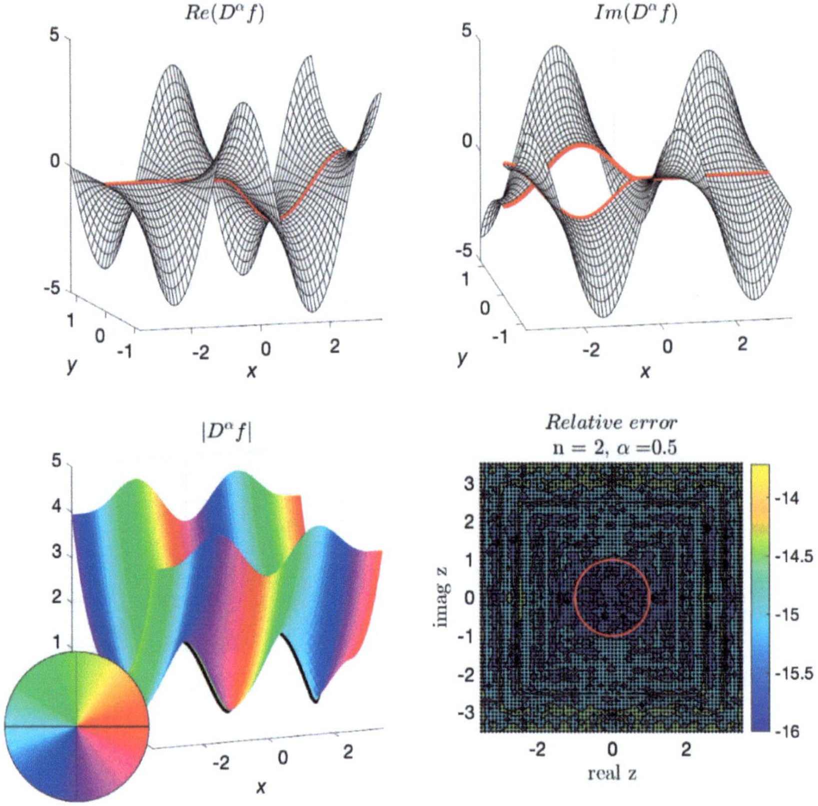


**FIGURE 6.9**. Plots of the real, imaginary, magnitude, phase portrait, and relative error of $D^\alpha \cos\left(\frac{\pi}{2}z\right)$ with $\alpha = \frac{1}{2}$ and $h = \frac{1}{10}$. The exact value is $D^{1/2}\cos\left(\frac{\pi}{2}z\right) = \sqrt{\pi}(\cos\left(\frac{\pi}{2}z\right) S(\sqrt{z}) - \sin\left(\frac{\pi}{2}z\right) C(\sqrt{z}))$, where $S(z)$ and $C(z)$ are respectively the Fresnel sine and cosine integral functions.

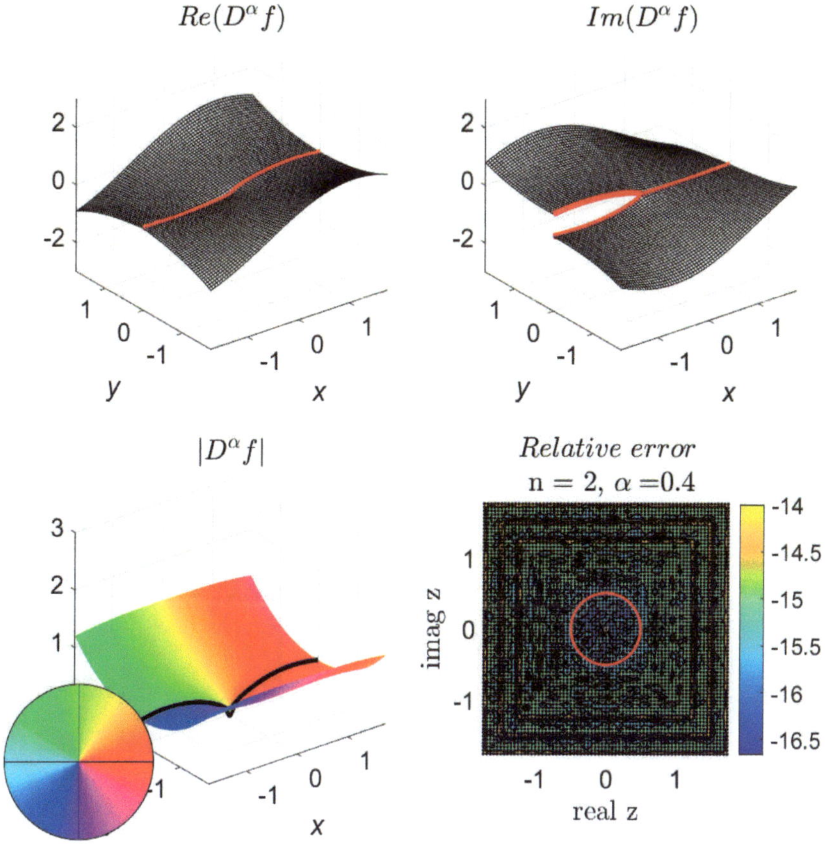


**FIGURE 6.10**. Plots of the real and imaginary parts, magnitude, phase portrait, and relative error of $D^\alpha J_0(z)$ where $J_0$ is the $J_0$ Bessel function, with $\alpha = 0.4$ and $h = 0.05$. The exact value is $D^\alpha J_0(z) = -\frac{1}{z^{2/5}\Gamma(3/5)}\left(1 - {}_1F_2(1/2; 3/10, 4/5; -z^2/4)\right)$.

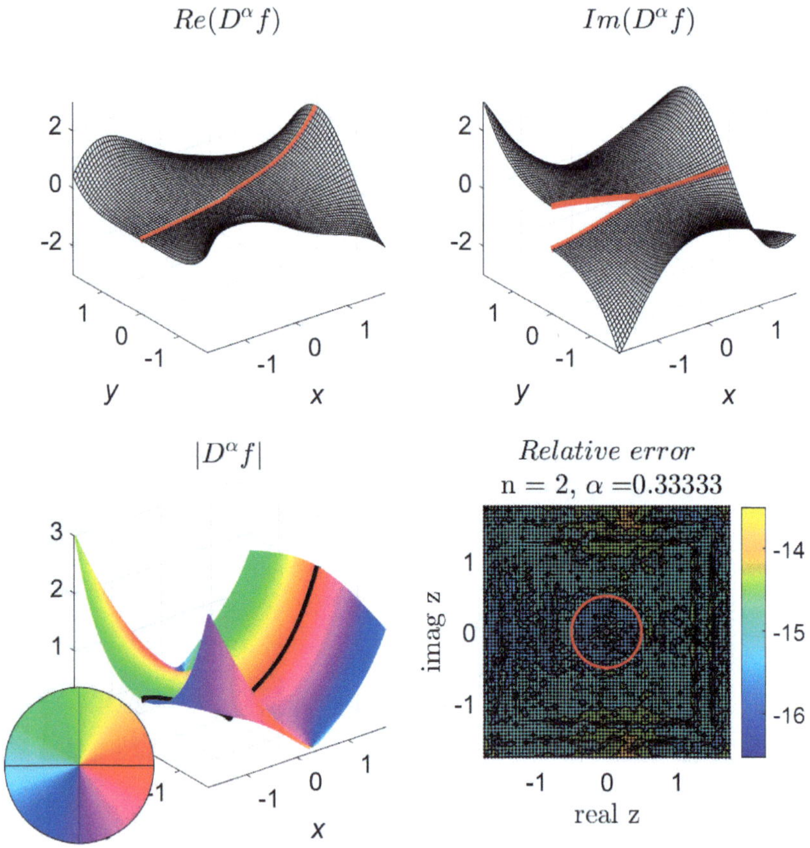


**FIGURE 6.11**. Plots of the real and imaginary parts, magnitude, phase portrait, and relative error of $D^\alpha \mathrm{Bi}(z)$ where Bi is the Airy Bi function, with $\alpha = 1/3$ and $h = 0.05$. The exact value is $D^{1/3}\mathrm{Bi}(z) = -\frac{z^{2/3}}{4\sqrt{3}\Gamma(-1/9)\Gamma(2/3)\Gamma(11/9)\Gamma(14/9)}$ $\times \left(20\Gamma(\frac{2}{3})^2 \,{}_2F_3(\frac{2}{3}, \frac{1}{5}; \frac{5}{9}, \frac{8}{9}, \frac{11}{9}; \frac{z^3}{9}) + 3^{7/6}\pi\, z^2 \,{}_2F_3(1, \frac{4}{3}; \frac{11}{9}, \frac{14}{9}, \frac{17}{9}; z^3/9)\right)$.

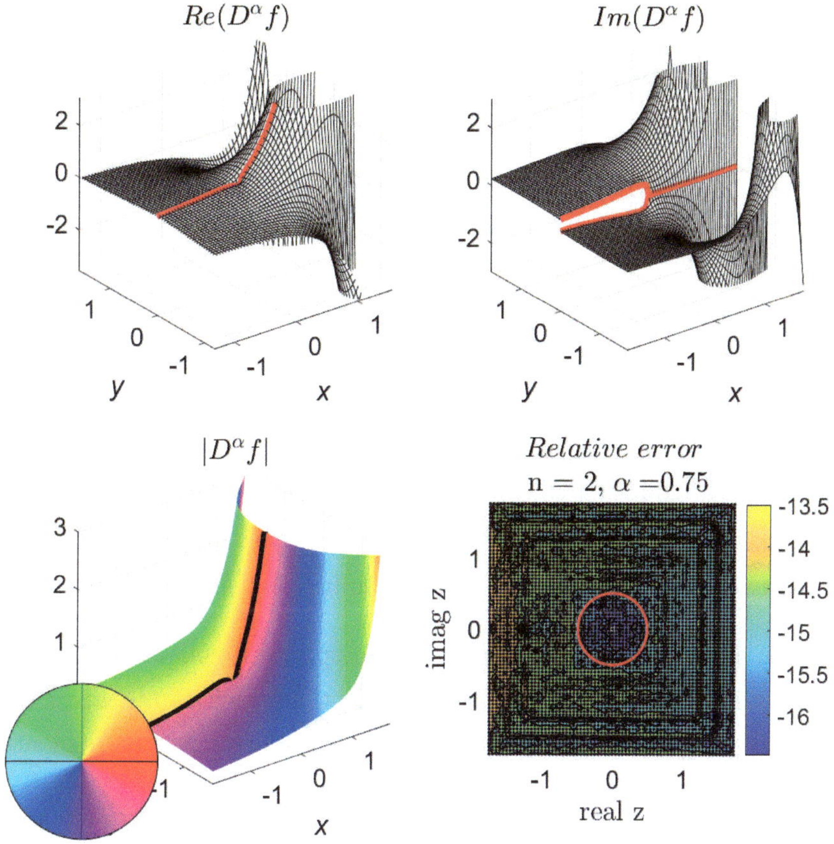


**FIGURE 6.12.** Plots of the real and imaginary parts, magnitude, phase portrait, and relative error of $D^\alpha E_{1/2}(z)$ where $E_{1/2}(z) = \operatorname{erfc}(-z)\, e^{z^2}$ is the Mittag-Leffler function, with $\alpha = 3/4$ and $h = 0.05$. The exact value is $D^\alpha E_{1/2}(z) = \frac{1}{\Gamma(1/4)}\left(\frac{8\,z^{1/4}}{\sqrt{\pi}} + \frac{32}{45}\,z^{5/4}\left(9\,{}_2F_2(1,\tfrac{3}{2};\tfrac{9}{8},\tfrac{13}{8};z^2) + \frac{16}{\sqrt{\pi}}z\,{}_2F_2(1,2;\tfrac{13}{8},\tfrac{17}{8};z^2)\right)\right)$.

## 6.4 Concluding Discussion

The approach described here for calculating fractional derivatives based on equispaced data (trapezoidal rule along grid lines together with end corrections) is both highly accurate and computationally very fast. For real-valued data along an interval (using only function values within the interval), it readily reaches convergence rates of around $O(h^{10})$, and for function values on a complex plane grid, convergence rates higher than $O(h^{20})$. While the Figures 6.6–6.12 were designed for visual display, the same grid density was, in all cases, sufficient for machine precision accuracy. Examples of MATLAB codes for the present algorithm can be found on GitHub [24]. Some current and future opportunities for investigations include:

i. Numerical evaluation of certain special functions. Cases listed in Table 17.1 in [20] include functions such as ${}_1F_1(a;c;z)$ and ${}_2F_1(a;b;c;z)$ expressed as fractional derivatives of elementary functions.

ii. Application of the real axis method to solving fractional order ODEs (initial and boundary value problems).

iii. Improve numerical stability by use of optimized under-determined Gregory-type quadrature, as described for non-singular integrals in [8].

# References


[1] Abdeljawad, T., (2015). On conformable fractional calculus. *J. Comp. Appl. Math., 279*, 57–66.

[2] Brockmann, D., Hufnagel, L., & Geisel, T., (2006). The scaling laws of human travel. *Nature 439*, 462–465.

[3] Caputo M., (1967). Linear model of dissipation whose Q is almost frequency independent-II. *Geophys. J. R. Astr. Soc., 13*(5), 529–539.

[4] Deng, Z.-Q., Lima, J. L. M. P. de, & Singh, V. P., (2006). A fractional dispersion model for overland solute transport. *Water Resources Research, 423*, Article Nr W03416, 14 pages.

[5] Driscoll, T. A., Hale, N., & Trefethen, L. N., (2014). *Chebfun Guide* , Pafnuty Press, Oxford, see also https://www.chebfun.org.

[6] Fornberg, B., (2021). Contour integrals of analytic functions given on a grid in the complex plane. *IMA Journal of Numerical Analysis 41*(2), 814–825.

[7] Fornberg, B., (2021). Generalizing the trapezoidal rule in the complex plane. *Numerical Algorithms, 87*(1), 187–202.

[8] Fornberg, B., (2021). Improving the accuracy of the trapezoidal rule. *SIAM Review 63*(1), 167–180.

[9] Fornberg, B., (2022). Finite difference formulas in the complex plane. *Numerical Algorithms 90*(3), 1305–1326.

[10] Fornberg, B., (2023). Infinite order accuracy limit of finite difference formulas in the complex plane. *IMA J. Num. Anal., 43*, 3055–3072.

[11] Fornberg, B., & Flyer, N., (2015). *A Primer on Radial Basis Functions with Applications to the Geosciences. CBMS-NSF, 87*, SIAM, Philadelphia.

[12] Fornberg, B., & Piret, C., (2020). *Complex Variables and Analytic Functions: An Illustrated Introduction*, SIAM, Philadelphia.

[13] Fornberg, B., & Piret, C., (2023). Computation of fractional derivatives of analytic functions. *Journal of Scientific Computing, 96.* Article ID 79.

[14] Gregory, J., (1841). Letter to J. Collins, 23 November 1670. *Oxford University Press*, 203–212. In Rigaud: Correspondence of Scientific Men.

[15] Guzmán, P. M., Langton, G., Bittencurt, L. M. Lugo Motta , Medina, J., Valdes, & J. E. Nápoles, (2018). A new definition of a fractional derivative of local type. *J. Math. Analysis, 9*(2), 88–98.

[16] Higgins, A., (2022). Numerical computations of fractional derivatives of analytic functions. *SIAM Undergraduate Research Online 15*, 511–525.

[17] Ishteva, M. K., (2005). *Properties and applications of the Caputo fractional operator.* Master's thesis, University of Karlsruhe.

[18] Jacobs, B. A., (2015). A new Grünwald–Letnikov derivative derived from a second-order scheme. *Abstract and Applied Analysis.* Article ID 952057.

[19] Khalil, R., Horani, M. Al., Yousef, A., & Sababheh, M., (2014). A new definition of fractional derivative. *J. Comp. Appl. Math., 264*, 65–70.

[20] Lavoie, J. L., Osler, T. J., & Tremblay, R., (1976). Fractional derivatives and special functions. *SIAM Review 18*(2), 240–268.

[21] Li, C., & Zeng, F., (2015). *Numerical Methods for Fractional Calculus.* CRC Press, Boca Raton, FL.

[22] Martensen, E., (1964). Optimale Fehlerschranken für die Qadraturformel von Gregory. *ZAMM 44*(4/5), 159–168.

[23] Olver, F. W.J., Lozier, D. W., Boisvert, R. F., & Clark, C. W., (2010). *NIST Handbook of Mathematical Functions.* Cambridge University Press.

[24] Piret, C., (2023). https://github.com/cmpiret/Fractional Derivatives.

[25] Podlubny, I., (1990). *Fractional Differential Equations.* Academic Press, San Diego.

[26] Scalas, E., Gorenflo, R., & Mainardi, F., (2000). Fractional calculus and continuous-time finance. *Physica A: Statistical Mechanics and its Applications, 284*(1–4), 376–384.

[27] Valério, D., Ortigueira, M. D., & Lopes, A. M., (2022). How many fractional derivatives are there? *Mathematics, 10*, Article ID 737.

# Chapter 7

## *Some Remarks on Caputo-Weighted Integrals and Applications*

**J. E. Nápoles-Valdés[1,2] and B. Bayraktar[3]**

[1] *UNNE, FaCENA Ave., Corrientes, Argentina*

[2] *UTN-FRRE, Resistencia, Chaco, Argentina*

[3] *Bursa Uludag University, Faculty of Education, Görukle Campus, Bursa, Turkey*

**Abstract**

In this chapter, we present weighted integrals as a versatile tool with various applications, from which new formulations of well-known integral inequalities are derived.

## Introduction

A function $f : I \to \mathbb{R}$, where $I := [a, b]$, is classified as convex if the inequality $f(tx + (1-t)y) \leq tf(x) + (1-t)f(y)$ holds true for all $x, y \in I$ and $t \in [0, 1]$. When the above-mentioned inequality is reversed, the function $f$ is considered concave on the interval $[a, b]$. Convex functions have been thoroughly investigated, and researchers have proposed various extensions of this concept. These include

$m$-convex functions, $n$-convex functions, $r$-convex functions, $h$-convex functions, $(h, m)$-convex functions, $s$-convex functions, and many others. For a comprehensive overview of current developments in this field, interested readers are encouraged to refer to [44].

The Hermite-Hadamard inequality has been a focal point for numerous experts in inequalities over recent decades and is regarded as one of the most significant inequalities.

$$f\left(\frac{a+b}{2}\right) \leq \frac{1}{b-a}\int_a^b f(x)\,dx \leq \frac{f(a)+f(b)}{2} \tag{7.0.1}$$

holds for any function $f$ convex on the interval $[a, b]$. The inequality was introduced by Hermite in 1883 and independently by Hadamard in 1893. It provides an estimate of the average value for a convex function, with the noteworthy aspect that it refines the Jensen inequality as well. The interested reader is referred to [1, 7, 11, 12, 25, 30, 33, 35, 36, 40–43, 51–53], and references therein for more information and other extensions of the Hermite–Hadamard inequality.

In the framework of non-conformable fractional integrals, Ali et al. [1] presented new Hermite-Hadamard-type inequalities for generalized $\phi$-convex functions. In [2], Aljaaidia and Pachpatte obtained some new fractional integral inequalities of Minkowski-type by using the Riemann-Liouville fractional integral operator. Atagana and Belanu in [3] defined a new fractional derivative with a non-local and non-singular kernel, presented some properties of this derivative, and gave applications. In [1, 7, 8], the authors, using Riemann-Liouville fractional integrals, obtained Hadamard's inequalities for functions whose second derivatives have convexity of the $(s, m)$, Godunov-Levin, and $(s, m_1, m_2)$ types.

Bohner et al., in [12], obtained some inequalities of Hadamard type for integrals arising in conformable fractional calculus and illustrated their findings via applications to modified Bessel functions, special means, and midpoint approximations. Bayraktar and Nápoles, in [9], for functions $(h, m, s)$-convex of the modified second type, obtained various extensions of the classic Hadamard inequality using Katugampola integrals.

In [11], the authors obtained several new generalized integral inequalities of Hadamard and Simpson type via fractional operators for functions whose first and third derivatives are either convex or satisfy the Lipschitz condition or the conditions of the Lagrange theorem. In [25, 40], the authors presented new Hermite-Hadamard inequalities for $h$-convex functions within the framework of the generalized integral operator. In [15], Butt et al. obtained several Hadamard-type inequalities using a new version of the Atangana-Baleanu integral operator. Cerone et al., in [19], obtained Ostrowski-type inequalities for twice differentiable functions whose derivatives are bounded and provided applications for special means.

In [22], Dragomir and Agarwal presented Hadamard's integral inequality for functions whose first derivative is convex and provided natural applications to special means of numbers. In [23], Du et al. established $k$-fractional integral inequalities with multiple parameters involving the first derivative and provided applications to random variables. In [27], the authors defined weighted fractional operators on some spaces and discussed their properties. Jarad et al., in [28], introduced new fractional integration and differentiation operators based on conformable derivatives. In [38], the authors derived new Hadamard-type inequalities using the Riemann-Liouville fractional integral operator for functions with convex and quasi-convex third derivatives, applying these findings to Bessel functions and special means.

Nápoles et al., in [41], obtained new Hermite-Hadamard-type integral inequalities for convex and quasi-convex functions in a generalized context. In [48], Sarıkaya and Ertugrul gave a new definition generalizing some important classical fractional integral operators and obtained new generalized Hermite-Hadamard-type inequalities through this operator. Vivas-Cortez et al., in [51], used a generalized fractional integral to construct a new Hermite-Hadamard-type integral inequality for functions of two independent variables that are semi-convex in coordinates. In [52], the authors obtained some inequalities of Hermite-Hadamard-Fejér-type for generalized integrals. Wu et al., in [54], established some new Hadamard-type inequalities via $k$-Riemann-Liouville fractional integrals and considered applications of the obtained results in error estimations of quadrature formulas. Xi and Qi, in [57], and Yildiz et al., in [58], obtained new integral inequalities for $s$-convex, $(\alpha, s)$-convex, and extended $s$-convex functions.

In our work, we will use the following notions of convexity ([8, 9]):

**Definition 7.0.1.** *A function* $f : I = [0, +\infty) \to [0, +\infty)$ *is considered a* $(h, m)$*-convex modified of the first type on* $I$ *if the following condition holds true: For all* $a, b \in I$ *and* $t \in [0, 1]$, *where* $h : [0, 1] \to (0, 1]$ *and* $m \in [0, 1], s \in (0, 1]$, *the inequality*

$$f(ta + m(1-t)b) \leq h^s(t)f(a) + m(1 - h^s(t))f(b)$$

*is satisfied.*

**Definition 7.0.2.** *A function* $f : I = [0, +\infty) \to [0, +\infty)$ *is considered a* $(h, m)$*-convex modified of the second type on* $I$ *if the following condition holds true: For all* $a, b \in I$ *and* $t \in [0, 1]$, *where* $h : [0, 1] \to (0, 1]$ *and* $m \in [0, 1], s \in [-1, 1]$, *the inequality*

$$f(ta + m(1-t)b) \leq h^s(t)f(a) + m(1-h(t))^s f(b)$$

*is satisfied.*

**Remark 7.0.1.** *Based on Definitions 7.0.1 and 7.0.2, we have the ability to establish* $N^s_{h,m}[a, b]$, *where* $a$ *and* $b \geq 0$, *as the collection of functions that are* $(h, m)-$*convex modified, with the condition that* $f(a) \geq 0$. *These functions can be distinguished or identified by the triple* $(h(t), m, s)$. *It is essential to observe that if:*

*i.* $(t, 0, 1)$, *then we get the starshaped functions ([14]).*

*ii.* $(t, 0, s)$, *then we get the* $s-$*starshaped functions ([14]).*

*iii.* $(h(t), 0, 0)$, *then we get the increasing functions ([14]).*

*iv.* $(t, 1, 1)$, *then* $f$ *is a convex function on* $[0, +\infty)$ *([14]).*

*v.* $(t, 1, s)$ *and* $s \in (0, 1]$, *then* $f$ *is a* $s-$*convex function on* $[0, +\infty)$ *([13, 26]).*

*vi.* $(t, m, 1)$, *then* $f$ *is a* $m-$*convex function on* $[0, +\infty)$ *([50]).*

*vii.* $(t, m, s)$ *and* $s \in (0, 1]$, *then* $f$ *is a* $(s, m)$*-convex function on* $[0, +\infty)$ *([45]).*

*viii.* $(t, 1, s)$ *and* $s \in [-1, 1]$, *then* $f$ *is a extended* $s-$*convex function on* $[0, +\infty)$ *([57]).*

*ix.* $(t^{\alpha}, m, s)$, *then* $f$ *is a* $s-(\alpha, m)$*-convex function on* $[0, +\infty)$ *([55]),*

*x.* $(t^{\alpha}, 1, s)$, *then* $f$ *is a* $(\alpha, s)$*-convex function on* $[0, +\infty)$ *([56]).*

*xi.* $(t^{\alpha}, m, 1)$, *then* $f$ *is a* $(\alpha, m)$*-convex function on* $[0, +\infty)$ *([36]),*

*where* $\alpha \in (0, 1]$.

**Remark 7.0.2.** *Within various notions of convexity, when the direction of the inequality is reversed, the concept is referred to as concave.*

In our study, we employ the Euler Gamma functions $\Gamma$ (see [47]) and $\Gamma_{\kappa}$ (see [21]):

$$\Gamma(z) = \int_0^{\infty} t^{z-1} e^{-t}\, dt, \qquad \operatorname{Re}(z) > 0,$$

$$\Gamma_{\kappa}(z) = \int_0^{\infty} t^{z-1} e^{-t^{\kappa}/\kappa}\, dt, \qquad \kappa > 0.$$

Here $\Gamma_{\kappa}(z+\kappa) = z\Gamma_{\kappa}(z)$, $\lim\limits_{\kappa \to 1} \Gamma_{\kappa}(z) = \Gamma(z)$ and $\Gamma_{\kappa}(z) = (\kappa)^{\frac{z}{\kappa}-1}\Gamma\left(\frac{z}{\kappa}\right)$.

In order to enhance understanding of the topic, we give the definition of the Riemann-Liouville fractional integral, considering the range $0 \leq a < t < b \leq \infty$.

**Definition 7.0.3.** *Let* $f \in L_1[a, b]$. *Then the Riemann-Liouville fractional integrals of order* $\alpha \in \mathbb{C}$, $\operatorname{Re}(\alpha) > 0$ *are defined by (right and left, respectively):*

$$^{\alpha}I_{a^+} f(x) = \frac{1}{\Gamma(\alpha)} \int_a^x (x-t)^{\alpha-1} f(t)dt, \qquad x > a$$

*and*

$$ {}^{\alpha}I_{b^-} f(x) = \frac{1}{\Gamma(\alpha)} \int_x^b (t-x)^{\alpha-1} f(t)dt, \qquad x < b. $$

In the realm of Fractional Calculus, the Caputo Derivative stands out as one of the most renowned differential operators. A notable feature of this derivative is that, when employing Caputo's definition to solve differential equations, there is no need to specify fractional order initial conditions. Researchers have widely applied this derivative in the analysis of numerous practical problems (refer to [16, 17]). The Caputo derivative of a differentiable function $f$ of order $0 < \alpha < 1$ is defined as

$$ {}^{C}D_t^{\alpha} f(t) = \frac{1}{\Gamma(1-\alpha)} \int_0^t \frac{f'(s)}{(t-s)^{\alpha}} ds. \tag{7.0.2} $$

An extension of ${}^{CF}D^{\alpha}$ is the so-called *Caputo-Fabrizio derivative* (see [4, 18]), given by:

$$ {}^{CF}D_t^{\alpha} f(t) = \frac{M(\alpha)}{1-\alpha} \int_a^t f'(s) \exp\left[-\frac{\alpha}{1-\alpha}(t-s)\right] ds, \tag{7.0.3} $$

where $M(\alpha)$ is a normalization function such that $M(0) = M(1) = 1$.

A more recent extension is the *Atangana-Baleanu derivative*, defined in [3] by

$$ {}_{0}^{ABC}D_t^{\alpha} f(t) = \frac{AB(\alpha)}{1-\alpha} \int_0^t f'(s) E_{\alpha}\left[-\frac{\alpha(t-s)^{\alpha}}{1-\alpha}\right] ds, $$

where

$$ E_{\alpha}(z) = \sum_{k=0}^{\infty} \frac{z^k}{\Gamma(\alpha k + 1)} $$

is the *Mittag-Leffler function.* Note that $E_1(z) = e^z$.

Some weighted fractional integral operators were defined in [8, 10, 53].

**Definition 7.0.4.** *([8]) Consider a function $f \in L([a,b])$ and non-negative $w \in C[0,1]$ with piecewise continuous first and second order derivatives on the interval $[0,1]$. Under these circumstances, the weighted fractional integrals are defined in the following manner(left and right, respectively):*

$$J^w_{a+}f(x) = \int_a^x w'\left(\frac{t-a}{x-a}\right) f(t)dt$$

*and*

$$J^w_{b-}f(x) = \int_x^b w'\left(\frac{b-t}{b-x}\right) f(t)dt,$$

*with $a < r \leq b$.*

**Definition 7.0.5.** *([10]) Consider a function $f \in L([a,b])$ and non-negative $w \in C[0,1]$ with piecewise continuous first and second order derivatives on the interval $[0,1]$, $w(0) = 0$ and $w(1) = 0$. Under these circumstances, the weighted fractional integrals are defined in the following manner(left and right, respectively):*

$$I_{a+}f(x) = \int_\rho^x w''\left(\frac{x-t}{\frac{a+b}{2}-a}\right) f(t)dt, \quad x > a$$

*and*

$$I_{b-}f(x) = \int_x^b w''\left(\frac{t-x}{b-\frac{a+b}{2}}\right) f(t)\,dt, \quad x < b.$$

**Definition 7.0.6.** *([53]) Consider a function $f \in L([a,b])$ and non-negative $w \in C[0,+\infty)$ with piecewise continuous first and second order derivatives on the interval $[0,+\infty)$. Under these circumstances, the weighted fractional integrals are defined in the following manner(left and right, respectively):*

$$^{(n+1)}I^{w}_{a+}f(x) = \int_{\rho}^{x} w'\left(\frac{(n+1)(x-t)}{b-a}\right) f(t)dt, \quad x > a$$

*and*

$$^{(n+1)}I^{w}_{b-}f(x) = \int_{x}^{b} w'\left(\frac{(n+1)(t-x)}{b-a}\right) f(t)\,dt, \quad x < b.$$

In our work, we will use the following weighted differential operators, of type Caputo:

**Definition 7.0.7.** *Let $f' \in L_1[a,b]$ and $w : I = [0,1] \to \mathbb{R}$ be a continuous and non-negative function, with first and second order derivatives integrable on $I$. Then the weighted differential operators, of type Caputo are defined by (right and left, respectively):*

$$D^{w}_{a+}f(x) = \int_{a}^{x} w'\left(\frac{t-a}{b-a}\right) f'(t)dt, \quad x > a$$

*and*

$$D^{w}_{b-}f(x) = \int_{x}^{b} w'\left(\frac{b-t}{b-a}\right) f'(t)dt, \quad x < b.$$

**Remark 7.0.3.** *Interested readers will have no difficulty in verifying that the Fractional Derivatives presented above are particular cases of our definition. For example: if we take $w(t) = \frac{(b-x)^{1-\alpha}(t-1)^{1-\alpha}}{(b-x)\Gamma(2-\alpha)}$, then* *we get right side Caputo fractional integral operator:*

$$^{C}D_{b-}^{\alpha}f(x) = \frac{1}{\Gamma(1-\alpha)}\int_{x}^{b}\frac{f'(t)}{(b-t)^{\alpha}}dt.$$

In this chapter, with the help of Caputo-type weighted derivatives, as defined in Definition 7.0.7, we obtain new variants of well-known integral inequalities within the framework of the $(h,m)$-convex functions of the second type.

---

## 7.1 Results

Next, we present a lemma that establishes an equality, which is fundamental for the rest of the chapter.

**Lemma 7.1.1.** *Suppose a function $f : I = [a,b] \to \mathbb{R}$ is twice differentiable on $(a,b)$ with $a < b$. If $f'' \in L_1[a,b]$, then for all $x \in (a,b)$ and $t \in [0,1]$, the following equality for Caputo-type Weighted Derivatives is fulfilled:*

$$\begin{aligned} & w(1)f'(x)\left[\frac{1}{b-x}+\frac{1}{x-a}\right] - w(0)\left[\frac{f'(b)}{b-x}+\frac{f'(a)}{x-a}\right] \\ & \quad - \frac{1}{(x-a)^2}\left(D_{a+}^{w}f\right)(t) - \frac{1}{(b-x)^2}\left(D_{b-}^{w}f\right)(t) \\ & = \int_0^1 w(t)\left|f''(tx+(1-t)b)\right|dt + \int_0^1 w(t)\left|f''(tx+(1-t)a)\right|dt. \end{aligned} \tag{7.1.4}$$

*Proof.* From the second member and properties of integral, we have:

$$\begin{aligned}
&\int_0^1 w(t)\left[f''(tx+(1-t)b)+f''(tx+(1-t)a)\right]dt \\
&=\int_0^1 w(t)f''(tx+(1-t)b)dt+\int_0^1 w(t)f''(tx+(1-t)a)dt \\
&=I_1+I_2.
\end{aligned}$$

To $I_1$ we have:

$$\begin{aligned}
&\int_0^1 w(t)f''(tx+(1-t)b)dt \qquad (7.1.5)\\
&=\frac{1}{x-b}\left[w(1)f'(x)-w(0)f'(b)\right]-\frac{1}{x-b}\int_0^1 w'(t)f'(tx+(1-t)b)dt \\
&=\frac{1}{x-b}\left[w(1)f'(x)-w(0)f'(b)\right]-\frac{1}{(b-x)^2}\int_x^b w'\left[\frac{b-u}{b-x}\right]f'(u)du.
\end{aligned}$$

The result was obtained after integrating by parts and making a change of variables in the resulting integral on the right-hand side. Similarly for integral $I_2$ we obtain:

$$\begin{aligned}
&\int_0^1 w(t)f''(tx+(1-t)a)dt \qquad (7.1.6)\\
&=\frac{1}{x-a}\left[w(1)f'(x)-w(0)f'(a)\right]-\frac{1}{(x-a)^2}\int_a^x w'\left[\frac{u-a}{x-a}\right]f'(u)du.
\end{aligned}$$

From (7.1.5) and (7.1.6) we obtain easily (7.1.4).

**Remark 7.1.1.** *Putt in $w(t)=t^{\delta+1}$, the above result extends the Lemma 1 of [29], obtained for AB-fractional integral.*

**Remark 7.1.2.** *If we take $w(t)=t^2$ and integrate by parts twice, we get Lemma 1 of [49].*

**Remark 7.1.3.** *If we take $w(t) = t^2$ and integrate by parts twice, we get Lemma 1 of [49].*

For simplicity, let's call

$$L = w(1)f'(x)\left[\frac{1}{b-x} + \frac{1}{x-a}\right] - w(0)\left[\frac{f'(b)}{b-x} + \frac{f'(a)}{x-a}\right]$$
$$- \frac{1}{(x-a)^2}\left(D^w_{a+}f\right)(t) - \frac{1}{(b-x)^2}\left(D^w_{b-}f\right)(t).$$

So we have

**Theorem 7.1.1.** *Let $f : I = [a,b] \to \mathbb{R}$ be a twice differentiable on $(a,b)$ with $a < b$ and $f'' \in L_1[a,b]$. If $|f''|$ is a $(h,m)$-convex of the second type, then the following inequality holds:*

$$|L| \leq 2\left|f''(x)\right| \int_0^1 w(t)h^s(t)dt$$
$$+ m\left(\left|f''(\frac{a}{m})\right| + \left|f''(\frac{b}{m})\right|\right) \int_0^1 w(t)\left(1-h(t)\right)^s dt. \tag{7.1.7}$$

*Proof.* From the $(h,m)$-convexity of $|f''|$ we have

$$\left|f''(tx+(1-t)b)\right| \leq \left|f''(x)\right| h^s(t) + m\left|f''(\frac{b}{m})\right|\left(1-h(t)\right)^s,$$
$$\left|f''(tx+(1-t)a)\right| \leq \left|f''(x)\right| h^s(t) + m\left|f''(\frac{a}{m})\right|\left(1-h(t)\right)^s.$$

Using these inequalities on the right-hand side of the Lemma 7.1.1, we obtain

$$|I_1| \leq \left|f''(x)\right| \int_0^1 w(t)h^s(t)dt + m\left|f''(\frac{b}{m})\right| \int_0^1 w(t)\left(1-h(t)\right)^s dt,$$
$$|I_2| \leq \left|f''(x)\right| \int_0^1 w(t)h^s(t)dt + m\left|f''(\frac{a}{m})\right| \int_0^1 w(t)\left(1-h(t)\right)^s dt.$$

$$|I_1|+|I_2| \leq 2\left|f''(x)\right|\int_0^1 w(t)h^s(t)dt$$
$$+m\left[\left|f''(\frac{b}{m})\right|+\left|f''(\frac{a}{m})\right|\right]\int_0^1 w(t)\left(1-h(t)\right)^s dt$$

Adding these two inequalities and grouping appropriately, we obtain (7.1.7).

**Remark 7.1.4.** *If we put $w(t) = t^{\delta+1}$, this result extends Theorem 1 of [29] obtained for integrals of the AB type of s-convex functions of the second type, to the case of weighted Caputo-type derivatives.*

The following result is a trivial consequence of the Theorem obtained before.

**Corollary 7.1.1.** *If we put $K = \sup_{[a,b]}|f''|$, from above Theorem 7.1.1 we obtain the following inequality:*

$$|L| \leq 2K\left(\int_0^1 w(t)h^s(t)dt + \int_0^1 w(t)\left(1-h(t)\right)^s dt\right)$$
$$= 2K\int_0^1 w(t)\left(h^s(t)+\left(1-h(t)\right)^s\right)dt.$$

**Remark 7.1.5.** *Considering $w(t) = t$, $m = 1$ and $h(t) = t$ we obtain the Corollary 2 of [49]. Additionally, if $s = 1$, i.e., $f''$ is a convex function, from the above result, we obtain the Theorem 2.1 of [19].*

By imposing additional conditions on $|f''|$, we can obtain various refinements of the result given in the Theorem 7.1.1. So, we have:

**Theorem 7.1.2.** *Let $f : I = [a, b] \to \mathbb{R}$ be a twice differentiable on $(a, b)$ with $a < b$ and $f'' \in L_1[a, b]$. If $|f''|^q$ is a $(h, m)$-convex of* *second type, $q > 1$ the following inequality:*

$$|L| \leq \left(\int_0^1 w(t)^p dt\right)^{\frac{1}{p}} \left\{ \left( \left|f''(x)\right|^q H_1 + m \left|f''(\frac{b}{m})\right|^q H_2 \right)^{\frac{1}{q}} \right. \tag{7.1.8}$$

$$\left. + \left( \left|f''(x)\right|^q H_1 + m \left|f''(\frac{a}{m})\right|^q H_2 \right)^{\frac{1}{q}} \right\}.$$

*holds, with $\frac{1}{p} + \frac{1}{q} = 1$, $H_1 = \int_0^1 h(t)^s dt$ and $H_2 = \int_0^1 (1 - h(t))^s dt$.*

*Proof.* From the Lemma 7.1.1, and using the well-known Hölder integral inequality we obtain:

$$\begin{aligned}
|L| &\leq \int_0^1 w(t) \left|f''(tx + (1-t)b)\right| dt + \int_0^1 w(t) \left|f''(tx + (1-t)a)\right| dt \\
&\leq \left(\int_0^1 w(t)^p dt\right)^{\frac{1}{p}} \left(\int_0^1 \left|f''(tx + (1-t)b)\right|^q dt\right)^{\frac{1}{q}} \\
&\quad + \left(\int_0^1 w(t)^p dt\right)^{\frac{1}{p}} \left(\int_0^1 \left|f''(tx + (1-t)a)\right|^q dt\right)^{\frac{1}{q}} \\
&\leq \left(\int_0^1 w(t)^p dt\right)^{\frac{1}{p}} \\
&\quad \times \left\{ \left(\int_0^1 \left|f''(tx + (1-t)b)\right|^q dt\right)^{\frac{1}{q}} + \left(\int_0^1 \left|f''(tx + (1-t)a)\right|^q dt\right)^{\frac{1}{q}} \right\}.
\end{aligned}$$

Since $|f''|^q$ is $(h, m)$-convex of second type, we have

$$\int_0^1 \left|f''(tx + (1-t)b)\right|^q dt \tag{7.1.9}$$

$$\leq \int_0^1 \left( \left|f''(x)\right|^q h^s(t) + m \left|f''(\frac{b}{m})\right|^q (1 - h(t))^s \right) dt$$

$$\int_0^1 \left|f''(tx + (1-t)a)\right|^q dt \tag{7.1.10}$$

$$\leq \int_0^1 \left( |f''(x)|^q h^s(t) + m \left| f''(\frac{a}{m}) \right|^q (1-h(t))^s \right) dt.$$

Using (7.1.9) and (7.1.10) we obtain

$$|L| \leq \left( \int_0^1 w(t)^p dt \right)^{\frac{1}{p}} \left\{ \left( |f''(x)|^q H_1 + m \left| f''(\frac{b}{m}) \right|^q H_2 \right)^{\frac{1}{q}} \right.$$
$$\left. + \left( |f''(x)|^q H_1 + m \left| f''(\frac{a}{m}) \right|^q H_2 \right)^{\frac{1}{q}} \right\}.$$

What is the desired inequality?

**Remark 7.1.6.** *As say before, putting* $w(t) = t^{\delta+1}$*, this result extends Theorem 2 of [29] obtained for integrals of the AB type of s-convex functions of the second type, to the case of weighted Caputo-type derivatives.In the case* $K = \sup_{[a,b]} |f''|$*, we can obtain an extension of the Corollary 2 of [29].*

**Remark 7.1.7.** *With* $w(t) = t^2$*,* $m = 1$ *and* $h(t) = t$*, this result cover Theorem 5 of [49]. Additionally, if* $K = \sup_{[a,b]} |f''|$*, and considering* $x = \frac{a+b}{2}$ *we extend the Corollary 4 and 5, respectively, of the above paper.*

The above remark can specify it by putting $x = a$ in $I_1$ and $x = b$ in $I_2$. Thus, we obtain the following result.

**Corollary 7.1.2.** *Under assumptions of Theorem 7.1.2, we have the following perturbed trapezoid-like inequality*

$$|L| \leq \left( \int_0^1 w(t)^p dt \right)^{\frac{1}{p}} \left\{ \left( |f''(a)|^q H_1 + m \left| f''(\frac{b}{m}) \right|^q H_2 \right)^{\frac{1}{q}} \right.$$
$$\left. + \left( |f''(b)|^q H_1 + m \left| f''(\frac{a}{m}) \right|^q H_2 \right)^{\frac{1}{q}} \right\}.$$

**Remark 7.1.8.** *With $w(t) = t^2$, $m = 1$ and $h(t) = t$, we can extend from the above result, the Corollary 6 of [49].*

We can obtain a slight variant of Theorem 7.1.2 if instead of using Hölder's Inequality, Young's Inequality establishes: Let $p, q > 1$ be conjugate exponents, that is, $\frac{1}{p} + \frac{1}{q} = 1$ and let $a, b \geq 0$. So that $ab \leq \frac{a^p}{p} + \frac{b^q}{q}$. Using this fact we have the following result.

**Theorem 7.1.3.** *Let $f : I = [a, b] \to \mathbb{R}$ be a twice differentiable on $(a, b)$ with $a < b$ and $f'' \in L_1[a, b]$.If $|f''|^q$ is a $(h, m)$-convex of second type, $q > 1$ the following inequality:*

$$|L| \leq \frac{2}{p}\left(\int_0^1 w(t)^p dt\right) \tag{7.1.11}$$
$$+\frac{1}{q}\left\{2\left|f''(x)\right|^q H_1 + m\left(\left|f''(\frac{b}{m})\right|^q + \left|f''(\frac{a}{m})\right|^q\right) H_2\right\}.$$

*holds, with $\frac{1}{p} + \frac{1}{q} = 1$, $H_1 = \int_0^1 h(t)^s dt$ and $H_2 = \int_0^1 (1 - h(t))^s dt$.*

*Proof.* It is enough to use, as we said, Young's Inequality instead of Hölder's and proceed as in the proof of Theorem 7.1.2.

**Theorem 7.1.4.** *Let $f : I = [a, b] \to \mathbb{R}$ be a twice differentiable on $(a, b)$ with $a < b$ and $f'' \in L_1[a, b]$.If $|f''|^q$ is a $(h, m)$-convex of second type, $q \geq 1$ the following inequality:*

$$|L| \leq \left(\int_0^1 w(t)dt\right)^{1-\frac{1}{q}} \left\{\left(\left|f''(x)\right|^q H_3 + m\left|f''(\frac{b}{m})\right|^q H_4\right)^{\frac{1}{q}}\right.$$
$$\left.+\left(\left|f''(x)\right|^q H_3 + m\left|f''(\frac{a}{m})\right|^q H_4\right)^{\frac{1}{q}}\right\}.$$

*holds, with $H_3 = \int_0^1 w(t)h(t)^s dt$ and $H_4 = \int_0^1 w(t)(1 - h(t))^s dt$.*

*Proof.* Using the power-mean integral inequality in the Lemma 7.1.1, we obtain:

$$|L| \leq \int_0^1 w(t) \left| f''(tx + (1-t)b) \right| dt + \int_0^1 w(t) \left| f''(tx + (1-t)a) \right| dt \tag{7.1.12}$$

$$\begin{aligned} &\leq \left( \int_0^1 w(t)dt \right)^{1-\frac{1}{q}} \left( \int_0^1 w(t) \left| f''(tx + (1-t)b) \right|^q dt \right)^{\frac{1}{q}} \\ &\qquad + \left( \int_0^1 w(t)dt \right)^{1-\frac{1}{q}} \left( \int_0^1 w(t) \left| f''(tx + (1-t)a) \right|^q dt \right)^{\frac{1}{q}} \\ &\leq \left( \int_0^1 w(t)dt \right)^{1-\frac{1}{p}} \left\{ \left( \int_0^1 w(t) \left| f''(tx + (1-t)b) \right|^q dt \right)^{\frac{1}{q}} \right. \\ &\qquad \left. + \left( \int_0^1 w(t) \left| f''(tx + (1-t)a) \right|^q dt \right)^{\frac{1}{q}} \right\}. \end{aligned}$$

Taking into account the $(h, m)$-convexity of the second type of $|f''|^q$ and proceeding as in the proof of the previous Theorem, we obtain the desired inequality.

**Remark 7.1.9.** *Like before, with* $w(t) = t^{\delta+1}$, *this result extends Theorem 3 of [29]. If additionally* $K = \sup_{[a,b]} |f''|$, *we can obtain an extension of the Corollary 3 of [29].*

**Remark 7.1.10.** *With* $w(t) = t^2$, $m = 1$ *and* $h(t) = t$, *this result cover Theorem 6 of [49]. As before if* $K = \sup_{[a,b]} |f''|$, *and considering* $x = \frac{a+b}{2}$ *we extend the Corollaries 7,8 and 9, of the cited above paper.*

## 7.2 Conclusions

In our chapter, we have obtained new versions of well-known integral inequalities within the framework of the Definition 7.0.7 operators. Following the ideas presented throughout the chapter, various inequalities can also be derived in terms of other integral operators, fractional or not. In [20], the following functional was considered:

$$T(f,g) = \frac{1}{b-a}\int_a^b f(x)g(x)dx - \left(\frac{1}{b-a}\int_a^b f(x)dx\right)\left(\frac{1}{b-a}\int_a^b g(x)dx\right).$$

In applied mathematics, inequalities and their applications are crucial, not only those presented above. For instance, using our weighted operators, generalized versions of the Grüss inequality ([24]) can be obtained. The latter, introduced in 1935, is closely linked to the Chebyshev inequality, making it highly significant. This includes its extension to the case of weighted Caputo derivatives. All of the above demonstrates the breadth and strength of our results.

# References

[1] Ali, M. A., Nápoles V, J. E., Kashuri, A., Zhang, Z., (2020). Fractional nonconformable Hermite-Hadamard inequalities for generalized $\phi$-convex functions. *Fasciculi Mathematici, Nr 64*, 5–16. DOI: 10.21008/j.0044–4413.2020.0007.

[2] Aljaaidi, T. A., & Pachpatte, D., (2021). A new generalization of reverse Minkowski's inequality for fractional integral. *Advances in the Theory of Nonlinear Analysis and its Applications, 5*(1), 72–81. https://doi.org/10.31197/atnaa.756605

[3] Atangana, A., & Baleanu, D., (2016). New fractional derivatives with nonlocal and non-singular kernel. *Theory and application to heat transfer model. Therm Sci., 20*(2), 763–769.

[4] Baleanu, D., Diethelm, K. et al., (2017). *Fractional Calculus: Models and Numerical Methods.* Vol. 5. Worls Scientific Publishing Co. Pte. Ltd., Series on Complexity, Nonlinearity and Chaos, Second Ed.

[5] Bayraktar, B., (2020). Some integral inequalities of Hermite-Hadamard type for differentiable $(s, m)$-convex functions via fractional integrals. *TWMS J. App. Eng. Math. 10*(3), 625–637.

[6] Bayraktar, B., (2020). Some New Inequalities of Hermite-Hadamard Type for Differentiable Godunova-Levin Functions via Fractional Integrals. *Konuralp Journal of Mathematics, 8*(1), 91–96.

[7] Bayraktar, B., Butt, S. I., Shaokat, Sh., & Nápoles Valdés, J. E., (2021). New Hadamard-type inequalities via $(s, m_1, m_2)$-convex functions. *Vestnik Udmurtskogo Universiteta. Matematika. Mekhanika, Kompyuternye Nauki, 31*(4), 597–612.

[8] Bayraktar, B., & Nápoles Valdes, J. E., (2022). New Generalized integral inequalities via $(h, m)$-convex modified functions. *Izvestiya Instituta Matematiki Informatiki Udmurtskogo Gosudarstvennogo Universiteta, 60*, 3–15. DO: 10.35634/2226–3594–2022–60—01.

[9] Bayraktar, B., & Nápoles, J. E., (2022). Integral inequalities for mappings whose derivatives are $(h, m, s)-$convex modified of second type via Katugampola integrals. *Annals of the Universityof Craiova, Mathematics and Computer Science Series, 49*(2), 371–383. DOI:10.52846/ami.v49i2.1596.

[10] Bayraktar, B., & Nápoles Valdes, J. E., (2022). Hermite-Hadamard Weighted integral inequalities for $(h, m)$-convex modified functions.*Fractional Differential Calculus, 12*(2), 235–248. doi:10.7153/fdc-2022–12–15

[11] Bayraktar, B., Nápoles, J. E., Rabossi, F., (2022). On generalizations of integral inequalities. *Probl. Anal. Issues Anal., 11*(29), 3–23. DOI: 10.15393/j3.art.2022.11190.

[12] Bohner, M., Kashuri, A., Mohammed, P. O., & Nápoles V., J. E., (2022). Hermite-Hadamard-type Inequalities for Integrals arising in Conformable Fractional Calculus. *Hacet. J. Math. Stat., 51*(3), 775–786. DOI:10.15672/hujms.946069.

[13] Breckner, W. W., (1978). Stetigkeitsaussagen fur eine Klasse verallgemeinerter konvexer funktionen in topologischen linearen Raumen. *Pupl. Inst. Math., 23*, 13–20.

[14] Bruckner, A. M., & Ostrow, E., (1962). Some Function Classes Related To The Class Of Convex Functions. *Pacific J. Math. 12*, 1203–1215.

[15] Butt, S. I., Yousaf, S., Akdemir, A. O., & Dokuyucu, M. A., (2021). New Hadamard-type integral inequalities via a general form of fractional integral operators. *Chaos Solitons Fractals, 148*, 111025.

[16] Caputo, M., (1967). Linear model of dissipation whose Q is almost frequency independent II. *Geophysical Journal International, 13(5)*, 529–539.

[17] Caputo, M., (1969). Elasticitá e Dissipazione, *Bologna: Zanichelli.*

[18] Caputo, M., & Fabrizio, M., (2015). A new definition of fractional derivative without singular kernel. *Natural Sciences Publishing Cor. Progress in Fractional Differentiation and Applications, 1*, 73–85.

[19] Cerone, P. , Dragomir, S. S., & Roumeliotis, J. , (1998). An Inequality Of Ostrowski Type For Mappings Whose Second Derivatives Are Bounded And Applications. *RGMIA Research Report Collection, 1*(1), 35–42. https://rgmia.org/papers/v1n1/3res98.pdf.

[20] Chebyshev, P. L., (1882). Sur les expressions approximatives des integrales definies par les autres prises entre les mêmes limites. *Proc. Math. Soc. Charkov, 2*, 93–98.

[21] Díaz, R., & Pariguan, E., (2007). On hypergeometric functionsand Pochhammer $k-$*symbol.. Divulg. Mat. 15*(2), 179–192.

[22] Dragomir, S. S., & Agarwal, R. P., (1998). Two inequalities for differentiable mappings and applications to special means of real numbers and to trapezoidal formula. *Appl. Math. Lett., 11*(5), 91–95. https://doi.org/10.1016/S0893–9659(98)00086-X

[23] Du, T. S., Wang, H., Latif, M. A., & Zhang, Y., (2018). Estimation type results associated to $k-$fractional integral inequalities with applications. *Journal of King Saud University-Science, 31*(4), 1083–1088. https://doi.org/10.1016/j.jksus.2018.09.010.

[24] Grüss, G., (1935). Uber das maximum des absoluten Betrages von? Formel nicht darstellbar. *Mathematische Zeitschrift 39*, 215–226. http://eudml.org/doc/168548.

[25] Galeano, J., Lloreda, J., Nápoles V., J. E., & Pérez, E. Erez, (2021). Certain Integral Inequalities Of Hermite-Hadamard Type For $h$-Convex Functions. *Journal of Mathematical Control Science and Applications, 7*(2), 129–140.

[26] Hudzik, H., & Maligranda, L., (1994). Some remarks on $s-$convex functions. *Aequationes Math. 48*1, 100–111.

[27] Jarad, F., Abdeljawad, T., & Shah, T., (2020). On The Weighted Fractional Operators Of A Function With Respect To Another Function. *Fractals, 28*(8), 2040011 (12 pages). DOI:10.1142/S0218348X20400113.

[28] Jarad, F., Ugurlu, E., Abdeljawad, T., & Baleanu, D., (2017). On a new class of fractional operators. *Adv. Differ. Equ.*, 247.

[29] Karim, M., Fahmi, A, Ullah, Z., Bhatti, M. Awais T., & Qayyum, A., (2023). On certain Ostrowski type integral inequalities for convex function via *AB*-fractional integral operator, *AIMS Mathematics, 8*(4), 9166–9184. DOI: 10.3934/math.2023459.

[30] Kashuri, A., Nápoles Valdés, Juan E., Ali, M. A., & Ud Din, G. M., (2022). New Integral Inequalities Using Quasi-Convex Functions Via Generalized Integral Operators And Their Applications. *Applied Mathematics E-Notes, 22*, 221–231.

[31] Khan, T. U., & Khan, M. A., (2019). Generalized conformable fractional integral operators. *J. Comput. Appl. Math. 346*, 378–389.

[32] Kirmaci, U. S., Bakula, M. K., Özdemir, M. E., & Pecaric, J., (2007). Hadamard-Type Inequalities for $s-$Convex Functions. *Applied Mathematics and Computation, 193*(1), 26–35. doi:10.1016/j.amc.03.030.

[33] Kórus, P., & Nápoles Valdés, J. E., (2023). On some integral inequalities for (h, m)-convex functions in a generalized framework. *Carpathian Journal of Mathematics, 15 (1)*, 137–149. doi:10.15330/cmp.15.1.137–149

[34] Luo, C., & Du, T., (2020). Generalized Simpson Type Inequalities Involving Riemann-Liouville Fractional Integrals and Their Applications.*Filomat 34(3)*, 751–760. https://doi.org/10.2298/FIL2003751L.

[35] Mehmood, S., Nápoles Valdés J. E., Fatima, N., & Aslam, W., (2022). Some integral inequalities via fractional derivatives. *Adv. Studies:Euro-Tbilisi Math. J. 15(3)*, 31–44. DOI: 10.32513/asetmj/19322008222.

[36] Mihesan, V. G., (1993). A generalization of convexity, In: Proceedings of the Seminar on Functional Equations, Approximation and Convexity. *Cluj-Napoca, Romania.*

[37] Mubeen, S., & Habibullah, G. M., (2012). $k-$fractional integrals and application. *Int. J. Contemp. Math. Sciences, 7*(2), 89–94.

[38] Nasir, J., Qaisar, S., Butt, S. I., Aydi, H., & Sen, M. De la, (2022). Hermite-Hadamard-like inequalities for fractional integral operator via convexity and quasi-convexity with their applications. *AIMS Mathematics, 7*(3), 3418–3439. doi: 10.3934/math.2022190.

[39] Nápoles Valdés, & Juan E., (2022). On the Hermite-Hadamard type inequalities involving generalized integrals, *Contrib. Math. 5*, 45–51. DOI: 10.47443/cm.2022.020.

[40] Nápoles Valdés, J. E., & Bayraktar, B., (2021). On The Generalized Inequalities Of The Hermite-Hadamard Type. *Filomat 35(14)*, 4917–4924. https://doi.org/10.2298/FIL2114917N.

[41] Nápoles Valdés, J. E., Bayraktar, B., & Butt, S. I., (2021). New integral inequalities of Hermite-Hadamard type in a generalized context. *Punjab University Journal of Mathematics, 53(11)*, 765–777. https://doi.org/10.52280/pujm.2021.531101.

[42] Nápoles Valdés, J. E., Rabossi, F., & Ahmad, H., (2021). Inequalities of the Hermite-Hadamard type, for functions $(h, m)$-convex modifoied of the second type. *Commun. Combin., Cryptogr. & Computer Sci., 1*, 33–43.

[43] Nápoles Valdés, J. E., Rodríguez, J. M., & Sigarreta, J. M., 2019. On Hermite-Hadamard type inequalities for non-conformable integral operators, *Symmetry, 11*, 1108.

[44] Nápoles, J. E., Rabossi, F., & Samaniego A. D., (2020). Convex functions: Ariadne's Thread or Charlotte's spiderweb? *Advanced Mathematical Models & Applications, 5(2)*, 176–191.

[45] Park, J., (2011). Generalization of Ostrowski-type inequalities for differentiable real $(s, m)-$convex mappings. *Far East J. Math. Sci., 49*, 157–171.

[46] Pearce, C. E. M., & Pecaric, J., (2000). Inequalities for differentiable mappings with application to special means and quadrature formulae. *Appl. Math. Lett. 13*, 51–55.

[47] Rainville, E. D., (1960). *Special Functions.* Macmillan Co., New York.

[48] Sarikaya, M. Z., & Ertugral, F., (2020). On the generalized Hermite-Hadamard inequalities. *Annals of the University of Craiova, Mathematics and Computer Science Series* , *47*(1), 193–213.

[49] Set, E., Sarikaya, M. Z., & Özdemir, M. E., (2014). Some Ostrowski's type inequalities for functions whose second derivatives are $s$-convex in the second sense. *Demonstr. Math., 47*, 37–47. https://doi.org/10.2478/dema-2014–0003.

[50] Toader, G., (1985). Some generalizations of the convexity. *Proceedings of the Colloquium on Approximation and Optimization, University Cluj-Napoca*, 329–338.

[51] Vivas-Cortez, M., Kermausuor, S., & Nápoles Valdés, J. E., (2022). *Hermite-Hadamard Type Inequalities for Coordinated Quasi-Convex Functions via Generalized Fractional Integrals*, in P. Debnath et al. (eds.), Fixed Point Theory and Fractional Calculus: Recent Advances and Applications, Forum for Interdisciplinary Mathematics, Springer Nature Singapore Pte Ltd. https://doi.org/10.1007/978–981–19–0668–8_16.

[52] Vivas-Cortez, M., Kórus, P., & Nápoles Valdés, J. E., (2021). Some generalized Hermite-Hadamard-Fejér inequality for convex functions. *Advances in Difference Equations, 199*. https://doi.org/10.1186/s13662–021–03351–7.

[53] Vivas-Cortez, M., Nápoles Valdés, J. E., & Guerrero, J. A., (2022). Some Hermite-Hadamard Weighted Integral Inequalities for $(h, m)$-Convex Modified Functions. *Appl. Math. Inf. Sci. 16*(1), 25–33. http://dx.doi.org/10.18576/amis/160103.

[54] Wu, S., Iqbal, S., Aamir, M., Samraiz, M., & Younus, A., (2021). On some Hermite–Hadamard inequalities involving $k-$fractional operators. *Journal of Inequalities and Applications, 32*. https://doi.org/10.1186/s13660–020–02527–1.

[55] Xi, B. Y., Gao, D. D., & Qi, F., (2020). Integral inequalities of Hermite-Hadamard type for $(\alpha, s)-$convex and $(\alpha, s, m)-$convex functions. *Journal of Pure And Applied Mathematics, 44*, 499–510.

[56] Xi, B. Y., Gao, D., & Qi, F., (2020). Integral inequalities of Hermite-Hadamard type for $(\alpha, s)$-convex and $(\alpha, s, m)$-convex functions. *Journal of Pure And Applied Mathematics, 44*, 499–510.

[57] Xi, B. Y., & Qi, F., (2015). Inequalities of Hermite–Hadamard type for extended $s-$convex functions and applications to means. *J. Nonlinear Convex. Anal. 16*(5), 873–890.

[58] Yildiz, C., Özdemir, M. E., & Kavurmaci, H., (2017). Fractional Integral Inequalities Via $s-$Convex Functions. *Turkish Journal of Analysis and Number Theory, 5*(1), 18–22.

# Chapter 8

## Periodicity Solutions of $k$-Symbol Fractional Differential Equations with Applications

**Rabha W. Ibrahim**[1–3]

[1] *Department of Mathematics, Mathematics Research Center, Nicosia/Mersin, Turkey*

[3] *Information and Communication Technology Research Group, Scientific Research Center, Al-Ayen University, Thi-Qar, Iraq*

[2] *Department of Computer Science and Mathematics, Lebanese American University, Beirut, Lebanon*

**Abstract**

The benefit of using integro-differential equations (I-DEs) is that they allow for the exploration of the full transition process, including its early, middle, and late stages. As a result, this approach can distinguish a growing element for an architecture that exhibits the same pattern of action over a long period of time but has unique characteristics at the beginning and middle stages of growth. An I-DE for diffusion is also used for assessing sub-diffusive and super-diffusive processes. Furthermore, techniques for computing I-DEs have been devised, and differential equations have an analytic framework in both force-free and linear force instances.

In this study, in view of the fractional settle operator in a non-compact set, we show the existence of solutions for fractional boundary value problems (BVPs) with multi-evolution. There are particular instances that are both specific and generic. A simple instance is offered to demonstrate the effectiveness of the abstraction theory method. The fixed-point model of non-compactness assessment serves as the cornerstone of our technique.

---

## Introduction

Fractional calculus is a branch of mathematics that generalizes the traditional calculus operators of differentiation and integration to non-integer orders. It is based on the idea of extending the concept of a derivative or integral to fractional orders, such as $\frac{1}{2}$, $\frac{3}{4}$, or even more general fractions.

In the context of fractional calculus, $k$-symbol fractional calculus refers to a particular generalization of the fractional derivative and fractional integral operators. In $k$-symbol fractional calculus, the fractional order is represented by a $k$-symbol, which is a sequence of non-negative integers. The $k$-symbol can be thought of as a kind of multi-index notation that generalizes the notion of a single index, which is used to represent the order of a derivative or integral in traditional calculus. The $k$-symbol allows for more flexibility in specifying the order of the fractional derivative or integral and can be used to capture a wider range of phenomena in science and engineering.

The $k$-symbol fractional calculus is based on a generalization of the well-known gamma function, called the $k$-symbol gamma function. This generalization was first introduced by Diaz and Pariguan in 2004 [4]. As a consequence of this definition, many fractional derivative and integral operators have been generalized

[6, 8]. These operators are defined in terms of the Riemann-Liouville fractional derivative and fractional integral operators, but with the order parameter generalized by the $k$-symbol [12, 15]. The $k$-symbol fractional calculus has applications in many areas of science and engineering, including signal processing, control theory, and fractional differential equations [13, 16]. It provides a powerful tool for modeling and analyzing complex systems that exhibit non-integer order behavior.

In this chapter, we proceed to develop and explore additional applications of this type of fractional calculus. The tendency to act without consideration is known as impulsive. To explain this pattern, we design an integro-differential equation (I-DE) in this endeavor. For a $k$-symbol fractional class of I-DEs with non-quick impulses, we investigate periodic boundary value problems in Banach spaces. Using the metric of non-compactness, the process of resolving domestically, and the fixed-point outcome, we establish various adequate conditions for the solvability of mild solutions of I-DEs. Finally, we provide examples to highlight the important results of the study. Our research was inspired by certain recent attempts in this approach.

---

## 8.1 Methods and Materials

### *8.1.1 k-symbol Fractional Calculus*

The motivating gamma function, sometimes referred as the $k$-symbol gamma function $\Gamma_k$, is expressed as follows [4]:

$$\Gamma_k(\chi) = \int_0^\infty \exp(\frac{-\nu^k}{k})\nu^{\chi-1}d\nu, \quad \mathrm{Re}\,(\chi) > 0$$
$$= \lim_{m\to\infty} \frac{m!k^m(m\,k)^{\frac{\chi}{k}-1}}{(\chi)_{m,k}},$$

where

$$(\chi)_{m,k} := \chi(\chi+k)(\chi+2k)...(\chi+(m-1)k)$$

and

$$(\chi)_{m,k} = \frac{\Gamma_k(\chi+m\,k)}{\Gamma_k(\chi)}.$$

### *8.1.2 Theory of Fixed Points*

Let

- $V = [0, T]$;
- $C[V, \Sigma] = \{\ell : \ell : V \to \Sigma\}$ the space of all continuous functions on $\Sigma$, where $\Sigma$ is a Banach space;
- $C_\wp[V, \Sigma] = \{\ell : \ell : (\wp_i, \tau_{i+1}] \to \Xi\}$ where there are two values $\ell(\tau_i^-)$ and $\ell(\tau_i^+)$ fulfilling $\ell(\tau_i^-) = \ell(\tau_i)$, $i = 1, ..., n$ with the sup-norm
$$\|\ell\|_{C_\wp} = \sup\{\|\ell(\tau)\| : \tau \in V\}.$$

#### *8.1.2.1 Arguments*

**Lemma 8.1.1.** *[8] Assume that $S \subset \Sigma$ is a bounded closed and convex set, and $\Sigma$ is a Banach space. Furthermore, consider that the function $G : S \to S$ belongs to the strict set contraction. Then $G \in S$ must have at least a fixed point.*

**Lemma 8.1.2.** *[14] Assume that $\Sigma$ is a Banach space and $S \subset C[V,\Sigma]$ is equicontinuous and bounded, then the closed convex hull of $S$ $\bar{C}oS \subset C[V,\Sigma]$ is bounded as well as equicontinuous.*

**Lemma 8.1.3.** *[15] Assume that $\Sigma$ is a Banach space, and $S \subset \Sigma$ is bounded, then there is a countable set $S_0 \subset S$ such that $\gamma(S) \leq 2\gamma(S_0)$, where $\gamma(S)$ is the Kuratowski measure of non-compactness of the bounded set $S \subset \Sigma$. Obviously, $0 < \gamma(S) < \infty$ and*

$$\gamma(S) = \inf\{\varepsilon > 0 : \bigcup_{i=0}^{n} v_i, \quad diam(v_i) \leq \varepsilon\}.$$

**Lemma 8.1.4.** *[6] Assume that $\Sigma$ is a Banach space, and let $S \subset C[V,\Sigma]$ is equicontinuous and bounded, then $\gamma(S(\tau))$ is continuous on $V$, and*

$$\gamma\left(\int_V G(\tau)d\tau\right) \leq \int_V \gamma\left(G(\tau)\right)d\tau, \quad \gamma(G(\tau)) = \max \gamma(G(\tau_i)).$$

### *8.1.3 k-symbol Fractional Calculus*

The $k$-symbol Riemann-Liouville fractional order integral is formulated by [17]:

$$\Upsilon_k^{\beta}\varphi(\tau) = \frac{1}{k\Gamma_k(\beta)}\int_0^{\tau} \varphi(s)(\tau - s)^{\beta/k-1}\, ds, \quad \beta > 0.$$

For a function $\varphi \in C^n[0,\infty)$, the $k$-symbol fractional differential operator in terms of the Caputo structure of fractional order $\beta \in (n, n+1]$ can be presented by:

$${}^{C}\Omega_k^{\beta}\varphi(\tau) = \frac{1}{k\Gamma_k(n-\beta)}\int_0^{\tau} \varphi^{(n)}(s)(\tau - s)^{n-\beta/k-1}\, ds, \quad \tau > 0,\, n \in \mathbb{N}.$$

Figure 8.1 shows the behavior of the generalized gamma function, where

$$\Gamma_{1.5}(\chi) = (2/3)^{1-(2\chi)/3}\Gamma((2\chi)/3),\ \Gamma_2(\chi)$$
$$= 2^{\chi/2-1}\Gamma(\chi/2),\ \Gamma_3(\chi) = 3^{\chi/3-1}\Gamma(\chi/3).$$

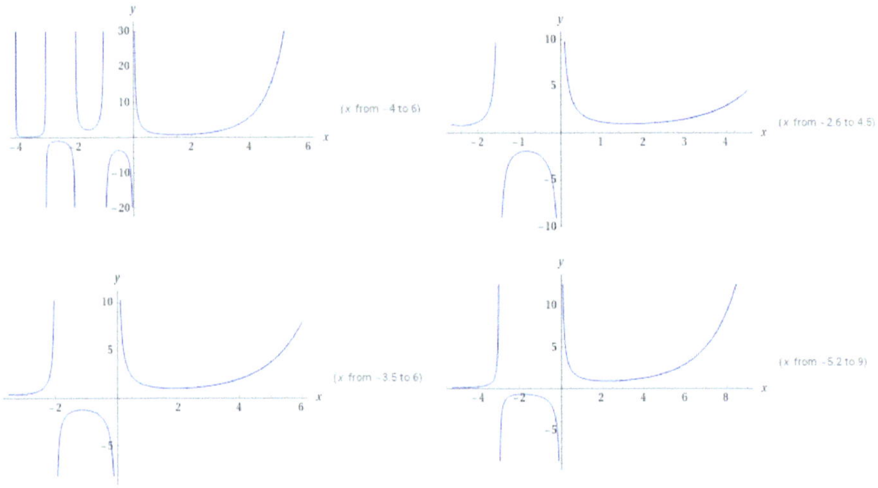


**FIGURE 8.1**. Plot of $\Gamma_1, \Gamma_{1.5}, \Gamma_2$ and $\Gamma_3$, respectively.

### *8.1.4 The Suggested Problem*

We investigate the periodicity of fractional multi-evolution equations (FME) via non-instantaneous impulses, which is created by the previous work:

$$\begin{cases} {}^{C}\Omega_k^{\beta}\varphi(\tau) = \Theta(\tau)\varphi(\tau) + \psi(\tau, s) \\ \qquad + \sum_{i=0}^{n} \int_0^{\tau} \varphi_i(\tau - s)\Psi_i(s, \varphi(s))d\,s & \tau \in (s_{i+1}, \tau_{i+1}], \\ \varphi(\tau) = p_i(\tau, \varphi(\tau))[q_\beta(\tau, \tau_i)]_k & \tau \in (\tau_i, \varsigma_i], \\ \varphi(0) = \varphi(\top), \end{cases} \tag{8.1.1}$$

where $i = 1, ..., n$, ${}^{C}\Omega_k^{\beta}$ is the $k$-symbol Caputo's fractional derivative of order $\beta \in (0, 1], \Theta$ is a linear operator, $[q_\beta]_k$ auxiliary function of fractional order and the numbers $\tau_i$ and $s_i$ satisfy $0 = s_0 < \tau_1 \leq s_1 < \tau_2 \leq .,. < \tau_{n+1} = \top$. Moreover, we have a set of continuous functions $\varphi_i : (\tau_i, s_i] \times \Sigma \to \Xi$, where $\Sigma$ indicates a Banach space, $[q_\beta(\tau, \tau_i)]_k : \mathbb{R}_+^2 \to \Sigma$ is the re-solvent operator generated by $\Theta$ such that $\{\lambda^{\beta/k} : \mathrm{Re}\,(\lambda) > 0\} \subset p(\Theta), p = \max p_i$ where $p_i$ are nonlinear functions in $\Sigma$. This work examines whether there are any moderate results for the FDEs (8.1.1) using the non-compact semi-group requirements, but the linear operator $\Theta$ is $\tau$-dependent. The results of this endeavor reinforce and enhance the key findings of several studies as a response of using a different technique.

If $\varphi \in C_{\wp}(V, \Sigma)$ satisfies the resulting equations, then it is considered to be a mild solution to (8.1.1):

$$\varphi(\tau) = [q_\beta(\tau, 0)]_k \Big[ [q_\beta(\top, \tau_n)]_k p_n(s_n, \varphi(s_n)) \\ + \int_{s_n}^{\top} [q_\beta(\tau, s)]_k \left( \psi(s, \varphi(s)) + \sum_{i=0}^{n} \int_0^{s} \varphi_i(\tau - s)^{\beta/k} \Psi_i(s, \varphi(s))\, ds \right) \Big]$$

$$+\int_0^{\tau}[q_\beta(\tau,s)]_k\left(\psi(s,\varphi(s))+\sum_{i=0}^{n}\int_0^{s}\varphi_i(\tau-s)^{\beta/k}\Psi_i(s,\varphi(s))\,ds\right)$$

$$\varphi(\tau)=[q_\beta(\tau,\tau_i)]_k p_i(\tau,\varphi(\tau)),\quad \tau\in(\tau_i,s_i]$$

$$\varphi(\tau)=[q_\beta(\tau,\tau_i)]_k p_i(s_i,\varphi(s_i))+\int_{s_i}^{\tau}[q_\beta(\tau,s)]_k$$

$$\times\left(\psi(s,\varphi(s))+\sum_{i=0}^{n}\int_0^{s}\varphi_i(\tau-s)^{\beta/k}\Psi_i(s,\varphi(s))\,ds\right).$$

$$\left(i=1,\ldots,n,\ \tau\in[0,\tau_1],\ \tau\in(s_i,\tau_{i+1}]\right)$$

---

## 8.2 Results

We have the following cases:

Formulate the following operator $\mathbb{P}:C_\wp[V,\Sigma]\to C_\wp[V,\Sigma]$, as follows:

$$(\mathbb{P}\varphi)(\tau)=[q_\beta(\tau,0)]_k\Big[[q_\beta(\top,\tau_n)]_k p_n(s_n,\varphi(s_n)) \tag{8.2.2}$$

$$+\int_{s_n}^{\top}[q_\beta(\tau,s)]_k\left(\psi(s,\varphi(s))+\sum_{i=0}^{n}\int_0^{s}\varphi_i(\tau-s)^{\beta/k}\Psi_i(s,\varphi(s))ds\right)\Big]$$

$$+\int_0^{\tau}[q_\beta(\tau,s)]_k\left(\psi(s,\varphi(s))+\sum_{i=0}^{n}\int_0^{s}\varphi_i(\tau-s)^{\beta/k}\Psi_i(s,\varphi(s))ds\right)$$

$$\varphi(\tau)=[q_\beta(\tau,\tau_i)]_k p_i(\tau,\varphi(\tau))$$

$$\varphi(\tau)=[q_\beta(\tau,\tau_i)]_k p_i(s_i,\varphi(s_i))+\int_{s_i}^{\tau}[q_\beta(\tau,s)]_k$$

$$\times \left( \psi(s, \varphi(s)) + \sum_{i=0}^{n} \int_0^s \varphi_i(\tau - s)^{\beta/k} \Psi_i(s, \varphi(s)) ds \right).$$

$$\Big( i = 1, ..., n, \ \tau \in [0, \tau_1], \ \tau \in (s_i, \tau_{i+1}] \Big)$$

We start with the following outcome:

**Theorem 8.2.1.** *Take into account these suppositions:*

*(S1) The functions* $\psi, \Psi_i, p_i \, (i = 1, ..., n) : V \times \Sigma \to \Sigma$ *are bounded and continuous in* $V \times \top_\iota$ *and*

$$\limsup_{\iota \to \infty} \frac{X(\iota)}{\iota} < \frac{1}{\varpi}, \tag{8.2.3}$$

*where*

$$\varpi = \max \Big\{ X^2 \left( 1 + (\top - s_n) + \int_{s_n}^{\top} \int_0^s \sum_{i=0}^{n} \varphi_i(s-t)^{\beta/k} dt ds \right)$$
$$+ X \left( \tau_1 + \int_0^{\tau} \int_0^s \sum_{i=0}^{n} \varphi_i(s-t)^{\beta/k} dt ds \right),$$
$$X \Big( 1 + (s_{i+1} - s_i) + \int_{s_i}^{\tau} \int_0^s \sum_{i=0}^{n} \varphi_i(s-t)^{\beta/k} \, dt \, ds,$$
$$i = 1, \ldots, n \Big), \ X \Big\},$$

*where*

$$X(\iota) = \sup \Big\{ \|\psi(\tau, \varphi)\|, \|\Psi_i(\tau, \varphi)\|, \|p_i(\tau, \varphi)\|,$$

$$i = 1, \ldots, n : (\tau, \varphi) \in V \times \top_\iota \Big\}.$$

*and*

$$\top_\iota = \{ \varphi \in \Sigma : \|\varphi\| \le \iota \}.$$

*The k-symbol resolvent operator* $[q_\beta(\tau, s)]_k$ *is non-compact for* $\tau, s > 0$, *where*

$$X = \max_{0 \le s < \tau \le \top} \|[q_\beta(\tau, s)]_k\| < \infty.$$

(S2) *There are Lebesgue integrator functions* $G_\psi, G_{\Psi_i}, G_{p_i} \in L^1(V, \mathbb{R}^+)(i = 1, 2, ..., n)$ *such that*

$$\begin{aligned}
\gamma\left(\psi(\tau, \delta)\right) &\le G_\psi(\tau)\gamma(\delta)\\
\gamma\left(\Psi_i(\tau, \delta)\right) &\le G_{\Psi_i}(\tau)\gamma(\delta)\\
\gamma\left(p_i(\tau, \delta)\right) &\le G_{p_i}(\tau)\gamma(\delta),
\end{aligned}$$

*where* $\delta \subset \Sigma$ *is equicontinuous and countable set. Consider the two sets*

$$\top_\iota = \{\varphi \in \Sigma : \|\varphi\| \le \iota, \quad \iota > 0\}$$

*and*

$$\begin{aligned}
\omega = \max\Big\{ & X^2 G_{p_n}(\tau) + X^2 \int_{s_n}^{\top} G_\phi(s)ds + X\int_0^\tau G_\psi(s)ds\\
& + (X + X^2)\int_0^\tau \int_0^s \sum_{i=0}^{n} \varphi_i(s-t)^{\beta/k} G_{\Phi_i}(\tau)dtd\varsigma,\\
& XG_{p_i}(\tau),\ G_{p_i}(\tau) + \int_{s_i}^{\tau} G_\psi(s)ds\\
& + \int_{s_i}^{\tau}\int_0^s \sum_{i=0}^{n} \varphi_i(s-t)^{\beta/k} G_{\Psi_i}(t)dtds\Big\}\\
& < 1.
\end{aligned}$$

$$\Big(i = 1, ..., n,\ \tau \in [0, \tau_1],\ \tau \in (s_i, \tau_{i+1}]\Big)$$

*Then the BVP* (8.1.1) *has a complete one mild solution* $\varphi \in C_\wp[V, \Sigma]$.

*Proof.* From (8.2.3), we obtain that there is a positive number $b \in (0, \frac{1}{\varpi})$ and an initial number $\iota_0 > 0$ with $\iota > \iota_0$ satisfying $X(\iota) < b\iota$. Corresponding to the initial value $\iota_0$, there exists $\iota^*$ such that $\iota^* \geq \iota_0$. Define a ball of radius $\iota^*$, as follows:

$$B_{\iota^*} = \{\varphi \in C_\wp[V, \Sigma] : \|\varphi\| \leq \iota^*\}.$$

We request $(\mathbb{P}\varphi) \in B_{\iota^*}$. We need the following steps:

Step 1: **Boundednees**

For $\tau \in [0, \tau_1]$, we have

$$\begin{aligned}
&\|(\mathbb{P}\varphi)\| \\
&\leq \|[q_\beta(\tau, 0)]_k\| \|[q_\beta(\top, \tau_n)]_k p_n(s_n, \varphi(s_n))\| \|[q_\beta(\tau, 0)]_k\| \\
&+ \left\| \int_{s_n}^{\top} [q_\beta(\tau, s)]_k \Big( \psi(s, \varphi(s)) + \sum_{i=0}^{n} \int_0^s \varphi_i(\tau - s)^{\beta/k} \Psi_i(s, \varphi(s)) ds \Big) \right\| \\
&+ \|[q_\beta(\tau, 0)]_k\| \left\| \int_0^{T} [q_\beta(\tau, s)]_k \Big( \psi(s, \varphi(s)) \right. \\
&\left. + \sum_{i=0}^{n} \int_0^s \varphi_i(\tau - s)^{\beta/k} \Psi_i(s, \varphi(s)) ds \Big) \right\| \\
&\leq X^2 b\iota^* \left( 1 + (\top - s_n) + \int_{s_n}^{\top} \int_0^s \sum_{i=0}^{n} \varphi_i(s - t)^{\beta/k} dt ds \right) \\
&+ X b\iota^* \left( \tau_1 + \int_0^{T} \int_0^s \sum_{i=0}^{n} \varphi_i(s - t)^{\beta/k} dt ds \right) \\
&\leq \iota^*.
\end{aligned}$$

Moreover, we get for $\tau \in (\tau_i, s_i]$ the following relation

$$\|(\mathbb{P}\varphi)\| \le \|[q_\beta(\top, \tau_n)]_k p_i(s_n, \varphi(s_n))\|$$

$$\le Xb\iota^* \le \iota^*.$$

For the interval $\tau \in (s_i, \tau_{i+1}]$, we get

$$\begin{aligned}
&\|(\mathbb{P}\varphi)\| \\
&\le \|[q_\beta(\top, \tau_n)]_k p_n(s_n, \varphi(s_n))\| \\
&+ \left\| \int_{s_n}^{\top} [q_\beta(\tau, s)]_k \left( \psi(s, \varphi(s)) + \sum_{i=0}^{n} \int_0^s \varphi_i(\tau - s)^{\beta/k} \Psi_i(s, \varphi(s)) ds \right) \right\| \\
&\le Xb\iota^* \left( 1 + (\tau_{i+1} - s_i) + \int_{s_i}^{\tau} \int_0^s \sum_{i=0}^{n} \int_0^s \varphi_i(\tau - s)^{\beta/k} \right) \\
&\le \iota^*.
\end{aligned}$$

Step 2: **Continuity** We request to prove that $(\mathbb{P}\varphi)B_{\iota^*} \to B_{\iota^*}$ is continuous. By the continuity of $\psi, \Psi_i$ and $p_i$, we have

$$\lim_{n\to\infty} \sup_{\tau\in V} \|\psi(\tau, \varphi_n(\tau)) - \psi(\tau, \varphi(\tau))\| = 0$$

$$\lim_{n\to\infty} \sup_{\tau\in V} \|\Psi_i(\tau, \varphi_n(\tau)) - \Psi_i(\tau, \varphi(\tau))\| = 0$$

$$\lim_{n\to\infty} \sup_{\tau\in V} \|p_i(\tau, \varphi_n(\tau)) - p_i(\tau, \varphi(\tau))\| = 0.$$

Now, when $\tau \in [0, \tau_1]$, this implies

$$\begin{aligned}
&\|(\mathbb{P}\varphi_n)(\tau) - (\mathbb{P}\varphi)(\tau)\| \\
&\le X^2 \sup_{\tau\in V} \|p_n(\tau, \varphi_n(\tau)) - p_n(\tau, \varphi(\tau))\|
\end{aligned}$$

$$+ X^2 \int_{s_n}^{\top} \|\phi(\tau, \varphi_n(\tau)) - \psi(\tau, \varphi(\tau))\| ds$$

$$+ X^2 \int_{s_n}^{\top} \int_0^s \sum_{i=0}^{n} \left( \|\Psi_i(\tau, \varphi_n(\tau)) - \Psi_i(\tau, \varphi(\tau))\| \varphi_i(s-t)^{\beta/k} \right) dtds$$

$$+ X\tau_1 \sup_{\tau \in V} \left( \|\psi(\tau, \varphi_n(\tau)) - \psi(\tau, \varphi(\tau))\| \right)$$

$$+ X \int_0^{\top} \int_0^s \sum_{i=0}^{n} \left( \|\Psi_i(\tau, \varphi_n(\tau)) - \Psi_i(\tau, \varphi(\tau))\| \varphi_i(s-t)^{\beta/k} \right) dtds.$$

We continue to establish the upper bound, whenever $\tau \in (\tau_i, s_i], i = 1, ..., n$

$$\|(\mathbb{P}\varphi_n)(\tau) - (\mathbb{P}\varphi)(\tau)\| \leq X \sup_{\tau \in V} \|p_i(\tau, \varphi_n(\tau)) - p_i(\tau, \varphi(\tau))\|.$$

And for $\tau \in (s_i, \tau_{i+1}], i = 1, ..., n$, we have

$$\|(\mathbb{P}\varphi_n)(\tau) - (\mathbb{P}\varphi)(\tau)\|$$

$$\leq X \sup_{\tau \in V} \|p_i(\tau, \varphi_n(\tau)) - p_i(\tau, \varphi(\tau))\|$$

$$+ X \int_{s_i}^{\top} \|\psi(\tau, \varphi_n(\tau)) - \psi(\tau, \varphi(\tau))\| ds$$

$$+ X \int_{s_i}^{\top} \int_0^s \sum_{i=0}^{n} \left( \|\Psi_i(\tau, \varphi_n(\tau)) - \Psi_i(\tau, \varphi(\tau))\| \varphi_i(s-t)^{\beta/k} \right) dtds.$$

Hence, $(\mathbb{P}\varphi)$ is continuous where

$$\lim_{n \to \infty} \|(\mathbb{P}\varphi_n)(\tau) - (\mathbb{P}\varphi)(\tau)\|_{C_\wp} = 0.$$

Step 3: **Equi-Continuity**

We deal with the following matters. The first one, $z_1, z_2 \in [0, \tau_1]$, where $z_1 < z_2$.

$$\begin{aligned}
&\|(\mathbb{P}\varphi_n)(z_2) - (\mathbb{P}\varphi)(z_1)\| \\
&\leq \|[q_\beta(z_2,0)]_k - [q_\beta(z_1,0)]_k\| \|[q_\beta(\top,\tau_i)]_k p_i(s_i,\varphi(s_i))\| \\
&+ \|(\mathbb{P}\varphi_n)(\eta_2) - (\mathbb{P}\varphi)(z_1)\| \\
&\times \Big\| \int_{s_n}^{\top} [q_\beta(\top,s)]_k \Big(\psi(s,\varphi(s)) \\
&\quad + \int_0^s \sum_{i=1}^{n} \varphi_i(s-t)^{\beta/k} \Psi_i(t,\varphi(t))\, dt\Big)\, ds\Big\| \\
&+ \sup_{s\in[0,\tau_1]} \|[q_\beta(z_2,0)]_k - [q_\beta(z_1,0)]_k\| \\
&\times \Big\| \int_0^{z_1} \left(\psi(s,\varphi(s)) + \int_0^s \sum_{i=1}^{n} \varphi_i(s-t)^{\beta/k} \Psi_i(t,\varphi(t))dt\right) ds\Big\| \\
&+ \Big\| \int_{z_1}^{z_2} [q_\beta(z_2,s)]_k \Big(\psi(s,\varphi(s)) \\
&\quad + \int_0^s \sum_{i=1}^{n} \varphi_i(s-t)^{\beta/k} \Psi_i(t,\varphi(t))\, dt\Big)\, ds\Big\|.
\end{aligned}$$

The second one is in the interval $(\tau_i, \varsigma_i]$, we obtain

$$\begin{aligned}
&\|(\mathbb{P}\varphi_n)(z_2) - (\mathbb{P}\varphi)(z_1)\| \\
&\leq \|[q_\beta(z_2,\tau_i)]_k p_i(z_2,\varphi(z_2)) - [q_\beta(z_1,\tau_i)]_k p_i(z_1,\varphi(\eta_1))\| \\
&\leq \|[q_\beta(z_2,z_1)]_k [q_\beta(z_1,\tau_i)]_k p_i(z_2,\varphi(z_2)) - [q_\beta(z_1,\tau_i)]_k p_i(z_1,\varphi(z_1))\| \\
&\leq X \|[q_\beta(z_2,z_1)]_k p_i(z_2,\varphi(z_2)) - p_i(z_1,\varphi(z_1))\|.
\end{aligned}$$

The third one is given in the interval $(s_i, \tau_{i+1}]$, which implies

$$\|(\mathbb{P}\varphi_n)(z_2) - (\mathbb{P}\varphi)(z_1)\|$$

$$
\begin{aligned}
&\leq \|[q_\beta(z_2,\tau_i)]_k - [q_\beta(z_1,\tau_i)]_k\| \|p_i(s_i,\varphi(s_i))\| \\
&+ \sup_{s\in[0,\tau_1]} \|[q_\beta(z_2,s)]_k - [q_\beta(z_1,s)]_k\| \\
&\times \left\| \int_{s_i}^{z_1} \left( \psi(s,\varphi(s)) + \int_0^s \sum_{i=1}^n \varphi_i (s-t)^{\beta/k} \Psi_i(t,\varphi(t))dt \right) ds \right\| \\
&+ \left\| \int_{z_1}^{z_2} [q_\beta(z_2,s)]_k \Big( \psi(s,\varphi(s)) \right. \\
&\quad \left. + \int_0^s \sum_{i=1}^n \varphi_i (s-t)^{\beta/k} \Psi_i(t,\varphi(t))\, dt \Big)\, ds \right\|.
\end{aligned}
$$

Obviously, $z_2 \to z_1$ implies

$$
\|(\mathbb{P}\varphi_n)(z_2) - (\mathbb{P}\varphi)(z_1)\| = 0.
$$

Thus, $(\mathbb{P}\varphi)$ is equicontinuous in $B_{\iota^*}$. As a result, Lemma 8.1.2 yields that $\bar{C}o\mathbb{P}(B_{\iota^*}) \subset B_{\iota^*}$ is bounded and equicontinuous.

Step 4: **Condensity**

We have to prove that

$$
\mathbb{P} : B_{\iota^*} \to B_{\iota^*}
$$

is a condensing operator. Lemma 8.1.3 implies that there is a countable set

$$
\Lambda_0 = \{\varphi_n\} \subset \Lambda \subset \bar{C}o\mathbb{P}(B_{\iota^*})
$$

such that

$$
\gamma\left(\mathbb{P}(\Lambda)\right) \leq 2\gamma\left(\mathbb{P}(\Lambda_0)\right).
$$

We indicate the next matters. In $[0, \tau_1]$ there is a set $\Lambda_0 \subset \Lambda \subset \bar{C}o\mathbb{P}(B_{\iota^*})$ with

$$
\begin{aligned}
&\gamma\left(\mathbb{P}(\Lambda_0)(\tau)\right)\\
&\leq X^2\, \gamma\left(p_n(s_n, (\Lambda_0)(s_n))\right)\\
&+ X^2\, \gamma\left(\int_{s_n}^{\top}\left(\psi(s,\Lambda_0(s)) + \int_0^s \sum_{i=1}^{n} \varphi_i(s-t)^{\beta/k}\Psi_i(t,\Lambda_0(t))dt\right) ds\right)\\
&+ X\, \gamma\left(\int_0^{\tau}\left(\psi(s,\Lambda_0(s)) + \int_0^s \sum_{i=1}^{n} \varphi_i(s-t)^{\beta/k}\Psi_i(t,\Lambda_0(t))dt\right) ds\right)\\
&\leq X^2 G_{p_n}\gamma(\Lambda) + X^2\left(\int_{s_n}^{\top}\left(G_{\psi}(s)\gamma(\Lambda_0(s))\right.\right.\\
&\left.\left.+ \int_0^s \sum_{i=1}^{n} \varphi_i(s-t)^{\beta/k} G_{\Psi_i}(t)\gamma(\Lambda_0(t))dt\right)ds\right)\\
&+ X\left(\int_0^{\tau}\left(G_{\psi}(s)\gamma(\Lambda_0(s))\right.\right.\\
&\qquad\left.\left.+ \int_0^s \sum_{i=1}^{n} \varphi_i(s-t)^{\beta/k} G_{\Psi_i}(t)\gamma(\Lambda_0(t))\, dt\right) ds\right).\\
&\leq \gamma(\Lambda)\left[X^2\left(G_{p_n} + \int_{s_n}^{\top} G_{\psi}(s)ds + \int_0^{\tau}\int_0^s \sum_{i=1}^{n} \varphi_i(s-t)^{\beta/k} G_{\Psi_i}(t)dtds\right)\right.\\
&\left.+ X\left(\int_0^{\top} G_{\psi}(s)ds + \int_0^{\tau}\int_0^s \sum_{i=1}^{n} \varphi_i(s-t)^{\beta/k} G_{\Psi_i}(t)dtds\right)\right].
\end{aligned}
$$

The second matter is displaced in the interval $(\tau_i, s_i], i = 1, ..., n$, which gives

$$\gamma\left(\mathbb{P}(\Lambda_0)(\tau)\right) \leq X G_{p_i}(\tau)\gamma(\Lambda).$$

But, the third fact is in the interval $(s_i, \tau_{i+1}], i = 1, ..., n$, which brings

$$\begin{aligned}
&\gamma\left(\mathbb{P}(\Lambda_0)(\tau)\right) \le XG_{p_i}(\tau)\gamma(\Lambda) \\
&\quad + X\Bigg(\int_{s_i}^{\top}\Big(G_{\phi}(s)\gamma(\Lambda_0(s)) \\
&\quad + \int_0^s \sum_{i=1}^{n} \varphi_i(s-t)^{\beta/k} G_{\Psi_i}(t)\gamma(\Lambda_0(t))\, dt\Big)\, ds\Bigg). \\
&\le X\gamma(\Lambda)\Bigg(G_{p_i}(\tau) + \int_{s_i}^{\top} G_{\psi}(s)\, ds \\
&\quad + \int_{s_i}^{\top}\int_0^s \sum_{i=1}^{n} \varphi_i(s-t)^{\beta/k} G_{\Psi_i}(t)\, dt\, ds\Bigg).
\end{aligned}$$

Lemma 8.1.4 yields

$$\gamma\left(\Lambda_0\right) \le \max_{\tau \in V} \gamma\left(\Lambda_0(t)\right),$$

which gives

$$\gamma\left(\Lambda\right) \le \omega\gamma\left(\Lambda\right), \quad \omega \in (0,1).$$

Hence, we proved that $\mathbb{P}$ is a strict contraction mapping in $\bar{C}o\mathbb{P}(B_{\iota^*})$. Lemma 8.1.1 implies that $\mathbb{P}$ admits a fixed point in $\bar{C}o\mathbb{P}(B_{\iota^*}) \subset C_{\wp}[\jmath, \Xi]$. Therefore, Eq. (8.1.1) admits a mild solution in $C[V, \Sigma]$.

Note that when $k = 1$, we obtain the result in [5]. Similarly, we have the following outcome introduces the maximum value of $\varphi_i$ and $\Psi_i$.

**Theorem 8.2.2.** *Assume that*

*(H3) The functions* $\psi, \Psi_i, p_i\,(i = 1, ..., n) : V \times \Sigma \to \Sigma$ *are bounded and continuous in* $V \times \top_\iota$ *with*

$$\limsup_{\iota\to\infty} \frac{X(\iota)}{\iota} < \frac{1}{\overline{\varpi}_n}, \tag{8.2.4}$$

*where*

$$\begin{aligned}\overline{\varpi}_n = \max\Big\{ & X^2 \left(1 + (\top - s_n) + n\int_{s_n}^{\top}\int_0^s \Phi(s-t)^{\beta/k} dtds\right) \\ & + X\left(\tau_1 + n\int_0^{\tau}\int_0^s \Phi(s-t)^{\beta/k} dtds\right), \\ & X\left(1 + (s_{i+1} - s_i) + n\int_{s_i}^{\tau}\int_0^s \Phi(s-t)^{\beta/k} dtds\right), \\ & i = 1, ..., n,\ X\Big\},\end{aligned}$$

*where* $\Phi := \max(\varphi_i(s-t)^{\beta/k})$ *and* $\Psi := \max(\Psi_i(\tau, \varphi))$,

$$X(\iota) = \sup\{\|\psi(\tau, \varphi)\|, \|\Psi(\tau, \varphi)\|, \|p_i(\tau, \varphi)\|,$$

$$i = 1, ..., n : (\tau, \varphi) \in V \times \top_\iota\}$$

*and*

$$\top_\iota = \{\varphi \in \Sigma : \|\varphi\| \le \iota\}.$$

*The resolvent operator* $[q_\beta(\tau, s)]_k$ *is non-compact for* $\tau, s > 0$, *where*

$$X = \max_{0 \le s < \tau \le \top} \|[q_\beta(\tau, s)]_k\| < \infty.$$

*(H4) Non-negative Lebesgue integrable functions do exist $G_\psi, G_\Psi, G_{p_i} \in G^1(V, \mathbb{R}^+)(i = 1, 2, ..., n)$ fulfilling the relation*

$$\gamma(\psi(\tau,\sigma)) \le G_\psi(\tau)\gamma(\sigma)$$
$$\gamma(\Psi(\tau,\sigma)) \le G_\Psi(\tau)\gamma(\sigma)$$
$$\gamma(p_i(\tau,\sigma)) \le G_{p_i}(\tau)\gamma(\sigma),$$

*where $\sigma \subset \Sigma$ is equicontinuous and countable set. Formulate a double set by*

$$\top_\iota = \{\varphi \in \Sigma : \|\varphi\| \le \iota, \quad \iota > 0\}$$

*and*

$$\begin{aligned}\overline{\omega}_n &= \max\Big\{X^2 G_{p_n}(\tau) + X^2 \int_{s_n}^{\top} G_\psi(s)ds + X\int_0^\tau G_\psi(s)ds \\ &+ n(X + X^2)\int_0^\tau \int_0^s \Phi(s-t)^{\beta/k} G_\Psi(\tau)dtds, \\ &XG_{p_i}(\tau),\ G_{p_i}(\tau) + \int_{s_i}^\tau G_\psi(s)d\varsigma \\ &+ n\int_{s_i}^\tau \int_0^s \Phi(s-t)^{\beta/k} G_\Psi(t)dtds\Big\} \\ &< 1.\end{aligned}$$

*Then* (8.1.1) *has a fully mild outcome $\varphi \in C_\wp[V, \Sigma]$.*

The following repercussion is stated in [21], where $k = 1$ and $n = 1$

**Corollary 8.2.1.** *Assume that*

(H5) *The functions* $\psi, \Psi, p_i\,(i = 1, ..., n) : V \times \Sigma \to \Sigma$ *are bounded and continuous in* $V \times \top_\iota$ *and*

$$\limsup_{\iota \to \infty} \frac{X(\iota)}{\iota} < \frac{1}{\overline{\varpi}}, \tag{8.2.5}$$

*where*

$$\begin{aligned}\overline{\varpi} = \max\Big\{&X^2\left(1 + (\top - s_n) + \int_{s_n}^{\top}\int_0^s \Phi(s-t)^\beta dtds\right)\\ &+ X\left(\tau_1 + \int_0^{\tau}\int_0^s \Phi(\varsigma - t)^\beta dtds\right),\\ &X\left(1 + (s_{i+1} - s_i) + \int_{s_i}^{\tau}\int_0^s \Phi(s-t)^\beta dtds\right), i = 1, ..., n,\ X\Big\},\end{aligned}$$

$$\begin{aligned}X(\iota) = \sup\Big\{&\|\psi(\tau,\varphi)\|, \|\Psi(\tau,\varphi)\|, \|p_i(\tau,\varphi)\|,\\ &i = 1, \ldots, n : (\tau,\varphi) \in V \times \top_\iota\Big\}.\end{aligned}$$

*and*

$$\top_\iota = \{\varphi \in \Sigma : \|\varphi\| \le \iota\}.$$

*The resolvent operator* $[q_\beta(\tau, s)]_k$ *is non-compact for* $\tau, \varsigma > 0$, *such that*

$$X = \max_{0 \le s < \tau \le \top} \|[q_\beta(\tau,\varsigma)]\| < \infty.$$

(H6) *There are integrable non-negative Lebesgue functions* $G_\psi, G_\Psi, G_{p_i} \in L^1(V, \mathbb{R}^+)(i = 1, 2, ..., n)$ *achieving the relations*

$$\begin{aligned}\gamma\left(\psi(\tau,\sigma)\right) &\le G_\psi(\tau)\gamma(\sigma)\\ \gamma\left(\Phi(\tau,\sigma)\right) &\le G_\Psi(\tau)\gamma(\sigma)\\ \gamma\left(p_i(\tau,\sigma)\right) &\le G_{\rho_i}(\tau)\gamma(\sigma),\end{aligned}$$

*where $\sigma \subset \Sigma$ is equicontinuous and countable set. Formulate the following sets:*

$$\top_\iota = \{\varphi \in \Sigma : \|\varphi\| \le \iota, \quad \iota > 0\}$$

*and*

$$\begin{aligned}\overline{\omega} = \max\Big\{ & X^2 G_{p_n}(\tau) + X^2 \int_{s_n}^{\top} G_\psi(s)ds + X \int_0^{\tau} G_\psi(s)ds \\ & + (X + X^2) \int_0^{\tau} \int_0^{s} \Phi(s-t) G_\Psi(\tau) dt ds, \\ & X G_{p_i}(\tau),\, G_{p_i}(\tau) + \int_{s_i}^{\tau} G_\psi(s)ds + \int_{s_i}^{\tau} \int_0^{s} \Phi(s-t) G_\Psi(t) dt ds, \Big\} \\ & i = 1, ..., n. < 1.\end{aligned}$$

*Then the BVP* (8.1.1) *provides a completely mild solution $\varphi \in C_\wp[V, \Sigma]$.*

**Corollary 8.2.2.** *Suppose that all the conditions of Theorem 8.2.2 are occurred. Then the maximum mild outcome $\varphi \in C_\wp(V, \Sigma)$ of Eq.*(8.1.1) *can be seen as follows:*

$$\begin{aligned}&\varphi(\tau) \\ &= [q_\beta(\tau, 0)]_k \Big[ [q_\beta(\top, \tau_n)]_k p_n(s_n, \varphi(s_n)) \\ &+ \int_{s_n}^{\top} [q_\beta(\tau, s)]_k \left( \psi(s, \varphi(s)) + n \int_0^{s} \Phi(\tau - s)^{\beta/k} \Psi(s, \varphi(s)) ds \right) \Big] \\ &+ \int_0^{\tau} [q_\beta(\tau, s)]_k \left( \psi(s, \varphi(s)) + n \int_0^{s} \Phi(\tau - s)^{\beta/k} \Psi(s, \varphi(s)) ds \right), \; \tau \in [0, \tau_1] \\ &\varphi(\tau) = [q_\beta(\tau, \tau_i)]_k p_i(\tau, \varphi(\tau)), \quad \tau \in (\tau_i, s_i] \\ &\varphi(\tau) = [q_\beta(\tau, \tau_i)]_k p_i(s_i, \varphi(s_i)) + \int_{s_i}^{\tau} [q_\beta(\tau, s)]_k\end{aligned}$$

$$\times \left(\psi(s,\varphi(s)) + n\int_0^s \Phi(\tau - s)^{\beta/k}\Psi(s,\varphi(s))ds\right)$$

$\tau \in (s_i, \tau_{i+1}]$,

*where $\Phi$ and $\Psi$ have the same sign.*

Theorem 8.2.1 can be enhanced into two dimensional parameters $(n, m)$ as follows:

**Theorem 8.2.3.** *Consider the following hypotheses:*

*(H7) The functions $\psi, \Psi_j, \rho_i$ $(i = 1, ..., n,\ j = 0, ..., m) : \jmath \times \Sigma \to \Sigma$ are bounded and continuous in $V \times \top_\iota$ and*

$$\limsup_{\iota \to \infty} \frac{X(\iota)}{\iota} < \frac{1}{\varpi_m}, \tag{8.2.6}$$

*where*

$$\varpi_m = \max\left\{X^2\left(1 + (\top - s_n) + \int_{s_n}^{\top}\int_0^s \sum_{j=0}^{m} \varphi_j(s-t)^{\beta/k} dtds\right)\right.$$
$$+ X\left(\tau_1 + \int_0^{\tau}\int_0^s \sum_{j=0}^{m} \varphi_j(s-t)dtds\right),$$
$$X\left(1 + (s_{i+1} - s_i) + \int_{s_i}^{\tau}\int_0^s \sum_{j=0}^{m} \varphi_j(s-t)^{\beta/k} dtds, i = 1, ..., n\right),$$
$$\left. X\right\},$$

*where*

$$X(\iota) = \sup\{\|\psi(\tau,\varphi)\|, \|\Psi_j(\tau,\varphi)\|, \|p_i(\tau,\varphi)\|,$$

$$i = 1, ..., n,\ j = 0, ..., m : (\tau,\varphi) \in V \times \top_\iota\}$$

*and*

$$\top_\iota = \{\varphi \in \Sigma : \|\varphi\| \leq \iota\}.$$

*The resolvent operator* $[q_\beta]_k(\tau, s)$ *is non-compact for* $\tau, s > 0$, *where*

$$X = \max_{0 \leq s < \tau \leq \top} \|[q_\beta]_k(\tau, s)\| < \infty.$$

*(H8) Non-negative Lebesgue integrable functions do exist, as follows:* $G_\psi, G_{\Psi_j}, G_{p_i} \in L^1(V, \mathbb{R}^+)(i = 1, 2, ..., n,\ j = 0, ..., m)$ *fulfilling the relations*

$$\begin{aligned}
&\gamma\left(\psi(\tau, \sigma)\right) \leq G_\psi(\tau)\gamma(\sigma)\\
&\gamma\left(\Psi_j(\tau, \sigma)\right) \leq G_{\Psi_j}(\tau)\gamma(\sigma)\\
&\gamma\left(p_i(\tau, \sigma)\right) \leq G_{p_i}(\tau)\gamma(\sigma),
\end{aligned}$$

*where* $p \subset \Sigma$ *is equicontinuous and countable set. Formulate a double sets, as follows:*

$$\top_\iota = \{\varphi \in \Sigma : \|\varphi\| \leq \iota, \quad \iota > 0\}$$

*and*

$$\begin{aligned}
\omega_m = \max\Big\{&X^2 G_{p_n}(\tau) + X^2 \int_{s_n}^{\top} G_\psi(s)ds + X \int_0^\tau G_\psi(s)ds\\
&+ (X + X^2) \int_0^\tau \int_0^s \sum_{j=0}^{m} \varphi_j(s-t)^{\beta/k} L_{\Phi_j}(\tau)dtds,\\
&X G_{p_i}(\tau),\ G_{p_i}(\tau) + \int_{s_i}^{\tau} G_\psi(s)ds\\
&+ \int_{s_i}^{\tau} \int_0^s \sum_{j=0}^{m} \varphi_j(s-t)^{\beta/k} G_{\Psi_j}(t)dtds\Big\} < 1.
\end{aligned}$$

*Then, we have*

$$\begin{cases} {}^{C}\Omega_k^{\beta}\varphi(\tau) = \Theta(\tau)\varphi(\tau) + \psi(\tau, s) \\ \qquad\qquad + \sum_{j=0}^{m} \int_0^{\tau} \varphi_j(\tau - s)^{\beta/k}\Psi_j(s, \varphi(s))d\,s & \tau \in (s_{i+1}, \tau_{i+1}] \\ \varphi(\tau) = p_i(\tau, \varphi(\tau))[q_\beta]_k(\tau, \tau_i) & \tau \in (\tau_i, s_i] \\ \varphi(0) = \varphi(\top), \end{cases} \tag{8.2.7}$$

*has at least one mild outcome* $\varphi \in C_{\wp}[V, \Sigma]$ *which is given by*

$$\begin{aligned} &\varphi(\tau) \\ &= [q_\beta(\tau, 0)]_k \Big[ [q_\beta]_k(\top, \tau_n)p_n(s_n, \varphi(s_n)) \\ &+ \int_{s_n}^{\top} [q_\beta]_k(\tau, s) \left( \psi(s, \varphi(s)) + \sum_{j=0}^{m} \int_0^s \varphi_j(\tau - s)^{\beta/k}\Psi_j(s, \varphi(s))ds \right) \Big] \\ &+ \int_0^{\tau} [q_\beta]_k(\tau, s) \left( \psi(s, \varphi(s)) + \sum_{j=0}^{m} \int_0^s \varphi_j(\tau - s)^{\beta/k}\Psi_j(s, \varphi(s))ds \right) \\ &\varphi(\tau) = [q_\beta]_k(\tau, \tau_i)p_i(\tau, \varphi(\tau)), \quad \tau \in (\tau_i, s_i],\ i = 1, ..., n \\ &\varphi(\tau) = [q_\beta]_k(\tau, \tau_i)p_i(s_i, \varphi(s_i)) + \int_{s_i}^{\tau} [q_\beta]_k(\tau, s) \\ &\times \left( \psi(s, \varphi(s)) + \sum_{j=0}^{m} \int_0^s \varphi_j(\tau - s)^{\beta/k}\Psi_j(s, \varphi(s))ds \right) \\ &\tau \in (s_i, \tau_{i+1}],\ i = 1, ..., n. \end{aligned}$$

The next result presents the maximum solvability of $\varphi_j$ and $\Psi_j$, $j = 0, ..., m$. The proof is likely to Theorem 8.2.3.

**Theorem 8.2.4.** *Take into account these theories:*

*(H9) The functions* $\psi, \Psi_j, p_i \, (i = 1, ..., n, \; j = 0, ..., m) : V \times \Sigma \to \Sigma$ *are bounded and continuous in* $V \times \top_\iota$ *and*

$$\limsup_{\iota \to \infty} \frac{X(\iota)}{\iota} < \frac{1}{\overline{\varpi}_m}, \tag{8.2.8}$$

*where*

$$\begin{aligned}\overline{\varpi}_m = \max\Big\{ & X^2 \left(1 + (\top - s_n) + m \int_{s_n}^{\top} \int_0^s \Phi(s-t)^{\beta/k} dtds \right) \\ & + X \left(\tau_1 + m \int_0^{\tau} \int_0^s \Phi(s-t)^{\beta/k} dtds \right), \\ & X \left(1 + (s_{i+1} - s_i) + m \int_{s_i}^{\tau} \int_0^s \Phi(s-t)^{\beta/k} dtds, i = 1, ..., n \right), \\ & X \Big\},\end{aligned}$$

*where* $\Phi := \max(\varphi_j(s-t)^{\beta/k})$ *and* $\Psi := \max(\Psi_j(\tau, \varphi))$

$$X(\iota) = \sup\{\|\psi(\tau, \varphi)\|, \|\Psi(\tau, \varphi)\|, \|p_i(\tau, \varphi)\|,$$

$$i = 1, ..., n : (\tau, \varphi) \in V \times \top_\iota\}$$

*and*

$$\top_\iota = \{\varphi \in \Sigma : \|\varphi\| \le \iota\}.$$

*The* $k$*-symbol resolvent operator* $[q_\beta](\tau, s)$ *is non-compact for* $\tau, s > 0$*, where*

$$X = \max_{0 \le s < \tau \le \top} \|[q_\beta]_k(\tau, s)\| < \infty.$$

*(H10) Non-negative Lebesgue integrable functions do exist* $G_{\psi}, G_{\Psi}, G_{p_i} \in L^1(V, \mathbb{R}^+)(i = 1, 2, ..., n)$ *achieving the relations*

$$\gamma\left(\psi(\tau,\sigma)\right) \leq G_{\psi}(\tau)\gamma(\sigma)$$
$$\gamma\left(\Psi(\tau,\sigma)\right) \leq G_{\Psi}(\tau)\gamma(\sigma)$$
$$\gamma\left(p_i(\tau,\sigma)\right) \leq G_{p_i}(\tau)\gamma(\sigma),$$

*where* $\sigma \subset \Sigma$ *is equicontinuous and countable set. Formulate the following double sets*

$$\top_{\iota} = \{\varphi \in \Sigma : \|\varphi\| \leq \iota, \quad \iota > 0\}$$

*and*

$$\begin{aligned}\overline{\omega}_m = \max\Big\{&X^2 G_{p_n}(\tau) + X^2 \int_{s_n}^{\top} G_{\psi}(s)ds \\ &+ X \int_0^{\mathcal{T}} G_{\psi}(s)ds \\ &+ m(X + X^2) \int_0^{\mathcal{T}} \int_0^s \Phi(s-t)^{\beta/k} G_{\Psi}(\tau)dtds, \\ &X G_{p_i}(\tau),\ G_{p_i}(\tau) + \int_{s_i}^{\mathcal{T}} G_{\psi}(s)ds \\ &+ m \int_{s_i}^{\mathcal{T}} \int_0^s \Phi(s-t)^{\beta/k} G_{\Psi}(t)dtds\Big\} \\ &< 1.\end{aligned}$$

*Then* (8.2.7) *has a complete mild resolution* $\varphi \in C_{\wp}[V, \Sigma]$.

**Corollary 8.2.3.** *Suppose that the conditions of Theorem 8.2.4 are satisfied. Then, the maximum mild outcome* $\varphi \in C_{\wp}(V, \Sigma)$ *of Eq.(8.2.7) is given by the structure*

$$\begin{aligned}
\varphi(\tau) &= [q_\beta]_k(\tau, 0)\Big[[q_\beta]_k(\top, \tau_n)p_n(s_n, \varphi(s_n)) \\
&+ \int_{s_n}^{\top} [q_\beta]_k(\tau, s)\left(\psi(s, \varphi(s)) + m\int_0^s \Phi(\tau - s)^{\beta/k}\Psi(s, \varphi(s))ds\right)\Big] \\
&+ \int_0^{\tau} [q_\beta]_k(\tau, s)\left(\psi(s, \varphi(s)) + m\int_0^{\varsigma} \Phi(\tau - s)^{\beta/k}\Psi(s, \varphi(s))ds\right) \\
\varphi(\tau) &= [q_\beta]_k(\tau, \tau_i)p_i(\tau, \varphi(\tau)), \quad \tau \in (\tau_i, s_i] \\
\varphi(\tau) &= [q_\beta]_k(\tau, \tau_i)p_i(s_i, \varphi(s_i)) + \int_{s_i}^{\tau} [q_\beta]_k(\tau, s) \\
&\times \left(\psi(s, \varphi(s)) + m\int_0^s \Phi(\tau - s)\Psi(s, \varphi(s))ds\right) \\
&\tau \in (s_i, \tau_{i+1}],
\end{aligned}$$

*where* $\Phi$ *and* $\Psi$ *have the same sign.*

---

## 8.3 Applications

The periodicity of mild solutions of differential equations has important applications in several areas of mathematics and science, some of which are:

- Oscillations: The periodicity of solutions of differential equations is closely related to the concept of oscillations in physical systems. For example, the simple harmonic motion of a mass-spring system or a pendulum is a well-known example of a periodic motion that can be described by a differential equation [7].

- Control Theory: The periodicity of solutions of differential equations is also used in control theory to design control systems that can stabilize oscillatory behavior in mechanical or electrical systems [10].

- Signal Processing: The periodicity of solutions of differential equations is used in signal processing to analyze periodic signals such as audio or radio signals. This analysis involves Fourier series and transforms that express periodic functions as a sum of sine and cosine functions [16].

- Vibrations: The periodicity of solutions of differential equations is used in the study of vibrations and resonance phenomena in structures and materials, such as bridges, buildings, and machines [3].

- Celestial Mechanics: The periodicity of solutions of differential equations is used in celestial mechanics to study the motion of planets, satellites, and other celestial bodies in our solar system. Kepler's laws of planetary motion and the study of the stability of the solar system are examples of this application [19].

- Biology: The periodicity of solutions of differential equations is used in the study of biological rhythms, such as the circadian rhythm that governs the sleep-wake cycle in humans and other animals. Mathematical models of biological systems often involve differential equations that exhibit periodic solutions [1].

- Finance: The periodicity of solutions of differential equations is used in financial mathematics to model the behavior of financial markets and to price derivative securities, such as options and futures contracts. The Black-Scholes model, which is widely used in finance, is based on a partial differential equation with periodic boundary conditions [2].

Thus, the periodicity of solutions of differential equations has a wide range of applications in many fields of science and engineering, where it is used to describe and analyze phenomena that exhibit periodic behavior.

We illustrate the following problems, which refer to one of the above applications:

**Example 8.3.1.** *Assume that*

$$\begin{cases} {}^{C}\Omega_k^{\beta}\varphi(\tau) = \tau\varphi(\tau) + \left(\dfrac{\tau}{1-\tau}\right)\tanh\left(\varphi(\tau)\right) & \\ + \int_0^{\tau}\exp(\tau - s)^{\beta/k}(\dfrac{1}{8} + \dfrac{1}{9}s)d\,s & \tau \in [0, \dfrac{1}{4}) \bigcup (\dfrac{1}{3}, 1] \\ \varphi(\tau) = \dfrac{1}{16}[q_{\beta}]_k(\tau, 1) & \tau \in (\dfrac{1}{4}, \dfrac{1}{3}], i = 1, ..., n \\ \varphi(0) = \varphi(1) = 0, & \end{cases} \tag{8.3.9}$$

*where, for example* $\beta = 0.5$ *and* $k = 1.5$*, we have*

$$\int_0^{\tau}\exp(\tau - s)^{0.5/1.5}(\frac{1}{8} + \frac{1}{9}s)d\,s = 1/8e^{\tau^{1/3}}(8\tau^{4/3}$$

$$+163\tau^{2/3} - 48\tau - 326\tau^{1/3} + 326) - (2\tau)/3 - 163/4.$$

*And for* $\beta = 0.5, k = 2$, *we have*

$$\int_0^\tau \exp(\tau - s)^{0.5/2}(\frac{1}{8} + \frac{1}{9}s)d\,s$$

$$= e^{\tau^{1/4}}(-16\tau^{5/4} + (16\tau^{3/2})/9 - (2237\tau^{3/4})/6 + (272\tau)/3$$

$$+(2237\sqrt{(\tau)})/2 - 2237\tau^{1/4} + 2237) + (8\tau)/3 - 2237.$$

*Since* $\Theta(\tau)\varphi(\tau) = \tau\varphi(\tau)$ *then* $\Theta$ *is generated the resolve operator* $[q_\beta]_k$, $\Sigma = [0,1], \beta \in (n, 1+n]$. *Define the set*

$$\Lambda := \{\varphi : \varphi(0) = \sigma(1)\} \subset \Sigma.$$

*Suppose that*

$$\psi(\tau) = \frac{\tau}{1-\tau}\tanh(\varphi(\tau))$$

$$\Psi_1(\tau) = \frac{1}{8}, \quad \Psi_2(\tau) = \frac{1}{9}\tau, \Phi_{1,2}(\tau - s)^{\beta/k} = \exp(\tau - s)^{\beta/k}$$

*achieve the conditions (H1) and (H2). Then Theorem 8.2.1 implies that* (8.3.9) *admits a mild solution.*

**Example 8.3.2.** *Consider the BVP*

$$\begin{cases} {}^C\Omega_k^\beta \varphi(\tau) = \tau\varphi(\tau) + \dfrac{\tau}{(1-\tau)^2}\tanh\varphi(\tau) & \\ + \int_0^\tau \exp(\tau - s)^{\beta/k}(1+s)d\,s & \tau \in [0, \frac{1}{4}) \bigcup (\frac{1}{3}, 1] \\ & \\ \varphi(\tau) = \dfrac{1}{16}[q_\beta]_k(\tau, 1) & \tau \in (\frac{1}{4}, \frac{1}{3}], i = 1, ..., n \\ & \\ \varphi(0) = \varphi(1) = 0, & \end{cases} \tag{8.3.10}$$

*where, for* $\beta = k = 1$, *we have*

$$\int_0^\tau \exp(\tau - s)^{\beta/k}(1+s)d\,s = (e^\tau - 1)(\tau + 1)$$

$$= \tau + \frac{(3\tau^2)}{2} + \frac{(2\tau^3)}{3} + \frac{(5\tau^4)}{24} + \frac{\tau^5}{20} + O(\tau^6).$$

*Also, for* $\beta = 0.5$ *and* $k = 2$, *we get*

$$\int_0^\tau \exp(\tau - s)^{0.5/2}(1+s)d\,s = 4\Big(e^{\tau^{1/4}}(-36\tau^{5/4} + 4\tau^{3/2} - 839\tau^{3/4}$$

$$+204\tau + 2517\sqrt{(\tau)} - 5034\tau^{1/4} + 5034) + 6(\tau - 839)\Big)$$

*Since* $\Theta(\tau)\varphi(\tau) = \tau\varphi(\tau)$ *then* $\Theta$ *is generated the resolve operator* $[q_\beta]_k$, $\Sigma = [0,1], \beta \in (n, 1+n]$. *Put the set*

$$\Lambda := \{\varphi : \varphi(0) = \varphi(1)\} \subset \Sigma.$$

*Suppose that*

$$\psi(\tau) = \frac{\tau}{(1-\tau)^2}\tanh\varphi(\tau)$$

$$\Psi_1(\tau) = 1, \quad \Psi_2(\tau) = \tau, \Phi_{1,2}(\tau - s)^{\beta/k} = \exp(\tau - s)^{\beta/k}$$

*fulfill the conditions (H1) and (H2) (H2). That leaves at least a single mild solution in light of Theorem 8.2.1.*

---

## 8.4 Conclusion

The advantage of employing integro-differential equations (I-DEs) lies in their ability to investigate the entire transition process, encompassing its early, middle, and late stages. This method allows for a distinction between an evolving element and a structure that exhibits consistent behavior over a long period of time but demonstrates different characteristics during the initial and intermediate stages of development. Additionally, an I-DE for diffusion is used to characterize the sub-diffusive and super-diffusive phases. Furthermore, methods for solving I-DEs have been developed, and analytic models for differential equations have been established for both force-free and linear force cases. In prior investigations involving the fractional resolve operator in a non-compact setting, we established the necessary conditions for the existence of a mild solution to fractional boundary value problems (BVPs). We proposed utilizing an equation with multiple evolutions. Certain scenarios are both specific and general. A simple illustration is provided to demonstrate the effectiveness of the abstraction-based theoretical approach. The fixed-point model for measuring non-compactness served as the foundation of our methodology. To achieve a mild periodic result for future studies, one could utilize a compact sample and establish the necessary criteria.

# References


[1] Abuasbeh, K., Shafqat, R., Alsinai, A., & Awadalla, M. (2023). Analysis of the mathematical modeling of COVID-19 by using a mild solution with delay Caputo operator. *Symmetry, 15*(2), 286.

[2] Alfar, A. J., Kumpamool, C., Nguyen, D. T., & Ahmed, R., (2023). The determinants of issuing central bank digital currencies. *Research in International Business and Finance, 64*, 101884.

[3] Ali, M. A., Eltohamy, F., Abd-Elrazek, A., & Hanafy, M. E., (2023). Assessment of micro-vibrations effect on the quality of remote sensing satellite images. *International Journal of Image and Data Fusion*, 1-18.

[4] AlSheikh, M. H., Al-Saidi, N. M., & Ibrahim, R. W., (2022). Dental X-ray identification system based on association rules extracted by k-symbol fractional Haar functions. *Fractal and Fractional, 6*(11), 669.

[5] Baleanu, D., & Ibrahim, R. W., (2023). Optical applications of a generalized fractional integro-differential equation with periodicity. *AIMS Mathematics, 8*(5), 11953–11972.

[6] Banas, J. (1980). On measures of noncompactness in Banach spaces. *Commentationes Mathematicae Universitatis Carolinae, 21*(1), 131–143.

[7] Bchatnia, A., Boukhatem, A., & El Mufti, K. (2023). Almost periodicity and stability for solutions to networks of beams with structural damping. *Discrete and Continuous Dynamical Systems-S, 16*(6), 1201–1215.

[8] Deimling, K., (2010). *Nonlinear functional analysis. Courier Corporation.* Dover Publications, Inc. New York.

[9] Diaz, R., & Pariguan, E., (2004). *On hypergeometric functions and Pochhammer k-symbol.* arXiv preprint math/0405596.

[10] Du, C., & Liu, C., (2023). Time periodic solution to a mechanochemical model in biological patterns. *Evolution Equations and Control Theory, 12*(2), 502–524.

[11] Hadid, S. B., & Ibrahim, R. W., (2022). Geometric Study of 2D-Wave Equations in View of K-Symbol Airy Functions. *Axioms, 11*(11), 590.

[12] Ibrahim, R. W., (2022). K-symbol fractional-order discrete-time models of Lozi system. *Journal of Difference Equations and Applications*, 1-20.

[13] Ledesma, C. E. T., & Nyamoradi, N., (2023). (k, $\psi$)-Hilfer impulsive variational problem. *Revista de la Real Academia de Ciencias Exactas, Fisicas y Naturales. Serie A. Matematicas, 117*(1), 42.

[14] Liu, L., (2000). Iterative method for solutions and coupled quasi-solutions of nonlinear integro-differential equations of mixed type in Banach spaces. *Nonlinear Analysis: Theory, Methods & Applications, 42*(4), 583–598.

[15] Liu, L., Guo, F., Wu, C., & Wu, Y., (2005). Existence theorems of global solutions for nonlinear Volterra type integral equations in Banach spaces. *Journal of Mathematical Analysis and Applications, 309*(2), 638–649.

[16] Miao, Y., Li, C., Shi, H., & Han, T., (2023). Deep network-based maximum correlated kurtosis deconvolution: A novel deep deconvolution for bearing fault diagnosis. *Mechanical Systems and Signal Processing, 189*, 110110.

[17] Mubeen, S., & Habibullah, G. M., (2012). k-Fractional integrals and application. *Int. J. Contemp. Math. Sci, 7*(2), 89–94.

[18] Suthar, D., Baleanu, D., Purohit, S., & Ucar, F., (2020). Certain k-fractional calculus operators and image formulas of k-Struve function. *AIMS Mathematics, 5*(3), 1–14.

[19] Xu, X., (2023). Determination of the doubly symmetric periodic orbits in the restricted three-body problem and Hill's lunar problem. *Celestial Mechanics and Dynamical Astronomy, 135*(2), 8.

[20] Yildiz, C., & Cotirla, L. I., (2023). Examining the Hermite-Hadamard Inequalities for k-Fractional Operators Using the Green Function. *Fractal and Fractional, 7*(2), 161.

[21] Zhu, B., & Liu, L, (2018). Periodic boundary value problems for fractional semilinear integro-differential equations with non-instantaneous impulses. *Boundary Value Problems, 2018*(1), 1–14.

# Chapter 9

## Gingerbread-man Discrete System Formulated by the k-Symbol Fractional Calculus

**Rabha W. Ibrahim**[1–3]

[1]*Department of Mathematics, Mathematics Research Center, Near East Boulevard, Nicosia/Mersin, Turkey*

[2]*Department of Computer Science and Mathematics, Lebanese American University, Beirut, Lebanon*

[3]*Information and Communication Technology Research Group, Scientific Research Center, Al-Ayen University, Thi-Qar, Iraq*

**Abstract**

In the present study, we extend the gingerbread-man map by utilizing the concept of a $k$-symbol Caputo derivative within the framework of fractional calculus. A two-dimensional chaotic map, dependent on the $k$-symbol gingerbread-man map and the fractional flow map (K-GM), is proposed. We investigate some of the important dynamics of these maps and analyze the conditions for the stability of variable fractional dynamic systems. As a result, we present the K-GM with stability solution specifications based on the stability of the system's generating polynomial. Additionally, the Hermite matrix-based predecessor of the system considers the symmetry stability criterion. Furthermore, to achieve system stabilization, we propose incorporating these maps into control processes. In this study, both one-dimensional (1D) and two-dimensional (2D) controller rules are assumed.

The symmetry and stability of the controller system alternatives are examined, and we also present asymmetric stability control systems. The linearization approach is employed to compute synchronization errors for the two- and four-dimensional controller laws. The primary findings of this investigation are demonstrated through mathematical models.

---

## Introduction

The gamma function is a special function in mathematics that extends the concept of factorials to complex and real numbers. It is denoted by the symbol $\Gamma(\zeta)$ and is defined as:

$$\Gamma(\zeta) = \int_0^\infty \xi^{\zeta-1} e^{-\xi}\, d\xi,$$

where $\zeta$ is a complex or real number.

The gamma function has many important properties and applications in mathematics, including its use in probability theory, number theory, and mathematical physics. One significant application of the gamma function is in fractional calculus, a branch of calculus that deals with derivatives and integrals of non-integer orders. The gamma function is closely related to fractional derivative and fractional integral operators, which are defined as:

$$D^{\wp}\varphi(\chi) = \frac{1}{\Gamma(n-\wp)}(d^n/d\chi^n)\int_a^\chi (\chi-\tau)^{n-\wp-1}\varphi(\tau)d\tau$$

$$I^{\wp}\varphi(\chi) = \frac{1}{\Gamma(\wp)}\int_a^\chi (\chi-\tau)^{\wp-1}\varphi(\tau)d\tau,$$

where $\wp$ is a non-integer number, $n$ is the smallest integer greater than $\wp$, and $\varphi$ is a function of $\chi$. These operators have numerous important applications in physics, engineering, and other fields. For example, they can be used to model non-local diffusion processes, such as the diffusion of heat or particles in a medium with long-range interactions.

In general, the gamma function and fractional calculus are both significant areas of mathematics with wide-ranging applications in many fields of science and engineering.

The gamma function has several generalizations using different parameters. One such generalization is obtained by utilizing the $k$-symbol gamma function to develop the $k$-symbol fractional calculus. The $k$-symbol fractional calculus is an extension of fractional calculus, where $k > 1$. It deals with differentiation and integration of non-integer orders. In traditional fractional calculus, the order of differentiation or integration is represented by a single real number, typically denoted by a fraction or a real number between 0 and 1 [4, 6].

The $k$-symbol fractional calculus is used in a variety of fields in science and engineering, including signal processing, control theory, and fractional differential equations. It provides a powerful tool for modeling and analyzing complex systems that exhibit non-integer order behavior [8]–[13].

The motivating gamma function, sometimes referred to as the $k$-symbol gamma function $\Gamma_k$, is expressed as follows [4] (see Figure 9.1):

$$\Gamma_k(\zeta) = \int_0^{\infty} \exp(\frac{-\nu^k}{k})\nu^{\zeta-1} d\nu, \quad \mathrm{Re}\,(\zeta) > 0$$

$$= \lim_{m \to \infty} \frac{m! k^m (m\,k)^{\frac{\zeta}{k}-1}}{(\zeta)_{m,k}},$$

where

$$(\zeta)_{m,k} := \zeta(\zeta + k)(\zeta + 2k)...(\zeta + (m-1)k)$$

and

$$(\zeta)_{m,k} = \frac{\Gamma_k(\zeta + m\,k)}{\Gamma_k(\zeta)}.$$

For example, when $k = 2$ we have $\Gamma_2(\zeta) = 2^{\zeta/2-1}\Gamma(\zeta/2) for Re(\zeta) > 0$ and for $k = 3$, we have $\Gamma_3(\zeta) = 3^{\zeta/3-1}\Gamma(\zeta/3) for Re(\zeta) > 0$.

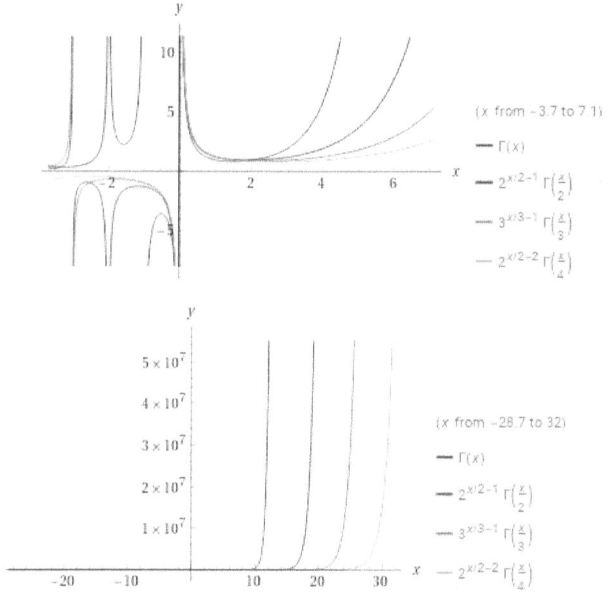


**FIGURE 9.1**. Plot of $\Gamma_k$ for $k = 1, 2, 3, 4$.

In accordance with the theory of dynamical systems, a two-dimensional chaotic map is a gingerbread-man map. The map is chaotic when certain initial conditions and initial conditions are fulfilled. The collection of chaotic solutions is organized, and the result is a map that resembles a gingerbread man. It is shown by the piecewise linear transformation [14]

$$\begin{aligned} F_{\imath+1} &= 1 - G_\imath + |F_\imath| \\ G_{\imath+1} &= F_\imath, \quad \imath \in \mathbb{N} \cup \{0\}. \end{aligned}$$

One dimensional parametric piecewise linear transformation is given by the formula [7]

$$\begin{aligned} F_{\imath+1} &= 1 - G_\imath + \alpha|F_\imath| \\ G_{\imath+1} &= F_\imath, \quad \imath \in \mathbb{N} \cup \{0\}. \end{aligned}$$

The two-dimensional parametric piecewise linear transformation described in the structure that follows is used to illustrate the symmetric performance

$$\begin{aligned} F_{\imath+1} &= 1 - G_\imath + \alpha|F_\imath| \\ G_{\imath+1} &= \beta F_\imath, \quad \imath \in \mathbb{N} \cup \{0\}. \end{aligned} \tag{9.0.1}$$

In this chapter, we suggest the convex parametric structure, as follows:

$$\begin{aligned} F_{\imath+1} &= 1 - G_\imath + \frac{\beta}{(1-\beta)}|F_\imath| \\ G_{\imath+1} &= F_\imath, \quad \imath \in \mathbb{N} \cup \{0\}, \, \beta \in (0,1). \end{aligned} \tag{9.0.2}$$

A convex parameter refers to a property of a system where any linear combination of two feasible outcomes is also a feasible outcome.

This is a fundamental characteristic of many optimization problems, implying that the set of feasible outcomes forms a convex set. A convex parametric discrete system is a mathematical model that describes a set of possible outcomes dependent on a set of parameters, where the system is both convex and discrete. Convex parametric discrete systems are widely used in optimization problems across various fields, including operations research, economics, and engineering. These systems are typically solved using mathematical programming techniques, such as linear programming or integer programming, which aim to determine the best possible outcome given the set of parameters and constraints. Different plots of System (9.0.2) are given in Figure 9.2.

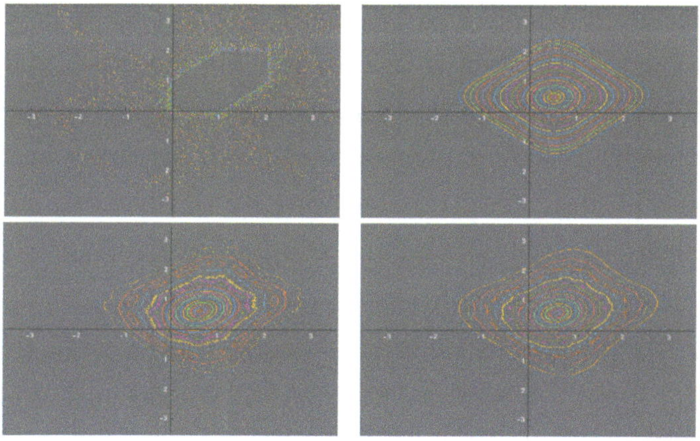

**FIGURE 9.2**. Plot of System (9.0.2), when $\beta = 0.5$(original case), 0.06, 0.3, 0.25.

Chaotic systems exhibit appealing features that are suitable for designing techniques for cryptography, in addition to ad hoc behavior. The chaotic maps have consistent qualities for the viewer while the system's imposed initial conditions are known in advance, but they display highly unpredictable traits when these parameters are uncertain. A chaotic map described as a "gingerbread man" has six separate polygonal parts and exhibits chaotic behavior over the entire sector (see [5, 11, 12]).

It has been noted that discrete fractional calculus (DFC) has advantages over traditional calculus for developing the dynamics of natural progressions. Diaz and Olserin first presented the fundamental illumination of DFC in 1974, and it is now acknowledged as one of the largest and most important contemporary divisions of calculus [3]. The fractional chaotic maps (FCMs) were initially shown in [17]. A fractional logistic map was presented by the study's authors in [18] along with a quantitative explanation of how chaotically it operates. A fractional Lorenz map is proposed and chaotic synchronization is examined in the context of stability findings. By utilizing several fractional calculi, especially the variable fractional order, the Lozi structure map of fractional order is adjusted (see [9, 10]).

In the present work, we use the notion of a $k$-symbol Caputo derivative of the fractional calculus form to expand the gingerbread-man map. We propose two different systems: the two-dimensional chaotic map based on the $k$-symbol gingerbread-man map and the fractional flow map (K-GM). We examine some key dynamics of these maps. We also examine the requirements for the stability of the variable fractional dynamic systems. As a result, we provide K-GM with specifications for stability solutions according to the stability of the system's defining polynomial. The system's Hermite matrix-based antecedent also takes into account the necessary criteria for symmetry stability.

In addition, we propose merging these maps into control procedures with the goal of system stabilization. The 1D and 2D controller rules are taken for granted in this research. We look at the symmetry and stability of the controller system solutions. Additionally, we offer asymmetric stability control systems. For two- and four-dimensional controller laws, the synchronization errors are calculated using the linearization procedure. Numerical simulations are also presented to illustrate the main findings of this investigation.

---

## 9.1 K-Symbol Gingerbread-Man Map (K-GM)

According to Caputo calculus, the fractional difference operator is formulated by the functional sum: [1]:

$$\begin{aligned} {}^{C}D^{\wp}\varphi(t) &= D^{-(u-\wp)}D^{u}\varphi(t) \\ &= \frac{1}{\Gamma(u-\wp)}\sum_{m=0}^{t-(u-\wp)}(t-\ell(m))^{u-\wp-1}D^{u}_{m}\varphi(t), \end{aligned} \tag{9.1.1}$$

where $\varphi : \mathbb{N}_m := \{m, m+1, ...\} \to \mathbb{R}, u = 1 + \lceil \wp \rceil$, $\wp \notin \mathbb{N}$ where $t \in \mathbb{N}_{m+u-\wp}$. In addition, the term $t^{(\wp)}$ is given by the formula

$$t^{(\wp)} = \frac{\Gamma(t+1)}{\Gamma(t+1-\wp)}, \quad \wp > 0,$$

and $D\varphi(t) = \varphi(1+t) - \varphi(t)$. The appropriate discrete integral equation can be created by

$$\varphi(t) = \varphi_0 + \frac{1}{\Gamma(\wp)} \sum_{m=0}^{t-\wp} (t - \ell(m))^{\wp-1} \varphi(\wp + \ell(m)).$$

By considering $\ell(m) = s + 1$, we get

$$\varphi(t) = \varphi_0 + \frac{1}{\Gamma(\wp)} \sum_{s=0}^{t-\wp} (t - s - 1)^{\wp-1} \varphi(\wp + s - 1), \tag{9.1.2}$$

where

$$\varphi_0 = \sum_{n=0}^{j-1} \frac{(t)^{(n)}}{\Gamma(n+1)} D^n \varphi(0).$$

The ordinary difference system according to (9.0.1) yields

$$\begin{aligned} DF &= 1 - G + \frac{\beta}{1-\beta}|F| - F \\ DG &= F - G. \end{aligned} \tag{9.1.3}$$

A rough resolution to the discrete system of (9.1.3) is as outlined below:

**Example 9.1.1.** *Suppose that*

$$\begin{aligned} DF &= 1 - G + (\frac{\beta}{1-\beta})F - F \\ DG &= F - G. \end{aligned} \tag{9.1.4}$$

*For symmetric stable solution, consider and* $\beta = 2/3 \rightarrow \beta/(1-\beta) = 2$, *we obtain the solution*

$$F(n) = -a_2\delta_{1-n} + a_1(1 - \vartheta(n-2)) + 2,$$
$$G(n) = a_1\delta_{1-n} + a_2(\vartheta(-n) - \vartheta(1-n)\vartheta(n-1)) + 1, \quad (a_1, a_2) \in \mathbb{Z}^2,$$

*where* $\delta$ *is the Kronecker Delta function and* $\theta$ *is the Heaviside step function.*

**Example 9.1.2.** *Consider the system*

$$DF = 1 - G - \frac{\beta}{1-\beta}F - F \tag{9.1.5}$$
$$DG = F - G.$$

*For symmetric stable solution, let* $\beta = 2/3$, *we get the outcome*

$$F(n) = a_1(-1)^n 2^{n-1}(n+2) + a_2(-1)^n 2^{n-1} n + 2/9,$$
$$G(n) = a_1(-1)^{n+1} 2^{n-1} n + a_2(-1)^{n+1} 2^{n-1}(n-2) + 1/9, \quad (c_1, a_2) \in \mathbb{Z}^2$$

While the fractional system turns into the following equations

$$^{C}D^{\wp}F(t) = 1 - G(t + F - 1) + \frac{\beta}{1-\beta}|F(t+\wp-1)| - F(t+\wp-1) \tag{9.1.6}$$
$$^{C}D^{\wp}G(t) = F(t+\wp-1) - G(t+\wp-1),$$

where $\wp \in (0,1], t \in \mathbb{N}_{1+\wp}$. Figure 9.3 indicates the solution of the fractional system (9.1.6).

It should be noticed that the fractional ordering for the two fractional differences is identical, resulting in what is commonly referred to as a symmetrical system. By utilizing the integral form (9.1.2), we get (see [17])

$$F(n) = F_0 + \frac{1}{\Gamma(\wp)} \sum_{\iota=1}^{n} \left( \frac{\Gamma(n-\iota+\alpha)}{\Gamma(n-\iota+1)} \right) (1 - G(\iota - 1) + \frac{\beta}{1-\beta} |F(\iota-1)| - F(\iota-1)) \tag{9.1.7}$$

$$G(n) = G_0 + \frac{1}{\Gamma(\wp)} \sum_{\iota=1}^{n} \left( \frac{\Gamma(n-\iota+\wp)}{\Gamma(n-\iota+1)} \right) (F(\iota-1) - G(\iota-1)).$$

The $k$-symbol discrete Caputo operator is given as follows:

$$^{C}D_k^{\wp}\varphi(t) = \frac{1}{k\Gamma_k(u-\wp)} \sum_{m=0}^{t-(u-\wp)} (t-\ell(m))^{(u-\wp-1)/k} D_m^u \varphi(t). \tag{9.1.8}$$

Thus, (9.1.6) becomes

$$^{C}D_k^{\wp}F(t) = 1 - G(t+F-1) + \frac{\beta}{1-\beta}|F(t+\wp-1)| - F(t+\wp-1) \tag{9.1.9}$$

$$^{C}D_k^{\wp}G(t) = F(t+\wp-1) - G(t+\wp-1),$$

And System (9.1.7) can be generalized in terms of $k$-symbol, as follows:

$$F(n) = F_0 + \frac{1}{k\Gamma_k(\wp)} \sum_{\iota=1}^{n} \left( \frac{\Gamma_k(n-\iota+\alpha)}{\Gamma_k(n-\iota+1)} \right) \left(1 - G(\iota-1) + \frac{\beta}{1-\beta}|F(\iota-1)| - F(\iota-1)\right) \tag{9.1.10}$$

$$G(n) = G_0 + \frac{1}{k\Gamma_k(\wp)} \sum_{\iota=1}^{n} \left( \frac{\Gamma_k(n-\iota+\wp)}{\Gamma_k(n-\iota+1)} \right) (F(\iota-1) - G(\iota-1)).$$

This calculus yields the significant contribution of an endless memory, as we just established. This is evident in (9.1.7), where the solution is dependent on all earlier values. Obviously, with the integer order scheme, this is not the case (9.1.3). Figures 9.3–9.5 show the behavior of solution of Systems (9.1.6), where $k = 1$ and (9.1.9) with $k = 1, 2, 3$ and 4.

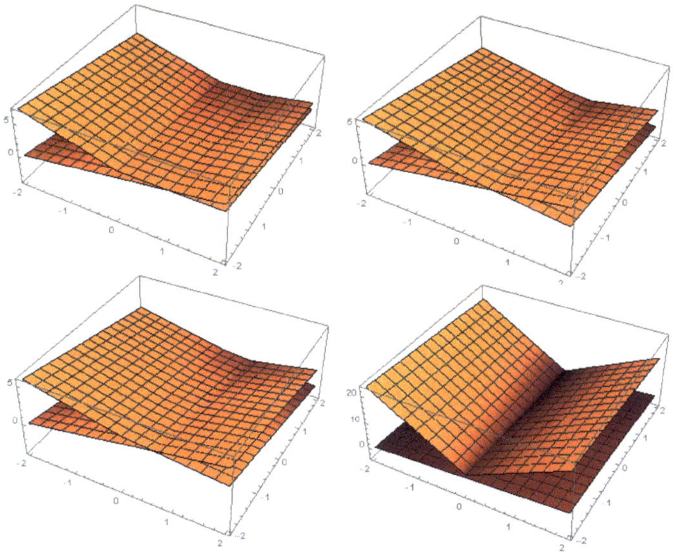

**FIGURE 9.3**. The solution behavior of System (9.1.6), for $\wp = 1, \beta = 0.5, 0.25, 0.75$, and 0.9, respectively.

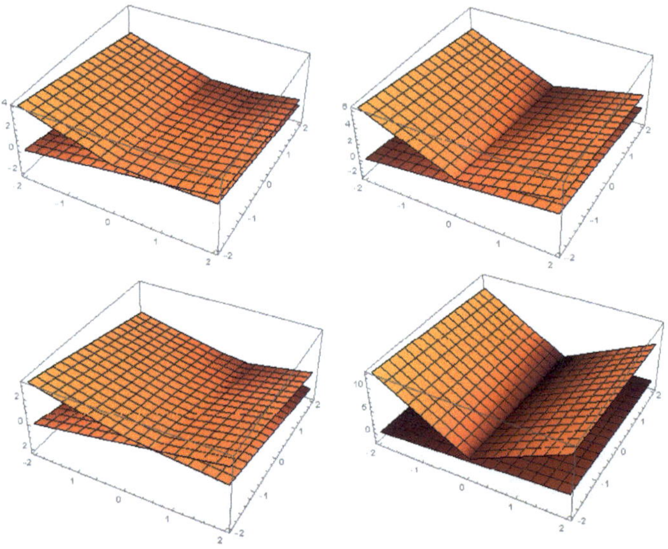

**FIGURE 9.4.** The solution behavior of System (9.1.6), for $\wp = 1/2, \beta = 0.5, 0.75, 0.25$, and $0.9$, respectively.

---

## 9.2 Steady of K-GM

This section examines the stable or steady state of the continuous form of K-GM

$$
\begin{aligned}
{}^{C}D_{k}^{\wp}F(t) &= \frac{\beta}{1-\beta}|F(t)| - F(t) - G(t) + 1 \qquad (9.2.1)\\
{}^{C}D_{k}^{\wp}G(t) &= F(t) - G(t),
\end{aligned}
$$

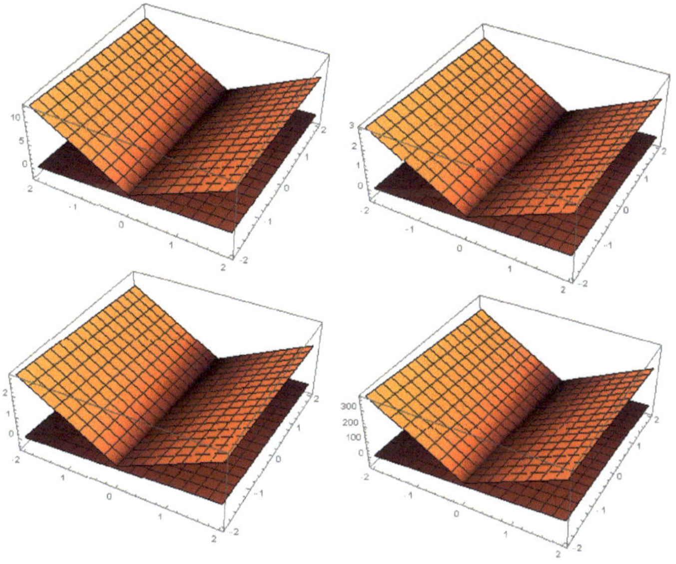

**FIGURE 9.5**. The solution behavior of System (9.1.9), for $\wp = 1/2, \beta = 0.9, k = 1, 2, 3$, and 4, respectively.

where System (9.2.1) can be structured as follows:

$$
\begin{aligned}
{}^{C}D_{k}^{\wp}F(t) &= (\frac{\beta}{1-\beta} - 1)F(t) - G(t) + 1 \\
{}^{C}D_{k}^{\wp}G(t) &= F(t) - G(t),
\end{aligned}
\qquad (9.2.2)
$$

and the equilibrium points $\Sigma_{\beta}\left(\frac{1}{(2-\frac{\beta}{1-\beta})}, \frac{1}{(2-\frac{\beta}{1-\beta})}\right)$, where $2 \neq \frac{\beta}{1-\beta}, \beta \in (0,1)$ and the fixed point

$\Phi_\beta\left(\dfrac{2}{(5-2\frac{\beta}{1-\beta})}, \dfrac{1}{(5-\frac{\beta}{1-\beta})}\right)$, where $5 \neq 2\frac{\beta}{1-\beta}$. Keep in mind that the equilibrium points exist in the initial system. $\Sigma_{0,1}\left(\frac{1}{2}, \frac{1}{2}\right)$, and the fixed point $\Phi\left(\frac{2}{5}, \frac{1}{5}\right)$.

Also, we have the system

$$
\begin{aligned}
{}^{C}D_k^{\wp}F(t) &= -(\frac{\beta}{1-\beta}+1)F(t) - G(t) + 1 \\
{}^{C}D^{\wp}G(t) &= F(t) - G(t),
\end{aligned}
\tag{9.2.3}
$$

with the equilibrium points $\Xi_\beta\left(\dfrac{1}{(2+\frac{\beta}{1-\beta})}, \dfrac{1}{(2+\frac{\beta}{1-\beta})}\right)$, where $2+\frac{\beta}{1-\beta} \neq 0$, and the fixed point $\Phi_\beta\left(\dfrac{2}{(5+\frac{\beta}{1-\beta})}, \dfrac{1}{(5+\frac{\beta}{1-\beta})}\right)$, where $5+\frac{\beta}{1-\beta} \neq 0$ and the original system has the equilibrium points $\Xi\left(\frac{1}{2}, \frac{1}{2}\right)$, with the fixed point $\Phi\left(\frac{2}{5}, \frac{1}{5}\right)$.

### *9.2.1 Stability of K-GM*

**Definition 9.2.1.** *A polynomial is considered to be stable if one or more of its zeros are on the open unit disk (complex-valued zeros) or all of them are situated in the open left half-plane (*$\mathrm{Re}(\lambda) < 0$*).*

The stability conditions of the System (9.2.2) is given in Table 1, where the characteristic polynomial is

$$\Theta_1(\lambda) = \lambda^2 + (2 - \frac{\beta}{1-\beta})\lambda + (2 - \frac{\beta}{1-\beta}).$$

And the characteristic polynomial of System (9.2.3) is

$$\Theta_2(\lambda) = \lambda^2 + (2 + \frac{\beta}{1-\beta})\lambda + (2 + \frac{\beta}{1-\beta}).$$

### *9.2.2 Symmetry of K-GM*

A symmetric stable solution refers to a solution that is both symmetric and stable. Symmetry means that the solution looks the same when viewed from different angles or directions. For example, a sphere is symmetric because it looks the same when viewed from any direction. Stability means that the solution is robust and resistant to small perturbations. In other words, if you slightly change the initial conditions or parameters of the system, the solution should not change too much. A symmetric stable solution is therefore a solution that is both robust and exhibits symmetry. Such solutions are often found in physics, engineering, and other fields where symmetry is an important characteristic of the system. Examples of symmetric stable solutions include the stable equilibrium point of a pendulum, which is symmetric about the vertical axis, and the stable equilibrium configuration of a hydrogen atom, which is symmetric about the nucleus (i.e., $\varphi(u) = \varphi(-u)$). Many fields of study, including physics, chemistry, and biology, employ the idea of a symmetric stable solution for describing systems that combine symmetry and stability.

**Definition 9.2.2.** *A matrix that contains real elements is symmetric if it is a Hermitian matrix. The Hermitian matrix $H$ is stable if and only if $\|H\| \leq 1$, that is, all the eigenvalues are in the unit close interval $|\lambda| \in [0, 1]$.*

For System (9.2.2), the eigenvalues of $H$ are:

$$\lambda_1 = 1/2\left(-\sqrt{(\frac{\beta}{1-\beta})^2 - 4} + \frac{\beta}{1-\beta} - 2\right),$$

$$\lambda_2 = 1/2\left(\sqrt{(\frac{\beta}{1-\beta})^2 - 4} + \frac{\beta}{1-\beta} - 2\right).$$

Thus, the symmetry of System (9.2.2) is valid when $2 \leq \beta/(1-\beta) < 5/2$. That is $2/3 \leq \beta < 1$. Now, for System (9.2.3) , the eigenvalues of $H$ are:

$$\lambda_1 = 1/2\left(-\sqrt{(\frac{\beta}{1-\beta})^2 - 4} - \frac{\beta}{1-\beta} - 2\right),$$

$$\lambda_2 = 1/2\left(\sqrt{(\frac{\beta}{1-\beta})^2 - 4} - \frac{\beta}{1-\beta} - 2\right).$$

Since there is no value of $\beta \in (0, 1)$ satisfies $|\lambda_{1,2}| \in [0, 1]$ then System (9.2.3) is anti-symmetry.

---

## 9.3 Stabilization of K-GM

Stabilization of a dynamic system refers to the process of designing a control system that ensures the system operates in a

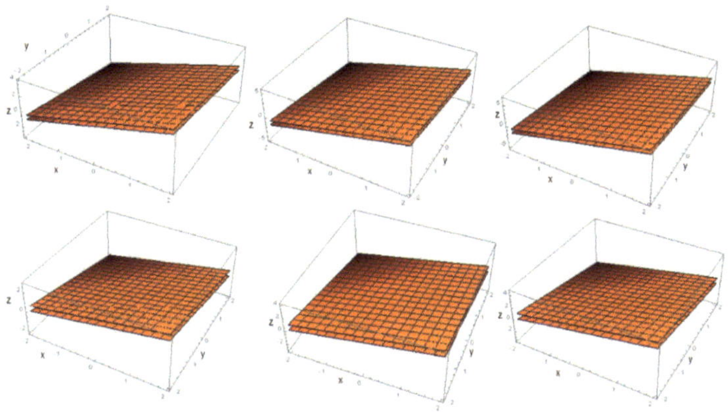

**FIGURE 9.6**. Various plots of the symmetric solution of System (9.2.2), for $k = 1, \beta = 1/2, \wp = 0.5, 0.75, 0.9$, respectively.

stable manner. A dynamic system can be any system that changes over time, such as a chemical reaction, a mechanical system, or an electrical circuit. The stabilization of a dynamic system involves designing a control system that can adjust the system's inputs or parameters to maintain stability in the face of disturbances. The control system can be designed using various techniques such as feedback control, adaptive control, and optimal control. Feedback control is a common technique used for stabilization, where the system's output is measured and compared to a desired reference value. The difference between the output and the reference value is fed back to the system's input, which is then adjusted to reduce the difference. Adaptive control is another technique used for stabilization, where the control system is designed to adjust its parameters based on the changes in the system's dynamics over time. This allows the control system to maintain stability even when the system's parameters change. Optimal control is a technique used for

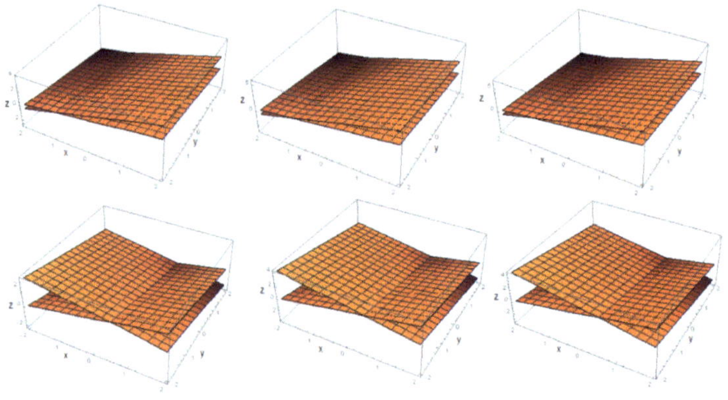

**FIGURE 9.7**. Different plots of the symmetric solution of System (9.2.3), for $k = 1, \beta = 1/2, \wp = 0.5, 0.75, 0.9$, respectively.

stabilization, where the control system is designed to minimize a cost function that represents the system's performance. The cost function can be defined based on various factors such as energy consumption, response time, or error rate. In general, the stabilization of a dynamic system is an important aspect of control engineering and is essential for ensuring that the system operates safely and reliably.

A one-dimensional controller function can use feedback control to control a discrete system. The controller function generates the appropriate control signal, such as a voltage or force, to be fed to the system in order to produce the intended output or behavior. It accepts data about the system's present state, such as its location or velocity. The system is then given this control signal, and the operation is subsequently repeated at specific time intervals. The fundamental principle of feedback regulation is to continuously monitor the state of the system and modify the control signal to produce what is needed.

**Theorem 9.3.1.** *System (9.1.9) can be controlled by the one-dimensional controller function*

$$U_1(G) := G(\wp - 1 + t) - 1,$$

*whenever* $\frac{\beta}{1-\beta} \in (0, 1)$.

*Proof.* System (9.1.6) becomes

$$\begin{aligned}
{}^C D_k^{\wp} F(t) &= 1 - G(F - 1 + t) + \frac{\beta}{1-\beta}|F(\wp - 1 + t)| \\
&\quad - F(t + \wp - 1) + U_1(F) \\
{}^C D_k^{\wp} G(t) &= F(t + \wp - 1) - G(t + \wp - 1).
\end{aligned}$$

Then, we have

$$\begin{aligned}
{}^C D_k^{\wp} F(t) &= \frac{\beta}{1-\beta}|F(\wp - 1 + t)| - F(\wp - 1 + t) \\
{}^C D_k^{\wp} G(t) &= F(\wp - 1 + t) - G(t + \wp - 1),
\end{aligned}$$

which can be seen by the two different systems

$$\begin{aligned}
{}^C D_k^{\wp} F(t) &= \frac{\beta}{1-\beta} F(\wp - 1 + t) - F(\wp - 1 + t) \\
{}^C D_k^{\wp} G(t) &= F(\wp - 1 + t) - G(\wp - 1 + t),
\end{aligned} \tag{9.3.1}$$

and

$$\begin{aligned}
{}^C D_k^{\wp} F(t) &= -\frac{\beta}{1-\beta} F(t + \wp - 1) - F(\wp - 1 + t) \\
{}^C D_k^{\wp} G(t) &= F(\wp - 1 + t) - G(t + \wp - 1).
\end{aligned} \tag{9.3.2}$$

The eigenvalues of (9.3.1) are $\lambda_1 = -1, \lambda_2 = \dfrac{\beta}{1-\beta} - 1$. The system is stable whenever $\dfrac{\beta}{1-\beta} < 1$ (Definition 9.2.1).

Now, the eigenvalues of (9.3.2) are $\lambda_1 = -1, \lambda_2 = -\dfrac{\beta}{1-\beta} - 1$. The system is stable whenever $\dfrac{\beta}{1-\beta} > -1$ $(0 < \beta < 1)$ (Definition 9.2.1).

**Theorem 9.3.2.** *System (9.1.6) can be controlled by the two-dimensional controller function*

$$U_1(G) := G(t + \wp - 1) - 1, \quad U_2(F) := -F(t + \wp - 1)$$

*whenever* $\dfrac{\beta}{1-\beta} \in (0,1)$. *Moreover, it admits a symmetric solution.*

*Proof.* System (9.1.6) becomes

$$\begin{aligned} CD_k^{\wp}F(t) &= 1 - G(t+\wp-1) + \frac{\beta}{1-\beta}|\aleph(t+\wp-1)| \\ &\quad -F(t+\wp-1) + U_1(G) \\ CD_k^{\wp}G(t) &= F(t+\wp-1) - G(t+\wp-1) + U_2(F). \end{aligned}$$

As a consequence, it yields that

$$\begin{aligned} {}^{C}D_k^{\wp}F(t) &= \frac{\beta}{1-\beta}|F(t+\wp-1)| - F(t+\wp-1) \\ {}^{C}D_k^{\wp}G(t) &= -G(t+\wp-1), \end{aligned}$$

which brings the system

$$\begin{aligned} {}^{C}D_k^{\wp}F(t) &= \frac{\beta}{1-\beta}F(t+\wp-1) - F(t+\wp-1) \\ {}^{C}D_k^{\wp}G(t) &= -G(t+\wp-1), \end{aligned} \tag{9.3.3}$$

and

$$
\begin{aligned}
{}^{C}D_{k}^{\wp}F(t) &= -\frac{\beta}{1-\beta}F(t+\wp-1) - F(t+\wp-1) \\
{}^{C}D_{k}^{\wp}G(t) &= -G(t+\wp-1).
\end{aligned}
\tag{9.3.4}
$$

The eigenvalues of (9.3.3) are $\lambda_1 = -1, \lambda_2 = \dfrac{\beta}{1-\beta} - 1$. The system is stable whenever $\dfrac{\beta}{1-\beta} < 1$. Moreover, the matrix $\begin{pmatrix} \dfrac{\beta}{1-\beta} - 1 & 0 \\ 0 & -1 \end{pmatrix}$ is not a Hermite matrix, but it can be replaced by a Hermite matrix, as follows: $\begin{pmatrix} \dfrac{1}{2}\left((\dfrac{\beta}{1-\beta})^* + (\dfrac{\beta}{1-\beta}) - 2\right) & 0 \\ 0 & -1 \end{pmatrix}$. Now the norm of the last matrix gives

$$
\begin{aligned}
\left\| \begin{pmatrix} \frac{1}{2}\left(2\frac{\beta}{1-\beta} - 2\right) & 0 \\ 0 & -1 \end{pmatrix} \right\| &= \max_{1\le j\le n} \sum_{i=1}^{m} |h_{ij}| \\
&= \max\left( |\frac{1}{2}\left(2\frac{\beta}{1-\beta} - 2\right)|, |-1| \right) \\
&= 1.
\end{aligned}
$$

Hence, it is symmetric.

Now, the eigenvalues of (9.3.4) are $\lambda_1 = -1, \lambda_2 = -\dfrac{\beta}{1-\beta} - 1$. The system is stable whenever $\dfrac{\beta}{1-\beta} > 0$. Moreover, the matrix $\begin{pmatrix} -(\dfrac{\beta}{1-\beta} + 1) & 0 \\ 0 & -1 \end{pmatrix}$ is not a Hermite matrix, but it can be

Hermited, as follows:

$$\begin{pmatrix} \frac{1}{2}\left(-(\frac{\beta}{1-\beta})^{*} - (\frac{\beta}{1-\beta}) - 2\right) & 0 \\ 0 & -1 \end{pmatrix}.$$

Now by taking the norm of the last matrix

$$\begin{aligned} &\left\| \begin{pmatrix} \frac{1}{2}\left(-(\frac{\beta}{1-\beta})^{*} - \frac{\beta}{1-\beta} - 2\right) & 0 \\ 0 & -1 \end{pmatrix} \right\| \\ &= \max_{1\le j\le n} \sum_{i=1}^{m} |h_{ij}| \\ &= \max\left( |\frac{1}{2}\left(-(\frac{\beta}{1-\beta})^{*} - \frac{\beta}{1-\beta} - 2\right)|, |-1| \right) \\ &= 1. \end{aligned}$$

Hence, it is symmetric.

**Theorem 9.3.3.** *System (9.1.9) can be controlled by the two-dimensional controller function*

$$U_3(F) := F(\wp - 1 + t) - 1, \quad U_2(G) := -F(\wp - 1 + t)$$

*whenever* $\frac{\beta}{1-\beta} \neq 0$. *Also, the system is not symmetric.*

*Proof.* System (9.1.9) becomes

$$\begin{aligned} {}^{C}D_{k}^{\wp}F(t) &= 1 - G(\wp - 1 + t) + \wp|F(\wp - 1 + t)| - F(\wp - 1 + t) + U_3(F, G) \\ {}^{C}D_{k}^{\wp}G(t) &= F(\wp - 1 + t) - G(\wp - 1 + t) + U_2(F). \end{aligned}$$

Accordingly, we get

$$
\begin{aligned}
{}^{C}D_k^{\wp}F(t) &= \frac{\beta}{1-\beta}|F(\wp-1+t)| - G(\wp-1+t)\\
{}^{C}D_k^{\wp}G(t) &= -G(\wp-1+t).
\end{aligned}
$$

A computation yields the system

$$
\begin{aligned}
{}^{C}D_k^{\wp}F(t) &= \frac{\beta}{1-\beta}F(\wp-1+t) - G(\wp-1+t) \qquad (9.3.5)\\
{}^{C}D_k^{\wp}G(t) &= -G(t+\wp-1),
\end{aligned}
$$

and

$$
\begin{aligned}
{}^{C}D_k^{\wp}F(t) &= -\frac{\beta}{1-\beta}F(\wp-1+t) - G(\wp-1+t) \qquad (9.3.6)\\
{}^{C}D_k^{\wp}G(t) &= -G(t+\wp-1).
\end{aligned}
$$

The eigenvalues of (9.3.5) are $\lambda_1 = -1, \lambda_2 = -\frac{\beta}{1-\beta}$. The system is stable whenever $\wp > 0$. Additionally, the matrix $\begin{pmatrix} \frac{\beta}{1-\beta} & -1 \\ 0 & -1 \end{pmatrix}$ is not a Hermite matrix, but it can be managed as a Hermite matrix, by

$$
\begin{pmatrix} \frac{1}{2}\left((\frac{\beta}{1-\beta})^* + (\frac{\beta}{1-\beta})\right) & -\frac{1}{2} \\ -\frac{1}{2} & -1 \end{pmatrix}.
$$

Next, by considering the norm of the last matrix, we get

$$
\left\| \begin{pmatrix} \frac{1}{2}\left((\frac{\beta}{1-\beta})^* + \frac{\beta}{1-\beta}\right) & -\frac{1}{2} \\ -\frac{1}{2} & -1 \end{pmatrix} \right\|
$$

$$= \max_{1\le j\le n} \sum_{i=1}^{m} |h_{ij}|$$
$$= \max\left( |\frac{1}{2}\left( (\frac{\beta}{1-\beta})^* + \frac{\beta}{1-\beta}\right)| + |-\frac{1}{2}|, |-\frac{1}{2}| + |-1|\right)$$
$$> 1.$$

Hence, the system is not symmetric.

Now, the eigenvalues of (9.3.4) are $\lambda_1 = -1, \lambda_2 = \frac{\beta}{1-\beta}$. The system is stable whenever $\frac{\beta}{1-\beta} > 0$. In addition, the matrix $\begin{pmatrix} -\frac{\beta}{1-\beta} & -1 \\ 0 & -1 \end{pmatrix}$ is not a Hermite matrix, but it can be Hermited, as follows: $\begin{pmatrix} \frac{1}{2}\left(-(\frac{\beta}{1-\beta})^* - \frac{\beta}{1-\beta}\right) & -\frac{1}{2} \\ -\frac{1}{2} & -1 \end{pmatrix}$. Now, by assuming the norm of the last matrix

$$\left\| \begin{pmatrix} \frac{1}{2}\left(-(\frac{\beta}{1-\beta})^* - \frac{\beta}{1-\beta}\right) & -\frac{1}{2} \\ -\frac{1}{2} & -1 \end{pmatrix} \right\| = \max_{1\le j\le n} \sum_{i=1}^{m} |h_{ij}|$$
$$= \max\left( |\frac{1}{2}\left(-(\frac{\beta}{1-\beta})^* - \frac{\beta}{1-\beta}\right)| + |-\frac{1}{2}|, |-\frac{1}{2}| + |-1|\right) > 1.$$

Hence, the system is not symmetric.

**Theorem 9.3.4.** *System (9.1.9) can be controlled by the two-dimensional controller function*

$$U_4(F) := -\frac{\beta}{1-\beta}|F(t+\wp-1)| - 1, \quad U_2(F) := -F(\wp - 1 + t).$$

*Moreover, the system is not symmetric.*

*Proof.* System (9.1.9) becomes

$$
\begin{aligned}
{}^{C}D_{k}^{\wp}F(t) &= 1 - G(\wp - 1 + t) + \wp|\aleph(t + \wp - 1)| - F(t + \wp - 1) + U_4(F, G)\\
{}^{C}D_{k}^{\wp}G(t) &= F(\wp - 1 + t) - G(\wp - 1 + t) + U_2(F).
\end{aligned}
$$

Consequently, we get

$$
\begin{aligned}
{}^{C}D_{k}^{\wp}F(t) &= -F(\wp - 1 + t) - G(\wp - 1 + t)\\
{}^{C}D_{k}^{\wp}G(t) &= -G(t + \wp - 1).
\end{aligned}
$$

The eigenvalues are $\lambda_1 = \lambda_2 = -1$. Therefore, the system is stable. Additionally, the matrix $\begin{pmatrix} -1 & -1 \\ 0 & -1 \end{pmatrix}$ is not a Hermite matrix, but it can be managed into a Hermite matrix, as follows: $\begin{pmatrix} -1 & -\frac{1}{2} \\ -\frac{1}{2} & -1 \end{pmatrix}$.
Now by considering the norm of the last matrix, we obtain

$$
\begin{aligned}
\left\| \begin{pmatrix} -1 & -\frac{1}{2} \\ -\frac{1}{2} & -1 \end{pmatrix} \right\| &= \max_{1 \le j \le n} \sum_{i=1}^{m} |h_{ij}|\\
&= \max \left( |-1| + |-\frac{1}{2}|, |-\frac{1}{2}| + |-1| \right)\\
&> 1.
\end{aligned}
$$

Hence, the system is not symmetric.

## 9.4 Synchronization of FGM

Consider the system (9.1.9), where $\frac{\beta}{1-\beta} \in (-1,1)$. Then the master system becomes

$$
\begin{aligned}
{}^{C}D_{k}^{\wp}F_m(t) &= 1 - G_m(\wp - 1 + t) + \frac{\beta}{1-\beta}|F_m(\wp - 1 + t)| - F_m(\wp - 1 + t) \qquad (9.4.1)\\
{}^{C}D_{k}^{\wp}G_m(t) &= F_m(\wp - 1 + t) - G_m(\wp - 1 + t).
\end{aligned}
$$

The slaves can be assumed by the system

$$
\begin{aligned}
{}^{C}D_{k}^{\wp}F_{\varsigma_1}(t) &= \gamma_1 \qquad (9.4.2)\\
&- G_{\varsigma_1}(\wp - 1 + t) - (1 + \frac{\beta}{1-\beta})F_{\varsigma_1}(\wp - 1 + t) + W_1(\wp - 1 + t)\\
{}^{C}D_{k}^{\wp}G_{\varsigma_1}(t) &= F_{\varsigma_1}(\wp - 1 + t) - G_{\varsigma_1}(\wp - 1 + t) + W_2(\wp - 1 + t),
\end{aligned}
$$

and

$$
\begin{aligned}
{}^{C}D_{k}^{\wp}F_{\varsigma_2}(t) &= \gamma_2 \qquad (9.4.3)\\
&- G_{\varsigma_2}(\wp - 1 + t) + (\frac{\beta}{1-\beta} - 1)F_{\varsigma_2}(\wp - 1 + t) + W_3(\wp - 1 + t)\\
{}^{C}D_{k}^{\wp}G_{\varsigma_2}(t) &= F_{\varsigma_2}(t + \wp - 1) - G_{\varsigma_2}(t + \wp - 1) + W_4(t + \wp - 1).
\end{aligned}
$$

In the systems mentioned above, where $W_i, i = 1, 2, 3, \&4$, are the synchronization controllers, we seek to identify the parameter values that are combination synchronized.

**Definition 9.4.1.** *If there are controllers that adhere to the following limitation, the master system and slave systems are said to be compositely synchronized.*

$$\lim_{t\to\infty}\left\|\begin{pmatrix}\mathsf{T}_1(t)\\ \mathsf{T}_2(t)\end{pmatrix}\right\| = \lim_{t\to\infty}\left\|\begin{pmatrix}F_{\varsigma_1}(t)+F_{\varsigma_2}(t)\\ G_{\varsigma_1}(t)+G_{\varsigma_2}(t)\end{pmatrix} - \begin{pmatrix}F_m(t)\\ G_m(t)\end{pmatrix}\right\| = 0$$

We have the following result

**Theorem 9.4.1.** *Assume the master and the slave systems* (9.4.1), (9.4.2) *and* (9.4.3). *By the following 2D-controllers*

$$W_1(t) = -\gamma_1 + \frac{1}{2} + \frac{\beta}{1-\beta}|F_m(t+\wp-1)| + \frac{\beta}{1-\beta}F_{\varsigma_1}(t+\wp-1)$$
$$W_2(t) = 0$$
$$W_3(t) = -\gamma_2 + \frac{1}{2} - \frac{\beta}{1-\beta}F_{\varsigma_2}(t+\wp-1)$$
$$W_4(t) = 0$$

*the master frame* (9.4.1) *and the frames* (9.4.2) *and* (9.4.3) *are combination synchronized.*

*Proof.* The error fractional system becomes

$$\begin{aligned}
{}^C D_k^{\wp}\mathsf{T}_1(t) &= (\gamma_1+\gamma_2-1) - [G_{\varsigma_1}\wp - 1 + t) + \flat_{\varsigma_2}(\wp-1+t)\\
&\quad - G_m(\wp-1+t)] + [-(1+\frac{\beta}{1-\beta})F_{\varsigma_1}(\wp-1+t)\\
&\quad + (\frac{\beta}{1-\beta}-1)F_{\varsigma_2}(\wp-1+t) - \frac{\beta}{1-\beta}|F_m(\wp-1+t)|\\
&+ F_m(t+\wp-1)] \qquad + W_1(\wp-1+t) + W_3(\wp-1+t)\\
{}^C D_k^{\wp}\mathsf{T}_2(t) &= [F_{\varsigma_1}(\wp-1+t) + F_{\varsigma_2}(\wp-1+t) - F_m(\wp-1+t)]
\end{aligned}$$

$$- [G_{\varsigma_1}(\wp - 1 + t) + G_{\varsigma_2}(\wp - 1 + t) - G_m(\wp - 1 + t)]$$
$$+ W_2(\wp - 1 + t) + W_4(\wp - 1 + t).$$

The decreased dynamics outcome of the aforesaid system with the controllers changed is as follows:

$$\begin{aligned}
{}^C D_k^{\alpha} \mathsf{T}_1(t) &= (\gamma_1 + \gamma_2 - 1) - [G_{\varsigma_1}(\wp - 1 + t) + G_{\varsigma_2}(\wp - 1 + t) \\
&- G_m(\wp - 1 + t)] + [-(1 + \frac{\beta}{1-\beta})F_{\varsigma_1}(\wp - 1 + t) \\
&+ (\frac{\beta}{1-\beta} - 1)F_{\varsigma_2}(\wp - 1 + t) - \frac{\beta}{1-\beta}|F_m(\wp - 1 + t)| \\
&+ F_m(\wp - 1 + t)] + (-\gamma_1 + \frac{1}{2} + \frac{\beta}{1-\beta}|F_m(\wp - 1 + t)| \\
+ \frac{\beta}{1-\beta}&F_{\varsigma_1}(\wp - 1 + t)) \\
&+ \left(-\gamma_2 + \frac{1}{2} - \frac{\beta}{1-\beta}F_{\varsigma_2}(t + \wp - 1)\right) \\
{}^C D_k^{\wp} \mathsf{T}_2(t) &= [F_{\varsigma_1}(t + \wp - 1) + F_{\varsigma_2}(t + \wp - 1) - \aleph_m(t + \wp - 1)] \\
&- [G_{\varsigma_1}(t + \wp - 1) + G_{\varsigma_2}(t + \wp - 1) - G_m(t + \wp - 1)].
\end{aligned}$$

As a result, we are able to create the error fractional system.

$$\begin{aligned}
{}^C D_k^{\wp} \mathsf{T}_1(t) &= -\mathsf{T}_1(t) - \mathsf{T}_2(t) \\
{}^C D_k^{\wp} \mathsf{T}_2(t) &= \mathsf{T}_1(t) - \mathsf{T}_2(t),
\end{aligned} \tag{9.4.4}$$

with the characteristic polynomial $p(\lambda) = 2 + 2\lambda + \lambda^2$. Thus, the solutions are $\lambda_1 = -i - 1$ and $\lambda_2 = i - 1$.

**Theorem 9.4.2.** *Suppose that the master and the slave systems* (9.4.1), (9.4.2) *and* (9.4.3) *are under the following 3D-controllers*

$$
\begin{aligned}
W_1(t) &= -\gamma_1 + \frac{1}{2} + \frac{\beta}{1-\beta}|F_m(t+\wp-1)| + \frac{\beta}{1-\beta}F_{\varsigma_1}(t+\wp-1) \\
W_2(t) &= \gamma[G_{\varsigma_1}(t+\wp-1) + G_{\varsigma_2}(t+\wp-1) - G_m(t+\wp-1)] \\
W_3(t) &= -\gamma_2 + \frac{1}{2} - \frac{\beta}{1-\beta}F_{\varsigma_2}(t+\wp-1) \\
W_4(t) &= 0.
\end{aligned}
$$

*Then the master frame* (9.4.1) *and the frames* (9.4.2) *and* (9.4.3) *are combination synchronized whenever* $\gamma \leq -2$.

*Proof.* The error fractional system is indicated by

$$
\begin{aligned}
{}^C D_k^{\wp} \mathsf{T}_1(t) &= (\gamma_1 + \gamma_2 - 1) - [G_{\varsigma_1}(t+\wp-1) + G_{\varsigma_2}(t+\wp-1) \\
&\quad - G_m(t+\wp-1)] + [-(1+\frac{\beta}{1-\beta})F_{\varsigma_1}(t+\wp-1) \\
&\quad + (\frac{\beta}{1-\beta} - 1)F_{\varsigma_2}(t+\wp-1) - \frac{\beta}{1-\beta}|F_m(t+\wp-1)| \\
&\quad + F_m(t+\wp-1)] + W_1(t+\wp-1) + W_3(t+\wp-1) \\
{}^C D_k^{\wp} \mathsf{T}_2(t) &= [F_{\varsigma_1}(t+\wp-1) + F_{\varsigma_2}(t+\wp-1) - F_m(t+\wp-1)] \\
&\quad - [G_{\varsigma_1}(t+\wp-1) + G_{\varsigma_2}(t+\wp-1) - G_m(t+\wp-1)] \\
&\quad + W_2(t+\wp-1) + W_4(t+\wp-1).
\end{aligned}
$$

When the controls are applied, the dynamics of the prior system diminish as follows:

$$
\begin{aligned}
{}^C D_k^{\wp} \mathsf{T}_1(t) &= (\gamma_1 + \gamma_2 - 1) \\
&\quad - [G_{\varsigma_1}(\wp-1+t) + G_{\varsigma_2}(\wp-1+t) - G_m(\wp-1+t)]
\end{aligned}
$$

$$+[-(1+\frac{\beta}{1-\beta})F_{\varsigma_1}(\wp-1+t)+(\frac{\beta}{1-\beta}-1)F_{\varsigma_2}(\wp-1+t)$$
$$-\frac{\beta}{1-\beta}|F_m(\wp-1+t)|+F_m(\wp-1+t)]$$
$$+\left(-\gamma_1+\frac{1}{2}+\frac{\beta}{1-\beta}|F_m(t+\wp-1)|+\frac{\beta}{1-\beta}F_{\varsigma_1}(t+\wp-1)\right)$$
$$+\left(-\gamma_2+\frac{1}{2}-\frac{\beta}{1-\beta}F_{\varsigma_2}(\wp-1+t)\right)$$
$${}^C D_k^{\wp}\top_2(t)=[F_{\varsigma_1}(\wp-1+t)+F_{\varsigma_2}(\wp-1+t)-F_m(\wp-1+t)]$$
$$-[G_{\varsigma_1}(\wp-1+t)+G_{\varsigma_2}(\wp-1+t)-G_m(\wp-1+t)]$$
$$+\gamma[G_{\varsigma_1}(\wp-1+t)+G_{\varsigma_2}(\wp-1+t)-G_m(\wp-1+t)]$$

Thus, we have the error fractional system.

$$\begin{aligned}{}^C D_k^{\wp}\top_1(t)&=-\top_1(t)-\top_2(t)\\ {}^C D_k^{\wp}\top_2(t)&=\top_1(t)-(1-\gamma)[\top_2(t)],\end{aligned} \tag{9.4.5}$$

with the characteristic polynomial $p(\lambda) = 2+(2-\gamma)\lambda+\lambda^2$. Consequently, the negative solutions are

$$\lambda_1=1/2(-\sqrt{\gamma^2-4}+\gamma-2),\quad \lambda_2=1/2(\sqrt{\gamma^2-4}+\gamma-2),$$

whenever $\gamma\le -2$.

**Theorem 9.4.3.** *Assume that the master and the slave systems* (9.4.1), (9.4.2) *and* (9.4.3) *are under the following 4D-controllers*

$$W_1(t)=-\gamma_1+\frac{1}{2}+\frac{\beta}{1-\beta}|F_m(t+\wp-1)|+\frac{\beta}{1-\beta}F_{\varsigma_1}(t+\wp-1)$$
$$W_2(t)=[G_{\varsigma_1}(t+\wp-1)-G_m(t+\wp-1)]$$

$$W_3(t) = -\gamma_2 + \frac{1}{2} - \frac{\beta}{1-\beta} F_{\varsigma_2}(t+\wp-1)$$

$$W_4(t) = [G_{\varsigma_2}(t+\wp-1)].$$

*Then the master frame (9.4.1) and the frames (9.4.2) and (9.4.3) are combination synchronized*

*Proof.* The error fractional system is as follows:

$$\begin{aligned}
{}^{C}D_{k}^{\wp}\mathsf{T}_1(t) &= (\gamma_1+\gamma_2-1)\\
&- [G_{\varsigma_1}(\wp-1+t) + G_{\varsigma_2}(\wp-1+t) - G_m(\wp-1+t)]\\
&+ [-(1+\frac{\beta}{1-\beta})F_{\varsigma_1}(t+\wp-1) + (\frac{\beta}{1-\beta}-1)F_{\varsigma_2}(\wp-1+t)\\
&- \frac{\beta}{1-\beta}|F_m(\wp-1+t)| + F_m(\wp-1+t)]\\
&+ W_1(\wp-1+t) + W_3(t+\wp-1)\\
{}^{C}D_{k}^{\wp}\mathsf{T}_2(t) &= [F_{\varsigma_1}(\wp-1+t) + F_{\varsigma_2}(\wp-1+t) - F_m(\wp-1+t)]\\
&- [G_{\varsigma_1}(t+\wp-1) + G_{\varsigma_2}(t+\wp-1) - G_m(\wp-1+t)]\\
&+ W_2(t+\wp-1) + W_4(\wp-1+t).
\end{aligned}$$

When the controllers are substituted into the above system, the reduced dynamics result is as follows:

$$\begin{aligned}
{}^{C}D_{k}^{\wp}\mathsf{T}_1(t) &= (\gamma_1+\gamma_2-1)\\
&- [G_{\varsigma_1}(\wp-1+t) + G_{\varsigma_2}(\wp-1+t) - G_m(\wp-1+t)]\\
&+ [-(1+\frac{\beta}{1-\beta})F_{\varsigma_1}(t+\wp-1) + (\frac{\beta}{1-\beta}-1)F_{\varsigma_2}(\wp-1+t)\\
&- \frac{\beta}{1-\beta}|F_m(\wp-1+t)| + F_m(\wp-1+t)]\\
&+ \left(-\gamma_1 + \frac{1}{2} + \frac{\beta}{1-\beta}|F_m(t+\wp-1)| + \frac{\beta}{1-\beta}F_{\varsigma_1}(\wp-1+t)\right)
\end{aligned}$$

$$+\left(-\gamma_2+\frac{1}{2}-\frac{\beta}{1-\beta}F_{\varsigma_2}(\wp-1+t)\right)$$

$$\begin{aligned}{}^{C}D_k^{\alpha}\top_2(t) &= [F_{\varsigma_1}(\wp-1+t)+F_{\varsigma_2}(\wp-1+t)-F_m(\wp-1+t)]\\ &\quad-[G_{\varsigma_1}(\wp-1+t)+G_{\varsigma_2}(\wp-1+t)-G_m(\wp-1+t)]\\ &\quad+[G_{\varsigma_1}(\wp-1+t)+G_{\varsigma_2}(\wp-1+t)-G_m(\wp-1+t)].\end{aligned}$$

As a result, we get the error fractional system

$$\begin{aligned}{}^{C}D_k^{\wp}\top_1(t) &= -\top_1(t)-\top_2(t)\\ {}^{C}D_k^{\wp}\top_2(t) &= \top_1(t),\end{aligned} \tag{9.4.6}$$

with the characteristic polynomial $p(\lambda) = 1+\lambda+\lambda^2$. Thus, the negative solutions are

$$\lambda_1 = 1/2(-1+i\sqrt{(3)}), \quad \lambda_2 = 1/2(-1-i\sqrt{(3)}).$$

---

## 9.5 Conclusion and Discussion

This study acknowledged the generalization of the gingerbread-man map using fractional calculus for the first time. We put forth two distinct systems: a dynamic fractional flow map (K-GM) and a two-dimensional chaotic map based on a fractional gingerbread-man map. We looked into some of these maps' most important dynamics. In addition, we investigated the stability requirements for the variable fractional dynamic systems.

As an outcome based on the stability of the system's defining polynomial, we provided K-GM with requirements for stability solutions. The relevant requirements for symmetry behavior are also taken into account by the system's Hermite grid-based follow-up. It was also suggested to incorporate these maps into control rules with system stabilization as the end goal.

The 1D and 2D controller laws were assumed in the above calculation. The symmetry and stability of the controller system solutions were examined. We also provided asymmetric stability control options. Asymptotically stable solutions of the Caputo difference equation can be studied when $-2^{\wp} < \lambda < 0$ (see Corollary 4.1 in [2]). For example, in Theorem 9.3.1, the stability is given when $\frac{\beta}{1-\beta} < 1$, then based on the above inequality, we have that the system is asymptotically stable satisfying the fractional power inequality.

$$\frac{\beta}{1-\beta} < 1, \quad \frac{\log(1 - \frac{\beta}{1-\beta})}{\log(2)} < \wp < 1.$$

Theorem 9.3.4 indicated the situation when $\lambda_{1,2} = -1$, which satisfy the condition of asymptotically stable whenever

$$\lambda := \lambda_{1,2} = -1, \quad \frac{\log(-\lambda)}{\log(2)} < \wp < 1.$$

Future works might address functional fractional power, quantum calculus, double quantum calculus, or any other sort of fractional calculus.

# References


[1] Abdeljawad, T., (2011). On Riemann and Caputo fractional differences. *Computers & Mathematics with Applications, 62*(3), 1602–1611.

[2] Cermak, J., Gyori, I., & Nechvatal, L., (2015). On explicit stability conditions for a linear fractional difference system. *Fractional Calculus and Applied Analysis, 18*, 651–672.

[3] Diaz, J. B., & Osler, T. J., (1974). Differences of fractional order. *Mathematics of Computation, 28*(125), 185–202.

[4] Diaz, R., & Pariguan, E., (2004). *On hypergeometric functions and Pochhammer k-symbol.* arXiv preprint math/0405596.

[5] Durafe, A., & Patidar, V., (2022). *Comparative analysis of chaotic image encryption using improved 2D Gingerbreadman map and Chirikov standard map.* In 2022 International Conference for Advancement in Technology (ICONAT) (pp. 1–6). IEEE.

[6] Hadid, S. B., & Ibrahim, R. W., (2022). Geometric Study of 2D-Wave Equations in View of K-Symbol Airy Functions. *Axioms, 11*(11), 590.

[7] Hao, R., & Ma, X., (2019). *Dynamical Analysis of Nose-Hoover Continuous Chaotic System Based on Gingerbreadman Discrete Chaotic Sequence.* In Green Energy and Networking: 6th EAI International Conference, GreeNets 2019, Dalian, China, May 4, 2019, Proceedings 6 (pp. 170–180). Springer International Publishing.

[8] Ibrahim, R. W., (2022). K-symbol fractional-order discrete-time models of Lozi system. *Journal of Difference Equations and Applications*, 1–20.

[9] Ibrahim, R. W., & Baleanu, D., (2022). Global stability of local fractional Henon-Lozi map using fixed point theory. *AIMS Math., 7*(6), 11399–11416.

[10] Ibrahim, R. W., (2022). K-symbol fractional-order discrete-time models of Lozi system. *Journal of Difference Equations and Applications*, 1–20.

[11] Khan, F. A., Ahmed, J., Khan, J. S., Ahmad, J., & Khan, M. A., (2017). A novel image encryption based on Lorenz equation, Gingerbreadman chaotic map and S 8 permutation. *Journal of Intelligent and Fuzzy Systems, 33*(6), 3753–3765.

[12] Khan, M., & Asghar, Z., (2018). A novel construction of substitution box for image encryption applications with Gingerbreadman chaotic map and S 8 permutation. *Neural computing and applications, 29*, 993–999.

[13] Ledesma, C. E. T., & Nyamoradi, N., (2023). (k, $\psi$)-Hilfer impulsive variational problem. *Revista de la Real Academia de Ciencias Exactas, Fisicas y Naturales. Serie A. Matematicas, 117*(1), 42.

[14] Peitgen, H. O., Jürgens, H., & Saupe, D., (2013). *Fractals for the classroom: part one introduction to fractals and chaos.* Springer Science & Business Media, New York.

[15] Suthar, D., Baleanu, D., Purohit, S., & Ucar, F., (2020). Certain k-fractional calculus operators and image formulas of k-Struve function. *AIMS Mathematics, 5*(3), 1–14.

[16] Yildiz, C., & Cotirla, L. I., (2023). Examining the Hermite-Hadamard Inequalities for k-Fractional Operators Using the Green Function. *Fractal and Fractional, 7*(2), 161.

[17] Wu, G. C., & Baleanu, D., (2014). Discrete fractional logistic map and its chaos. *Nonlinear Dynamics, 75*, 283–287.

[18] Wu, G. C., Baleanu, D., Xie, H. P., & Chen, F. L., (2016). Chaos synchronization of fractional chaotic maps based on the stability condition. *Physica A: Statistical Mechanics and Its Applications, 460*, 374–383.

# Chapter 10

## An Improved Numerical Algorithm for Solving Multi-Order Fractional Differential Equations

**Zulfiqar Ahamad Noor,[1] Imran Talib,[2] and Muhammad Bilal Riaz[1,3]**

[1]*Department of Mathematics, University of Management and Technology, Lahore, Pakistan*

[2]*Nonlinear Analysis Group (NAG), Mathematics Department, Virtual University of Pakistan, Pakistan*

[3]*IT4Innovations, VŠB–Technical University of Ostrava, Ostrava, Poruba, Czech Republic*

**Abstract**

This chapter extends a highly efficient computing technique based on newly developed operational matrices involving fractional integral and derivative operators of Vieta-Lucas function vectors (VLFVs). In contrast to the spectral collocation approach, our method does not depend on specific collocation point choices for solving fractional differential equations (FDEs). Additionally, it eliminates the need to expand residual functions as orthogonal polynomial series, a requirement in the spectral Tau method. Consequently, our proposed method demonstrates superior efficiency when compared to other methods documented in existing literature. Notably, we introduce a novel integral operational matrix for the Vieta-Lucas polynomial in the Riemann-Liouville context as a significant contribution to this work. Our computational approach facilitates the conversion of

FDEs into a system of Sylvester-type matrix equations, efficiently solvable using available software like MATLAB. To validate the accuracy of our proposed numerical method, we compare it with exact solutions and achieve promising results in comparison with other numerical solvers introduced in existing literature.

---

## Introduction

Fractional calculus (FC) explores derivatives and integrals of arbitrary order. In its early stages, FC was regarded as an abstract mathematical concept with few practical applications. However, the current perception of FC has drastically changed, as it has become a highly valuable and significant topic within the scientific community. Fractional derivatives have proven to be effective in describing various phenomena in science and engineering, such as diffusion models, earthquake dynamics, viscoelastic systems, biological systems, chaos, and wave propagation [2, 7, 8, 15, 25, 29, 30].

The study of classical (integer-order) and fractional-order differential equations, incorporating nonhomogeneous initial conditions and generalized source terms, presents a challenge when seeking their analytical solutions. Consequently, researchers have turned to reliable numerical methods for tackling these complex problems. For instance, in [28, 35], the authors employed numerical techniques to address Caputo-type tempered fractional integrodifferential equations. A noteworthy contribution by Zeidan [33] introduced the characteristic Adomian decomposition method for solving various types of Riemann problems. Additionally, Goncalves [10] explored cavitation models with thermodynamic effects through the development of a Navier-Stokes solver. In [16], computational examinations of the flow model were studied.

Spectral collocation and spectral Tau methods in conjunction with the operational matrices of orthogonal polynomials have been effectively utilized to solve a wide class of FDEs with various initial and boundary conditions. For example, Saadatmandi and Dehghan combined operational matrices with spectral collocation and Tau methods to address multi-order FDEs (MOFDEs) [26]. Doha et al. [9] utilized the Chebyshev operational matrix with spectral methods to solve MOFDEs. Similarly, Atabakzadeh [4] reduced MOFDEs to a system of FDEs and employed operational matrices with spectral methods for solving them. Zhang [34] employed the derivative operational matrix of FLFVs in the Caputo context in conjunction with spectral Tau methods to address MOFDEs. Additionally, Bhrawy and Zaky [5] developed the integral operational matrix within the Riemann-Liouville framework and utilized it alongside spectral Tau methods to numerically solve MOFDEs.

In this work, we have employed a numerical approach exclusively based on the operational matrices of VLFVs. This method operates independently of collocation point selection and the expansion of the residual function into orthogonal polynomials. Consequently, it yields highly efficient numerical results compared to existing techniques such as spectral Tau methods [26], Bessel collocation methods [32], and stochastic techniques [23]. Additionally, our article introduces a novel fractional-order integral operational matrix in the Riemann-Liouville sense. To demonstrate the method's efficacy, we apply it to solve various MOFDEs with both homogeneous and nonhomogeneous initial conditions, considering various source terms. Our approach entails the transformation of MOFDEs into a set of matrix equations resembling Sylvester-type equations, which can be efficiently tackled using MATLAB. The subsequent sections of this chapter are structured as follows:

Preliminary definitions of FC are covered in Section 10.1. Section 10.2 reviews analytical expressions and useful properties of Vieta-Lucas Polynomials (VLPs), along with important properties of VLFVs and their analytical expressions. Section 10.3 includes the statement of the derivative operational matrix in the Caputo sense. Moreover, Section 10.3 includes the derivation of the newly developed fractional-order integral operational matrix in the Riemann-Liouville sense. Section 10.4 presents a numerical method based on the results presented in the previous sections. Section 10.5 examines the efficiency and validity of the proposed numerical method while also providing a comparative analysis with alternative methods published in the existing literature.

---

## 10.1 Basic Definitions

This section lists fundamental definitions of fractional-order operators that will be utilized throughout this chapter.

**Definition 10.1.1.** *The Riemann–Liouville fractional-order integral extends the notion of integration to include integrals of functions with non-integer orders. It's defined as:*

$$ {}_{\mathrm{RL}}J^{\alpha}_{b^+} f(\xi) = \frac{1}{\Gamma(\alpha)} \int_b^{\xi} (\xi - \vartheta)^{\alpha-1} f(\vartheta)\, d\vartheta, \quad \xi > b, \quad \alpha > 0. \tag{10.1.1} $$

**Definition 10.1.2.** *The fractional-order derivatives operators in Riemann–Liouville's sense and Caputo's sense are defined in the following way:*

$$ \begin{aligned} {}_{\mathrm{RL}}\mathrm{D}^{\alpha}_{b^+} f(\xi) &= D^{m}_{\mathrm{RL}} J^{m-\alpha}_{b^+} f(\xi) \\ &= \frac{1}{\Gamma(m-\alpha)} \frac{\mathrm{d}^{\mathrm{m}}}{\mathrm{d}\xi^{m}} \int_b^{\xi} (\xi - \vartheta)^{m-\alpha-1} f(\vartheta)\, \mathrm{d}\vartheta, \quad \xi > b, \\ {}_{\mathrm{C}}\mathrm{D}^{\alpha}_{b^+} f(\xi) &= {}_{\mathrm{RL}} J^{m-\alpha}_{b^+} D^{m} f(\xi) \\ &= \frac{1}{\Gamma(m-\alpha)} \int_b^{\xi} (\xi - \vartheta)^{m-\alpha-1} f^{(m)}(\vartheta)\, \mathrm{d}\vartheta, \quad \xi > b, \end{aligned} \tag{10.1.2} $$

*where* $m-1 < \alpha < m,\ m \in \mathbb{N}$, *and* $\alpha > 0$.

## 10.2 Vieta-Lucas Polynomials

Vieta-Lucas Polynomials (VLPs) are among the new types of orthogonal polynomials that had been proposed by Vladimir P. Agarwal in 2020 [1]. It is defined as:

$$\mathrm{VL}_m(\xi) = 2\cos(m\theta), \quad \theta = \cos^{-1}\left(\frac{\xi}{2}\right), \quad \theta \in [0,\pi]. \tag{10.2.3}$$

The following recurrence relation can be used to represent VLPs:

$$\mathrm{VL}_m(\xi) = \xi\,\mathrm{VL}_{m-1}(\xi) - \mathrm{VL}_{m-2}(\xi), \quad m = 2,3,\ldots,$$
$$\mathrm{VL}_0(\xi) = 2, \quad \mathrm{VL}_1(\xi) = \xi.$$

Also, $\mathrm{VL}_m(\xi)$ can be written as the power series expression:

$$\mathrm{VL}_m(\xi) = \sum_{r=0}^{\lceil \frac{m}{2} \rceil}(-1)^r \frac{m\Gamma(m-r)}{\Gamma(r+1)\Gamma(m+1-2r)}\xi^{m-2r}, \quad m = \{2,3,\ldots\}. \tag{10.2.4}$$

where, $\lceil . \rceil$ is the ceiling function.

Shifted VLPs, $\mathrm{VL}^*_m(\xi)$ of degree $m$, which is defined on the interval $[0,1]$ can also be obtained as follows:

$$\mathrm{VL}^*_m(\xi) = \mathrm{VL}_m(4\xi - 2) = \mathrm{VL}_{2m}(2\sqrt{\xi}). \tag{10.2.5}$$

Also, $\mathrm{VL}^*_m(\xi)$ can be expressed as recurrence relation:

$$\mathrm{VL}^*_{m+1}(\xi) = (4\xi - 2)\,\mathrm{VL}^*_m(\xi) - \mathrm{VL}^*_{m-1}(\xi), \quad m = 1,2,\ldots, \tag{10.2.6}$$

with the starting values

$$\mathrm{VL}^*_0(\xi) = 2, \; \mathrm{VL}^*_1(\xi) = 4\xi - 2. \tag{10.2.7}$$

Moreover, analytically $\mathrm{VL}^*_m(\xi)$ can be expressed as:

$$\mathrm{VL}_m^*(\xi) = 2n\sum_{j=0}^{m}(-1)^j \frac{4^{m-j}\Gamma(2m-j)}{\Gamma(j+1)\Gamma(2m-2j+1)}\xi^{m-j}, \quad m = \{2,3,\ldots\}. \tag{10.2.8}$$

Consider a square-integrable function $u(\xi)$ defined on the interval $[0,1]$, which can be represented in relation to $\mathrm{VL}_m(\xi)$ as:

$$u(\xi) = \sum_{j=0}^{\infty} c_j\,\mathrm{VL}_j^*(\xi), \qquad j = 0,1,2,\ldots,n \tag{10.2.9}$$

where the coefficients, $c_j$ can be evaluated as follows:

$$c_j = \frac{1}{\delta_j \pi}\int_{-2}^{2} \frac{u(\frac{\xi+2}{4})\,\mathrm{VL}_j(\xi)}{\sqrt{4-\xi^2}}\,d\xi, \tag{10.2.10}$$

or

$$c_j = \frac{1}{\delta_j \pi}\int_{0}^{1} \frac{u(\xi)\,\mathrm{VL}_j^*(\xi)}{\sqrt{\xi-\xi^2}}\,d\xi, \tag{10.2.11}$$

where

$$\delta_j = \begin{cases} 4, & j = 0, \\ 2, & j = \{1,2,\ldots,m\}. \end{cases} \tag{10.2.12}$$

If we take the first $m+1$ terms of the series, then $u(\xi)$ can be written as:

$$u_m(\xi) \simeq \sum_{j=0}^{m} c_j\,\mathrm{VL}_j^*(\xi) = C^T\Psi(\xi), \tag{10.2.13}$$

### *10.2.1 Vieta-Lucas Approximation of Functions*

In this section, we discuss the framework for approximating the elements of $L^2[0,1]$ spaces (space of square-integrable functions) via the bases of Vieta-Lucas polynomials.

Any $f \in L^2[0,1]$ can be approximated by using the bases of Vieta-Lucas polynomials as:

$$f(\xi) \simeq \sum_{q=0}^{M} a_q \, \mathrm{VL}_q^*(\xi). \tag{10.2.14}$$

The series coefficients $a_q$ can be easily computed by exploiting (10.2.11) as:

$$a_q = (2q+1) \int_0^1 f(\xi) \, \mathrm{VL}_q^*(\xi) d\xi, \; q = 0, 1, \cdots, M. \tag{10.2.15}$$

Equation (10.2.14) can also be expressed in vector form as:

$$f(\xi) = \mathbf{A}^T \Psi(\xi), \tag{10.2.16}$$

where, $\mathbf{A}^T = [a_0, a_1, \cdots, a_M]$ and $\Psi(\xi) = [\mathrm{VL}_0^*, \mathrm{VL}_1^*, \cdots, \mathrm{VL}_M^*]^T$.

For example, if $f(\xi) = \exp(\xi)$, then using (10.2.15), we have the following series coefficients for various $M$

$$\mathbf{A}^T = \begin{cases} [\; 0.4366 \quad 3.0000 \;], & M = 1, \\ [\; 0.3377 \quad 1.2258 \quad 3.5914 \;], & M = 2, \\ [\; 0.2498 \quad 0.8511 \quad 1.8282 \quad 3.9441 \;], & M = 3. \end{cases}$$

## 10.3 Vieta-Lucas Operational Matrix of Differentiation and Integration

In this section, we introduce the Vieta-Lucas operational matrix of differentiation in Caputo and integration in the Riemann–Liouville sense.

**Theorem 10.3.1.** *The fractional derivative of order $\alpha$ of $VL_i^*(x)$ can be expressed as*

$${}_{C}\mathrm{D}^{\alpha}(VL_i^*(x)) \simeq \mathbf{D}_{M\times M}^{\alpha}\Psi(x), \quad M = m+1,$$

$$\mathbf{D}^{\alpha}=\begin{pmatrix} 0 & 0 & \cdots & 0 \\ \vdots & \vdots & \cdots & \vdots \\ 0 & 0 & \cdots & 0 \\ \sum\limits_{s=0}^{l-\lceil\alpha\rceil}\xi_{l,0,s} & \sum\limits_{s=0}^{l-\lceil\alpha\rceil}\xi_{l,1,s} & \cdots & \sum\limits_{s=0}^{l-\lceil\alpha\rceil}\xi_{l,m,s} \\ \vdots & \vdots & \cdots & \vdots \\ \sum\limits_{s=0}^{m-\lceil\alpha\rceil}\xi_{m,0,s} & \sum\limits_{s=0}^{m-\lceil\alpha\rceil}\xi_{m,1,s} & \cdots & \sum\limits_{s=0}^{m-\lceil\alpha\rceil}\xi_{m,m,s} \end{pmatrix},$$

*and,* $\xi_{l,j,s}$ *is given by*

$$\xi_{l,j,s}=\begin{cases} l\sum_{s=0}^{l-\lceil\alpha\rceil}(-1)^{s}\frac{4^{l-s}\Gamma(2l-s)\Gamma(l-s+1)\Gamma(l-s-\alpha+1/2)}{\sqrt{\pi}\Gamma(s+1)\Gamma(2l-2s+1)\Gamma(l-s-\alpha+1)^{2}}, & j=0, \\ 2l\sum_{s=0}^{l-\lceil\alpha\rceil}\sum_{r=0}^{j}\frac{(-1)^{s+r}}{\sqrt{\pi}}\frac{4^{l-s}\Gamma(2l-s)\Gamma(l-s+1)}{\Gamma(s+1)\Gamma(2l-2s+1)\Gamma(l-s-\alpha+1)} \\ \qquad\times\frac{4^{j-r}\Gamma(2j-r)\Gamma(l+j-s-r-\alpha+1/2)}{\Gamma(r+1)\Gamma(2j-2r+1)\Gamma(l+j+s-r-\alpha+1)}, \\ & j=1,2,3,\ldots. \end{cases} \tag{10.3.17}$$

*Proof.* Proof of this theorem can be found in [18].

**Theorem 10.3.2.** *The integral of order* $\beta$ *of* $VL_i^*(x)$*, can be expressed as*

$$_{\mathrm{RL}}J_{0+}^{\beta}(VL_i^*(x))\simeq\mathbf{H}^{(\beta)}\Psi(x), \tag{10.3.18}$$

*where* $\Psi(x)$ *is the VLFVs defined in* (10.2.16) *and* $\mathbf{H}^{(\beta)}$ *is the integral operational matrix of order* $\beta\in\mathbb{R}_+$*, which is computed as follows:*

$$\mathbf{H}^{(\beta)}=\begin{pmatrix} \xi_{0,0,0} & \xi_{0,1,0} & \cdots & \xi_{0,j,0} \\ \vdots & \vdots & \cdots & \vdots \\ \sum_{s=0}^{i}\xi_{i,0,s} & \sum_{s=0}^{i}\xi_{i,1,s} & \cdots & \sum_{s=0}^{i}\xi_{i,l,s} \\ \vdots & \vdots & \cdots & \vdots \\ \sum_{s=0}^{l}\xi_{l,0,s} & \sum_{s=0}^{l}\xi_{l,1,s} & \cdots & \sum_{s=0}^{l}\xi_{l,l,s} \end{pmatrix},$$

*and,* $\xi_{i,j,s}$ *is given by*

$$\xi_{i,j,s}=\begin{cases} i\sum_{s=0}^{i}(-1)^s\frac{4^{i-s}\Gamma(2i-s)\Gamma(i-s+1)\Gamma(i-s+\beta+1/2)}{\sqrt{\pi}\Gamma(s+1)\Gamma(2i-2s+1)\Gamma(i-s+\beta+1)^2}, & j=0, \\ 2i\sum_{s=0}^{i}\sum_{r=0}^{j}\frac{(-1)^{s+r}}{\sqrt{\pi}}\frac{4^{i-s}\Gamma(2i-s)\Gamma(i-s+1)}{\Gamma(s+1)\Gamma(2i-2s+1)\Gamma(i-s+\beta+1)} \\ \qquad\times\frac{4^{j-r}\Gamma(2j-r)\Gamma(i+j-s-r+\beta+1/2)}{\Gamma(r+1)\Gamma(2j-2r+1)\Gamma(i+j+s-r+\beta+1)}, \\ & j=1,2,3,\ldots. \end{cases} \tag{10.3.19}$$

*Proof.* Applying the Riemann–Liouville fractional-order integral to (10.2.8), we have

$${}_{\mathrm{RL}}J_{0+}^{\beta}(VL_i^*(x)) = {}_{\mathrm{RL}}J_{0+}^{\beta}\left(\sum_{s=0}^{i}(-1)^s\frac{4^{i-s}2i\Gamma(2i-s)}{\Gamma(s+1)\Gamma(2i-2s+1)}x^{i-s}\right). \tag{10.3.20}$$

Applying the linearity of the Riemann–Liouville integral, we have

$${}_{\mathrm{RL}}J_{0+}^{\beta}(VL_i^*(x)) = \sum_{s=0}^{i}(-1)^s\frac{4^{i-s}2i\Gamma(2i-s)\Gamma(i-s+1)}{\Gamma(s+1)\Gamma(2i-2s+1)\Gamma(i-s+1-\alpha)}x^{i-s+\beta}. \tag{10.3.21}$$

Approximate $x^{i-s+\beta}$ by $(l+1)$ terms of Vieta-Lucas series, as

$$x^{i-s+\beta} \simeq \sum_{j=0}^{l} c_{sj}\,\mathrm{VL}_j^*(x), \tag{10.3.22}$$

where

$$c_{sj} = \frac{1}{\delta_j \pi}\int_0^1 \frac{u(x)\,\mathrm{VL}_j^*(x)}{\sqrt{x-x^2}}dx, \tag{10.3.23}$$

and, $u(x) = x^{i-s+\beta}$.

Now putting the value of $u(x)$ and $\mathrm{VL}_j^*(x)$ in (10.3.23), we get

$$c_{sj} = \begin{cases} \frac{1}{4\pi}\int_0^1 \frac{2x^{i-s-\alpha}}{\sqrt{x-x^2}}dx, & j=0, \\ \frac{1}{2\pi}\int_0^1 \frac{x^{i-s+\beta}}{\sqrt{x-x^2}}\sum_{r=0}^{j}\frac{(-1)^r 2j4^{j-r}\Gamma(2j-r)}{\Gamma(r+1)\Gamma(2j-2r+1)}x^{j-r}dx, & j=1,2,3,..., \end{cases}$$

$$= \begin{cases} \frac{1}{2\pi}\int_0^1 \frac{x^{i-s-\alpha}}{\sqrt{x-x^2}}dx, & j=0, \\ \frac{1}{\pi}\int_0^1 \sum_{r=0}^{j}\frac{(-1)^r j4^{j-r}\Gamma(2j-r)}{\Gamma(r+1)\Gamma(2j-2r+1)}\frac{x^{i+j-s-r+\beta}}{\sqrt{x-x^2}}dx, & j=1,2,3,..., \end{cases}$$

$$= \begin{cases} \frac{1}{2\sqrt{\pi}}\frac{\Gamma(i-s+\beta+1/2)}{\Gamma(i-s+\beta+1)}, & j=0, \\ \frac{1}{\sqrt{\pi}}\sum_{r=0}^{j}\frac{(-1)^r j4^{j-r}\Gamma(2j-r)\Gamma(i+j-s-r+\beta+1/2)}{\Gamma(r+1)\Gamma(2j-2r+1)\Gamma(i+j-s-r+\beta+1)}, & j=1,2,3,..., \end{cases} \tag{10.3.24}$$

by putting the value of $x^{i-s+\beta}$ in (10.3.21), we get

$$_{\mathrm{RL}}J_{0+}^{\beta}(VL_i^*(x)) \simeq \sum_{j=0}^{l} S_v(i,j)VL_j^*(x)), \tag{10.3.25}$$

where, $S_v(i,j) = \sum_{s=0}^{i}\xi_{i,j,s}$, and $\xi_{i,j,s}$ is given in (10.3.19).

## 10.4 Solution of Multi-Order FDEs

In this section, we introduce a numerical method that directly transforms the FDEs into a system of matrix equations without implementing the framework of spectral Tau and spectral collocation methods. Consequently, our proposed approaches offer the potential to be more efficient and user-friendliness when compared to alternative spectral methods and other numerical approaches found in existing literature.

Suppose we have the following generalized MOFDEs

$$\begin{gathered} {}_{\mathrm{C}}\mathrm{D}^{\alpha}\upsilon(x) = g(x, \upsilon(x), {}_{\mathrm{C}}\mathrm{D}^{\beta_1}\upsilon(x), {}_{\mathrm{C}}\mathrm{D}^{\beta_2}\upsilon(x), ..., {}_{\mathrm{C}}\mathrm{D}^{\beta_m}\upsilon(x)), \\ \upsilon^{(s)}(0) = h_s, s = 0, 1, ..., m. \end{gathered} \tag{10.4.26}$$

where $\alpha$ lies between $m$ and $m-1$, and $0 < \beta_i < \beta_m < \alpha$, where $i = 1, 2, ..m$

.

Now the left hand side of (10.4.26) can be approximated as:

$${}_{\mathrm{C}}\mathrm{D}^{\alpha}\upsilon(x) = \Lambda^T \Psi(x). \tag{10.4.27}$$

Now integrating (10.4.27) and using the initial conditions $\upsilon^{(s)}(0) = h_s, s = 0, 1, ..., m$, we get

$$\upsilon(x) = \Lambda^T_{\mathrm{RL}} J^{\alpha}_{0^+} \Psi(x) + \sum_{s=0}^{m} h_s x^s. \tag{10.4.28}$$

By employing (10.3.18), (10.4.28) can also be written as

$$\upsilon(x) = \Lambda^T \mathbf{H}^{(\alpha)}_{(m+1,m+1)} \Psi(x) + \sum_{s=0}^{m} h_s x^s. \tag{10.4.29}$$

We also approximate the term $\sum_{s=0}^{m} h_s x^s$ using VLFVs. So (10.4.29) can be expressed as

$$\upsilon(x) = \Lambda^T \mathbf{H}^{(\alpha)}_{(m+1,m+1)} \Psi(x) + M^T \Psi(x). \tag{10.4.30}$$

Now using Theorem 10.3.1, (10.4.26) can be expressed as

$$\begin{aligned}
{}_{C}\mathrm{D}^{\beta_0}v(x) &= \Lambda^T \mathbf{H}^{\alpha}_{(m+1,m+1)}\mathbf{D}^{\beta_0}_{(m+1,m+1)}\Psi(x) + M^T\mathbf{D}^{\beta_0}_{(m+1,m+1)}\Psi(x),\\
{}_{C}\mathrm{D}^{\beta_1}v(x) &= \Lambda^T \mathbf{H}^{\alpha}_{(m+1,m+1)}\mathbf{D}^{\beta_1}_{(m+1,m+1)}\Psi(x) + M^T\mathbf{D}^{\beta_1}_{(m+1,m+1)}\Psi(x),\\
&\vdots\\
{}_{C}\mathrm{D}^{\beta_m}v(x) &= \Lambda^T \mathbf{H}^{\alpha}_{(m+1,m+1)}\mathbf{D}^{\beta_m}_{(m+1,m+1)}\Psi(x) + M^T\mathbf{D}^{\beta_m}_{(m+1,m+1)}\Psi(x).
\end{aligned} \tag{10.4.31}$$

Now inserting (10.4.27) and (10.4.31) into (10.4.26), we have

$$\begin{aligned}
\Lambda^T\Psi(x) =& \Lambda^T\mathbf{H}^{\alpha}_{(m+1,m+1)}(\mathbf{D}^{\beta_0}_{(m+1,m+1)} + \mathbf{D}^{\beta_1}_{(m+1,m+1)} + \\
& \cdots + \mathbf{D}^{\beta_m}_{(m+1,m+1)})\Psi(x) + M^T(\mathbf{D}^{\beta_0}_{(m+1,m+1)} + \mathbf{D}^{\beta_1}_{(m+1,m+1)} + \\
& \cdots + \mathbf{D}^{\beta_m}_{(m+1,m+1)}) + G^T\Psi(x)
\end{aligned}$$

After simplification, we have

$$\Lambda^T - \Lambda^T\mathbf{H}^{\alpha}_{(m+1,m+1)}\widehat{Q} = M^T\widehat{Q} + G^T \tag{10.4.32}$$

where $\widehat{Q} = (\mathbf{D}^{\beta_0}_{(m+1,m+1)} + \mathbf{D}^{\beta_1}_{(m+1,m+1)} + \cdots + \mathbf{D}^{\beta_m}_{(m+1,m+1)})$.

Equation (10.4.32) is a Sylvester-type matrix equation that can be easily solved for the unknown vector $\Lambda^T$ which, when inserted into (10.4.28), provides the solution to the given problem.

## 10.5 Examples

This section is devoted to demonstrating the applicability and numerical accuracy of the proposed algorithm by solving various test examples. The results obtained by our proposed methodology are compared with the results of other methods in the literature.

**Example 10.5.1.** *Examine the non-homogeneous Bagley–Torvik FDEs, provided below:*

$$ {}_{\mathrm{C}}\mathrm{D}^{2}\upsilon(x) + {}_{\mathrm{C}}\mathrm{D}^{\frac{3}{2}}\upsilon(x) + \upsilon(x) = 1 + x, \;\; x \in [0,1], \tag{10.5.33} $$

*subject to the following initial conditions:*

$$ \upsilon(0) = 1 = \upsilon'(0). \tag{10.5.34} $$

*The exact solution is given under*

$$ \upsilon(x) = 1 + x. \tag{10.5.35} $$

*Here, we have*

$$ G^{T} = \left[\frac{3}{4}, \frac{1}{4}, 0\right], \quad \mathrm{M}^{T} = \left[\tfrac{3}{4}, \tfrac{1}{4}, 0\right], \quad \Lambda^{T} = [0, 0, 0], \tag{10.5.36} $$

$$ \mathbf{H}^{(\frac{3}{2})} = \begin{bmatrix} 0 & 0 & 0 \\ 0 & 0 & 0 \\ 0.1823 & 0.2917 & 0.1458 \end{bmatrix}, \tag{10.5.37} $$

$$ \mathbf{D}^{2} = \begin{bmatrix} 0 & 0 & 0 \\ 0 & 0 & 0 \\ 11.4936 & 7.6624 & -1.5325 \end{bmatrix}. $$

*Using* (10.4.32), (10.5.36), *and* (10.5.37), *we compute the approximate solution of Example 10.5.1, that coincides with its exact solution provided in* (10.5.35).

*In this work, we compare the acquired results against both the precise analytical solution and a numerical solution computed through alternative numerical techniques, as illustrated in Tables 10.1-10.2 and Fig. 10.1. Furthermore, Fig. 10.1 presents a graphical contrast between the exact solution and our approximate solution for Example 10.5.1. This analysis demonstrates a noteworthy observation, indicating that the numerical outcomes by our PNM significantly outperform those achieved through other methods documented in the literature for addressing Example 10.5.1.*

**TABLE 10.1**. Comparing the Approximation Solution (AS) for Example 10.5.1, Achieved Through Our PNM, with the AS Obtained Through Alternative Numerical Solver as Described in [11]

| $x$ | $v(x)$ | Our method ($n=3$) | Method in [11] ($n=10$) |
|---|---|---|---|
| 0.0 | 1.00 | 1.00 | 1.024862 |
| 0.2 | 1.20 | 1.20 | 1.220821 |
| 0.4 | 1.40 | 1.40 | 1.426952 |
| 0.6 | 1.60 | 1.60 | 1.634569 |
| 0.8 | 1.80 | 1.80 | 1.828738 |
| 1.0 | 2.00 | 2.00 | 1.985057 |

**TABLE 10.2**. Comparing the Absolute Errors of Example 10.5.1 Calculated Using Our PNM and the Approach Detailed in [11]

| $x$ | Method in [11] | Our method |
|---|---|---|
| 0.0 | 2.30e-2 | 0.00 |
| 0.2 | 3.13e-2 | 0.00 |
| 0.4 | 3.45e-2 | 0.00 |
| 0.6 | 1.36e-2 | 0.00 |
| 0.8 | 2.30e-2 | 0.00 |
| 1.0 | 3.13e-2 | 0.00 |

**Example 10.5.2.** *Examine the homogeneous MOFDEs, provided below:*

$$\begin{aligned} c\mathrm{D}^{\alpha}v(x) &= a_C\mathrm{D}^{\beta_0} + b_C\mathrm{D}^{\beta_1}v(x) + c_C\mathrm{D}^{\beta_2}v(x) \\ &+d_C\mathrm{D}^{\beta_3}v(x) + G(x), \ x \in [0,1], \ \ 1 < \alpha < 2, \end{aligned} \tag{10.5.38}$$

*with the given initial conditions*

$$v(0) = 0, \ v^{'}(0) = 0, \tag{10.5.39}$$

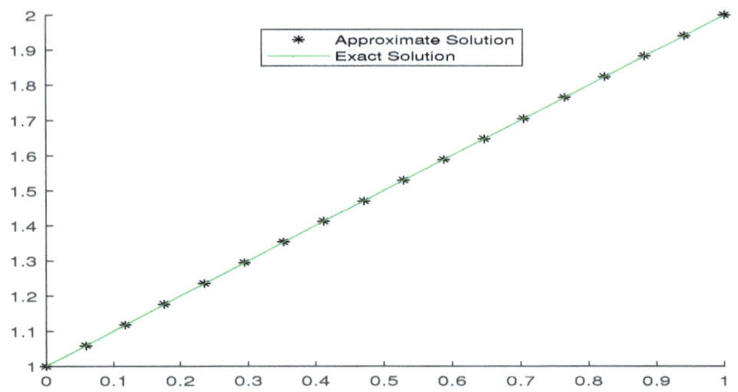


**FIGURE 10.1**. Graphical comparison of the results of Example 10.5.1 computed by using PNM with the exact solution.

*and the source term*

$$G(x) = x^7 + \frac{2048}{429\sqrt{\pi}}x^{6.5} - 14x^6 + 42x^5 - x^2$$

$$-\frac{8}{3\sqrt{\pi}}x^{1.5} + 4x - 2. \tag{10.5.40}$$

*If we take* $\alpha = 2, \beta_0 = 0, \beta_1 = 1, \beta_2 = \frac{1}{2}$, *and* $a = c = -1, b = 2, d = 0$, *then the exact solution of Example 10.5.2 is given below:*

$$v(x) = x^7 - x^2.$$

*We investigate Example 10.5.2 by comparing the approximate outcomes obtained through our PNM with those achieved using the methodologies outlined in [12]. Based on our observations, it is evident that our method yields superior approximations and enhanced precision in the solution, as depicted in Table 10.3. The effectiveness of our PNM is assessed across multiple scales, uncovering a striking resemblance between the approximate and exact solutions as the scale level increases. Refer to Fig. 10.2, Fig.*

*10.3 for visual representations. Furthermore, when $n = 7$, the approximated solution precisely matches the exact solution, $v(x) = x^7 - x^2$, as indicated in Table 10.3. Absolute error quantifications are also carried out at scale levels $n = 4$, 6, and 7. A comprehensive analysis showcases a substantial reduction in absolute errors with increasing scale level, underscoring this trend in Table 10.3, Figures 10.2 and 10.3. Notably, leveraging only a few terms of orthogonal VLFVs yields a remarkable congruence between the approximate and exact solutions.*

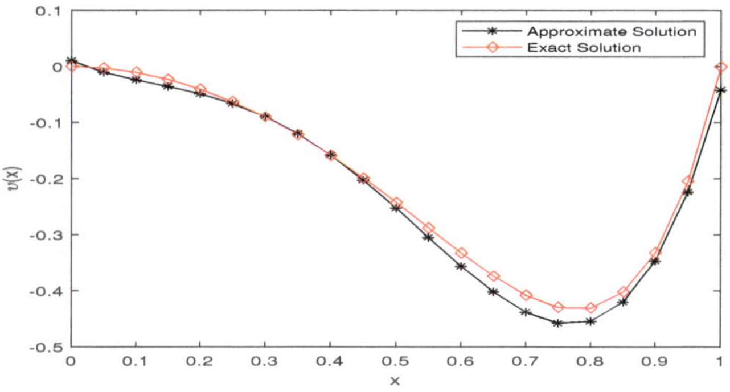


**FIGURE 10.2**. Plot comparing the exact solution and the approximation in Example 10.5.2 when $n = 4$.

**TABLE 10.3**. Comparison of Absolute Errors of Example 10.5.2 at $n = 4, 6$, and 7

| $x$ | [12]($n = 4$) | PNM | [12]($n = 6$) | PNM | [12]($n = 7$) | |
|---|---|---|---|---|---|---|
| 0.2 | 0.0844 | 0.0085 | 0.0044 | $6.949e-5$ | $2.8103e-15$ | 0 |
| 0.4 | 0.3501 | 0.0001 | 0.0079 | 0.0002 | $6.6336e-15$ | 0 |
| 0.6 | 0.6734 | 0.0240 | 0.0143 | $7.361e-5$ | $3.2752e-15$ | 0 |
| 0.8 | 1.0234 | 0.0243 | 0.0214 | 0.0004 | $4.2577e-14$ | 0 |
| 1.0 | 1.6700 | 0.0416 | 0.0280 | 0.0005 | $2.4382e-13$ | 0 |

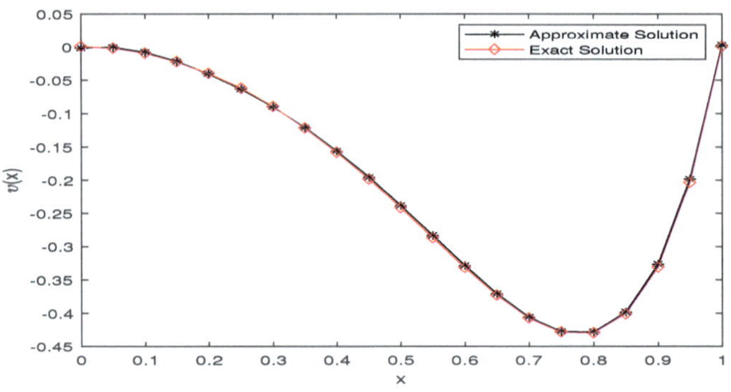


**FIGURE 10.3**. Plot comparing the exact solution and the approximation in Example 10.5.2 when $n = 5$.

**Example 10.5.3.** *Examine the homogeneous MOFDEs, provided below:*

$$
\begin{aligned}
{}_C\mathrm{D}^{\alpha}\upsilon(x) &= a_{cC}\mathrm{D}^{\beta_0}\upsilon(x) + b_C\mathrm{D}^{\beta_1}\upsilon(x) + c_C\mathrm{D}^{\beta_2}\upsilon(x) \\
&\quad + d_C\mathrm{D}^{\beta_3}\upsilon(x) + G(x), \quad x \in [0,1], \quad 0 < \alpha < 2,
\end{aligned} \tag{10.5.41}
$$

*subject to the given initial conditions*

$$\upsilon(0) = 0, \quad \upsilon^{'}(0) = 0, \tag{10.5.42}$$

*and the source term*

$$G(x) = 6x + x^3 - \frac{12}{\Gamma(\frac{7}{3})}x^{\frac{4}{3}} + \frac{6}{\Gamma(\frac{10}{3})}x^{\frac{7}{3}}. \tag{10.5.43}$$

*Below is the exact solution for the parameters:* $\alpha = 2$, $a = c = -1$, $b = 0$, $d = 2$, $\beta_0 = 0$, $\beta_2 = \frac{2}{3}$ *(within the range 0 to 1), and* $\beta_3 = \frac{5}{3}$ *(within the range 1 to 2):*

$$\upsilon(x) = x^3.$$

*Here, we have,*

$$\begin{aligned} G^T &= [-0.2118, -0.3032, 0.0653], \\ H^T &= [0, 0, 0, 0], \\ \Lambda^T &= [1.4991, 1.4991, -0, -0], \end{aligned} \quad . \tag{10.5.44}$$

$$\mathbf{D}^{\frac{2}{3}} = \begin{bmatrix} 0 & 0 & 0 & 0 \\ -1.5973 & -0.7986 & 0.2282 & -0.1141 \\ 0.3993 & -3.6509 & -1.5973 & 0.5617 \\ -2.6404 & 0 & -4.2653 & -2.3104 \end{bmatrix},$$

$$\mathbf{D}^{\frac{5}{3}} = \begin{bmatrix} 0 & 0 & 0 & 0 \\ 0 & 0 & 0 & 0 \\ 25.5567 & 12.7783 & --3.6510 & 1.8255 \\ -9.5837 & 87.6228 & 38.3350 & -13.4804 \end{bmatrix}, \tag{10.5.45}$$

$$\mathbf{H}^2 = \begin{bmatrix} 0.1875 & 0.25 & 0.0625 & 0 \\ -0.0833 & -0.0938 & 0 & 0.01041 \\ -0.0469 & -0.0833 & -0.0417 & 0 \\ 0.05 & 0.0625 & 0 & -0.0156 \end{bmatrix}.$$

*Using* (10.5.44) *and* (10.5.45) *in* (10.4.30)*, we get the exact solution,* $\upsilon(x) = x^3$.

*In Example 10.5.3, we compare the outcomes derived from our algorithmic approach with those obtained through the methodologies outlined in [12]. It is evident from our analysis that our method yields superior results, characterized by enhanced precision and accuracy in the approximated solution. see Table 10.4 and Figure 10.4.*

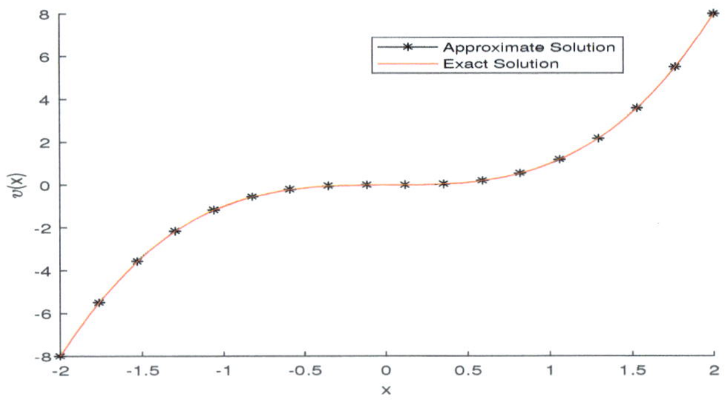


**FIGURE 10.4**. Graph of the exact solution and approximate solution of Example 10.5.3 at $n = 3$.

**TABLE 10.4**. Comparison of the Absolute Error of Example 10.5.3 As Calculated Through Our PNM Method with the Absolute Error Obtained Using the Approach Presented in [12] for the Cases Where $n = 2$ and $n = 3$

| $x$ | $\upsilon(x)$ | [12]$(n = 2)$ | PNM$(n = 2)$ | [12]$(n = 3)$ | PNM$(n = 3)$ |
|---|---|---|---|---|---|
| 0.2 | 0.0080 | 0.0614 | 0.0197 | $2.5500e - 16$ | 0 |
| 0.4 | 0.0640 | 0.1541 | 0.0558 | $2.7756e - 16$ | 0 |
| 0.6 | 0.2160 | 0.3339 | 0.0590 | $3.3307e - 16$ | 0 |
| 0.8 | 0.5120 | 0.6488 | 0.0190 | $4.4409e - 16$ | 0 |
| 1.0 | 1.0000 | 1.1468 | 0.2258 | $6.6613e - 16$ | 0 |

**Example 10.5.4.** *Examine the homogeneous MOFDEs, provided below:*

$$a_1 \mathrm{D}^{\beta_0} \upsilon(x) + a_2 \mathrm{D}^{\beta_1} \upsilon(x) + a_3 \upsilon(x) = g(x), \qquad (10.5.46)$$

$$x \in [0, 1], \ 1 < \beta_1 < 2,$$

*with the given initial conditions*

$$\upsilon(0) = d_1, \ \upsilon^{'}(0) = d_2 \qquad (10.5.47)$$

*where* $\beta_0 = 2, d_1 = d_2 = 0$, *and* $g(x) = 8$. *The analytical solution at* $a_1 = a_2 = a_3 = 1$ *is computed in [19].*

*We solve Example 10.5.4 using our proposed methodology by considering two choices of the coefficients, one is* $a_1 = a_2 = a_3 = 1$ *and the other is* $a_2 = a_3 = 0.5$ $a_1 = 1$. *For both cases* $\beta_1 = 1.5$. *The obtained results are compared with the exact solution and with the numerical solution computed by using other numerical methods, see Table 10.5 and Table 10.6.*

*For the case* $a_1 = 1$, $a_2 = a_3 = 0.5$, *the results obtained by using GA-PS [23], PSO-PS [21], NN-IPA [24] are compared with the result of our method and is given in Table 10.5.*

*In Table 10.6, we show the case when* $a_1 = a_2 = a_3 = 1$, *and we compare the result obtained by our proposed method at different scales with results obtained by [27] and [6].*

*From both tables, it is observed that the numerical results obtained by using our PNM are much better than those obtained by using other methods proposed in the literature.*

**TABLE 10.5.** Absolute Error Comparison of the Results of Example 10.5.4 Computed by PNM with the Results Obtained Otherwise in the Literature at $a_2 = a_3 = 0.5$, $a_1 = 1$, and $\beta_1 = 1.5$

| $x$ | Exact solution | GA-PS [23] | PSO-PS [21] | NN-IPA [24] | PNM ($n = 2$) |
|---|---|---|---|---|---|
| 0.0 | 1.0 | $1.60 \times 10^{-2}$ | $2.84 \times 10^{-3}$ | $4.90 \times 10^{-10}$ | 0.0 |
| 0.1 | 1.1 | $4.73 \times 10^{-3}$ | $3.40 \times 10^{-3}$ | $7.97 \times 10^{-10}$ | 0.0 |
| 0.2 | 1.2 | $1.95 \times 10^{-4}$ | $4.52 \times 10^{-3}$ | $7.42 \times 10^{-10}$ | 0.0 |
| 0.3 | 1.3 | $6.66 \times 10^{-4}$ | $2.02 \times 10^{-3}$ | $6.70 \times 10^{-10}$ | 0.0 |
| 0.4 | 1.4 | $1.62 \times 10^{-3}$ | $2.55 \times 10^{-3}$ | $8.56 \times 10^{-10}$ | 0.0 |
| 0.5 | 1.5 | $4.97 \times 10^{-3}$ | $7.59 \times 10^{-3}$ | $1.37 \times 10^{-9}$ | 0.0 |
| 0.6 | 1.6 | $7.42 \times 10^{-3}$ | $1.14 \times 10^{-2}$ | $2.08 \times 10^{-9}$ | 0.0 |
| 0.7 | 1.7 | $6.70 \times 10^{-3}$ | $1.24 \times 10^{-2}$ | $2.69 \times 10^{-9}$ | 0.0 |
| 0.8 | 1.8 | $1.27 \times 10^{-5}$ | $8.72 \times 10^{-3}$ | $2.90 \times 10^{-9}$ | 0.0 |
| 0.9 | 1.9 | $1.62 \times 10^{-2}$ | $1.29 \times 10^{-3}$ | $2.61 \times 10^{-9}$ | 0.0 |
| 1.0 | 2.0 | $4.62 \times 10^{-2}$ | $1.95 \times 10^{-2}$ | $2.14 \times 10^{-9}$ | 0.0 |

**TABLE 10.6**. Numerical Comparison of the Results of Example 10.5.4 Computed by PNM with the Results Obtained Otherwise in the Literature at $\beta_1 = 1.5$, $\beta_0 = 2$, $a_1 = a_2 = a_3 = 1$

| $x$ | Exact solution | ADM [27] | TMM [6] | PNM ($n = 16$) | PNM ($n = 20$) |
|---|---|---|---|---|---|
| 0.0 | 0.000000 | 0.000000 | 0.000000 | 0.000000 | 0.000000 |
| 0.2 | 0.125221 | 0.140640 | 0.125254 | 0.125221 | 0.125221 |
| 0.4 | 0.455435 | 0.533284 | 0.455468 | 0.455436 | 0.455435 |
| 0.6 | 0.950392 | 1.148840 | 0.950398 | 0.950393 | 0.950393 |
| 0.8 | 1.579557 | 1.963033 | 1.579689 | 1.579558 | 1.579557 |
| 1.0 | 2.315526 | 2.952567 | 2.315589 | 2.315527 | 2.315526 |

## 10.6 Conclusion

The proposed study showcases a research endeavor that introduces and applies an efficient computational technique based on operational matrices of fractional-order integral and derivative operators of VLFVs. The approach is presented as an efficient method for solving FODEs without the dependence on specific collocation point choices or the need for expanding residual functions as orthogonal polynomial series.

The key contributions highlighted in the results include the development of a new integral operational matrix for the Vieta-Lucas polynomial in the Riemann-Liouville context and the demonstration of the method's efficiency compared to existing techniques documented in the literature.

The results section further validates the proposed numerical method through comparisons with exact solutions and alternative numerical techniques across various examples. The results indicate that the PNM, excels in terms of precision and accuracy when solving the given fractional differential equations, surpassing other existing methods.

In conclusion, the research presents a promising and efficient computational method for addressing fractional differential equations, offering advantages over existing approaches. The comparisons with exact solutions and alternative numerical methods support the effectiveness of the proposed technique across multiple examples, emphasizing its potential applicability and superiority in solving fractional differential equations.

# References


[1] Agarwal, P., & El-Sayed, (2020). A. Vieta–Lucas polynomials for solving a fractional-order mathematical physics model. *Advances In Difference Equations*, 1–18.

[2] Ahmed, E., & Elgazzar, (2007). A. On fractional order differential equations model for nonlocal epidemics. *Physica A: Statistical Mechanics And Its Applications, 379*, 607–614.

[3] Al-Sharif, M., Ahmed, A., & Salim, M., (2020). An integral operational matrix of fractional-order Chelyshkov functions and its applications. *Symmetry, 12*(11), 1755.

[4] Atabakzadeh, M., Akrami, M., & Erjaee, G., (2013), Chebyshev operational matrix method for solving multi-order fractional ordinary differential equations. *Applied Mathematical Modelling, 37*, 8903–8911.

[5] Bhrawy, A., & Alofi, A., (2013). The operational matrix of fractional integration for shifted Chebyshev polynomials. *Applied Mathematics Letters, 26*, 25–31.

[6] Çenesiz, Y., Keskin, Y., & Kurnaz, A., (2010). The solution of the Bagley–Torvik equation with the generalized Taylor collocation method. *Journal of The Franklin Institute, 347*, 452–466.

[7] Chen, W., Sun, H., Zhang, X. and Korošak, D., (2010). Anomalous diffusion modeling by fractal and fractional derivatives. *Computers and Mathematics With Applications, 59*, 1754–1758.

[8] Chen, W., (2006). A speculative study of 2/3-order fractional Laplacian modeling of turbulence: Some thoughts and conjectures. *Chaos: An Interdisciplinary Journal of Nonlinear Science, 16.*

[9] Doha, E., Bhrawy, A., & Ezz-Eldien, S., (2011). A Chebyshev spectral method based on operational matrix for initial and boundary

value problems of fractional order. *Computers and Mathematics With Applications, 62*, 2364–2373.

[10] Goncalves, E., & Zeidan, D., (2017). Numerical study of turbulent cavitating flows in thermal regime. *International Journal of Numerical Methods For Heat and Fluid Flow, 27*, 1487–1503.

[11] Gülsu, M., Öztürk, Y., & Anapali, A., (2017). Numerical solution the fractional Bagley–Torvik equation arising in fluid mechanics. *International Journal of Computer Mathematics, 94*, 173–184.

[12] Han, W., Chen, Y., Liu, D., Li, X., & Boutat, D., (2018). Numerical solution for a class of multi-order fractional differential equations with error correction and convergence analysis. *Advances In Difference Equations*, 1–22.

[13] Kazem, S., Abbasbandy, S., & Kumar, S., (2013). Fractional-order Legendre functions for solving fractional-order differential equations. *Appl. Math. Model., 37*, 5498–5510, doi.org/10.1016/j.apm.2012.10.026.

[14] Kazem, S., (2013). An integral operational matrix based on Jacobi polynomials for solving fractional-order differential equations. *Applied Mathematical Modelling, 37*, 1126–1136.

[15] Kilbas, A. A., Srivastava, H. M., Trujillo, J. J., (2006). *Theory and applications of fractional differential equations* (Vol. 204). elsevier.

[16] Kuila, S., Raja Sekhar, T., & Zeidan, D., (2016). On the Riemann problem simulation for the drift-flux equations of two-phase flows. *International Journal of Computational Methods, 13*, 1650009.

[17] Mohammadi, F., & Hassani, H., (2022). Numerical solution of time-fractional Telegraph equation by using a new class of orthogonal polynomials. *Boletim Da Sociedade Paranaense De Matemática, 40*, 1–13.

[18] Noor, Z., Talib, I., Abdeljawad, T., & Alqudah, M., (2022). Numerical Study of Caputo Fractional-Order Differential Equations by Developing New Operational Matrices of Vieta–Lucas Polynomials. *Fractal And Fractional, 6*(79).

[19] Ray, S., & Bera, R., (2005). Analytical solution of the Bagley Torvik equation by Adomian decomposition method. *Applied Mathematics And Computation, 168*, 398–410.

[20] Raja, M., Khan, J., & Qureshi, I., (2011). Solution of fractional order system of Bagley-Torvik equation using evolutionary computational intelligence. *Mathematical Problems In Engineering.*

[21] Raja, M., Khan, J., & Qureshi, I., (2011). Swarm intelligence optimized neural networks in solving fractional system of Bagley-Torvik equation. *Engineering Intelligent Systems, 19*, 41–51.

[22] Raja, M., Khan, J., Qureshi, I., (2011). Solution of fractional order system of Bagley-Torvik equation using evolutionary computational intelligence. *Mathematical Problems In Engineering*, 2011.

[23] Raja, M., Khan, J., Qureshi, I., (2011). Solution of fractional order system of Bagley-Torvik equation using evolutionary computational intelligence. *Mathematical Problems in Engineering.*

[24] Raja, M., Samar, R., Manzar, M., & Shah, S., (2017). Design of unsupervised fractional neural network model optimized with interior point algorithm for solving Bagley–Torvik equation. *Mathematics And Computers In Simulation, 132*, 139–158.

[25] Rossikhin, Y., & Shitikova, M., (1997). Application of fractional derivatives to the analysis of damped vibrations of viscoelastic single mass systems. *Acta Mechanica, 120*, 109–125.

[26] Saadatmandi, A., & Dehghan, M., (2010). A new operational matrix for solving fractional-order differential equations. *Computers and Mathematics With Applications, 59*, 1326–1336.

[27] Setia, A., Liu, Y., & Vatsala, A., (2014). The solution of the Bagley-Torvik equation by using second kind Chebyshev wavelet. *2014 11th International Conference On Information Technology: New Generations.* pp. 443–446.

[28] Sultana, F., (2020). Singh, D., Pandey, R., & Zeidan, D. Numerical schemes for a class of tempered fractional integro-differential equations. *Applied Numerical Mathematics, 157*, 110–134.

[29] Sun, H., Chen, D., Zhang, Y., & Chen, L., (2015). Understanding partial bed-load transport: Experiments and stochastic model analysis. *Journal of Hydrology, 521*, 196–204.

[30] Sun, H., Chen, W., & Chen, Y., (2009). Variable-order fractional differential operators in anomalous diffusion modeling. *Physica A: Statistical Mechanics And Its Applications, 388*, 4586–4592.

[31] Talib, I., & Bohner, M., (2022).Numerical study of generalized modified Caputo fractional differential equations. *International Journal of Computer Mathematics.* 1–24.

[32] Yüzbaşı, Ş., (2013). Numerical solution of the Bagley–Torvik equation by the Bessel collocation method. *Mathematical Methods In The Applied Sciences, 36*, 300–312.

[33] Zeidan, D., Chau, C., & Lu, T., (2021). On the characteristic Adomian decomposition method for the Riemann problem. *Mathematical Methods In The Applied Sciences, 44*, 8097–8112.

[34] Zhang, H., Jiang, X., & Yang, X., (2018). A time-space spectral method for the time-space fractional Fokker–Planck equation and its inverse problem. *Applied Mathematics And Computation, 320*, pp. 302–318.

[35] Zheng, X., & Wang, H., (2020). An optimal-order numerical approximation to variable-order space-fractional diffusion equations on uniform or graded meshes. *SIAM Journal On Numerical Analysis, 58*, 330–352.

# Chapter 11

## Existence Results for Implicit FFDEs with Tempered Ξ-HFD

**R. Vivek,[1] K. Kanagarajan,[1] and D. Vivek[2]**
[1]*Department of Mathematics, Sri Ramakrishna Mission Vidyalaya College of Arts and Science, Coimbatore, India*
[2]*Department of Mathematics, PSG College of Arts and Science, Coimbatore, India*

**Abstract**
The foremost goal of this chapter is to study the existence and uniqueness of solutions for implicit fuzzy differential equations with tempered Ξ-Hilfer fractional derivatives. Then the uniqueness and continuous dependence of the $\delta$-approximate solutions are obtained using the generalized Gronwall inequality. Finally, an example is presented to illustrate the main results.

## Introduction

In recent years, fractional calculus has drawn more attention due to its applications in various applicable fields such as physics, mechanics, chemistry, engineering, etc. For more details on the applications of fractional calculus, the reader can refer to the books of Podlubny [26], Kilbas et al. [18], and Hilfer [11].

In the same context, Kilbas et al., in [18], proposed the properties of fractional integrals and fractional derivatives with respect to another function. Recently in 2017, Sousa and Oliviera [29] proposed an interpolator of $\psi$-RL and $\psi$-Caputo fractional derivatives in Hilfer [11] sense of the definition and named $\psi$-Hilfer fractional derivative. In [30], Vivek et al. established the existence and stability results for Hilfer fractional implicit differential equations with nonlocal conditions. For details, we refer the reader to [2, 3, 8, 17], and references therein.

Tempered fractional calculus is an extension of fractional calculus characterized by integrals, in which the kernel begins with a fractional power function multiplied by an exponential factor. The properties of tempered fractional derivative and integral operators are discussed in [10]. The Hadamard-type fractional integrals and derivatives with respect to functions are proposed in [9]. In their work cited as [19], Kucche et al., made substantial advancements in the field of tempered fractional integrals and derivatives. They introduced a new framework for calculating these integrals and derivatives and presented a comprehensive set of related properties and results. Continuing their research in [22], the same team extended the theory and explored tempered fractional calculus with respect to functions, introducing the tempered Hilfer-type operator. For more relevant applications concerning tempered fractional calculus, we refer the reader to [20, 24, 25, 28].

On the other hand, in recent years, the investigation of fuzzy calculus and FDEs has been picking up much attention from researchers. This is due to the fact that fuzzy calculus and FDEs have various applications in many mathematical or computer models of some deterministic real-world phenomena affected by uncertainty. Bede et al., in [2, 5–7], presented a great number of applications of this calculus. Various approaches and methods, based on Hukuhara differentiability or generalized Hukuhara differentiability, were then considered to investigate fuzzy fractional differential equations (FFDEs) that can be found in literature [1, 13, 14, 21]. Authors in [15] presented a few basic theories of different IVP of implicit FFDEs involving different fuzzy fractional derivatives. Recently, Vivek et al., [32], considered the stability of solutions for quaternion FFDEs with Ξ-Hilfer fractional derivative (Ξ-HFD) in the case of the generalized regular fuzzy function space. In [33], Vivek et al. further studied the existence and finite-time stability of the solution to the IVP of tempered-regularized Hilfer fractional derivative. In [31], Vivek et al., proved some existence and stability results of Ξ-Hilfer type FFDE with time delay. Relevant works can be found in [4, 12, 16, 27] and their references.

Inspired by the aforementioned works, in the present chapter, we will study the existing results for the IVP of a tempered Ξ-Hilfer type implicit FFDE of the form:

$$\begin{cases} {}^{TH}_{a^+}\mathfrak{D}^{\zeta_1,\zeta_2,\mu}_{\Xi(t)} w(t) = g(t, w(t), {}^{TH}_{a^+}\mathfrak{D}^{\zeta_1,\zeta_2,\mu}_{\Xi(t)} w(t)), \mu > 0, t \in [a, b], \\ {}^{T}_{a^+}\mathfrak{I}^{1-\vartheta,\mu}_{\Xi(t)} w(a) = w_0, \quad \vartheta = \zeta_1 + \zeta_2 - \zeta_1\zeta_2, \end{cases} \tag{11.0.1}$$

where ${}^{TH}_{a^+}\mathfrak{D}^{\zeta_1,\zeta_2,\mu}_{\Xi(t)}$ and ${}^{T}_{a^+}\mathfrak{I}^{1-\vartheta,\mu}_{\Xi(t)}$ are the tempered Ξ-HFD of order $\zeta_1 \in (0, 1)$ and type $\zeta_2 \in [0, 1]$ and tempered Ξ-RL fractional integral of order $1 - \vartheta$, respectively. Here $g : [a, b] \times E \times E \to E$ is a given continuous fuzzy function.

The rest of this chapter is outlined as follows: In Chapter 11.1, we introduce some known definitions and lemmas, which are used in our works. In Chapter 11.2, we prove the existence and uniqueness of solutions for problem (11.0.1). In Chapter 11.3, we discuss the $\delta$-approximate solution of the problem (11.0.1). In Chapter 11.4, we give an example to highlight the feasibility of our main results.

---

## 11.1 Preliminaries

Let $C([a, b], E)$, $AC([a, b], E)$ be the spaces of all continuous fuzzy functions on the interval $[a, b]$ with values in $E$. Let $L([a, b], E)$ be the space of all fuzzy functions $w : [a, b] \to E$ such that the functions $t \mapsto D_0[w(t), \widehat{0}]$

belongs to $L[a,b]$ (the space of all Lebesgue integrable real-valued functions on the bounded interval $[a,b]$).

For $0<\vartheta\le 1$, we need the following weighted spaces of continuous fuzzy functions:

$$C_{\vartheta,\Xi}([a,b],E)=\{w:(a,b]\to E:(\Xi(t)-\Xi(a))^{\vartheta}w(t)\in C([a,b],E)\},$$

with the norm

$$D_{C_{\vartheta,\Xi}}[w,\widehat{0}]=\sup_{t\in[a,b]}D_0[(\Xi(t)-\Xi(a))^{\vartheta}w(t),\widehat{0}].$$

We defined the weighted space of all bounded and continuous functions defined by:

$$BC_{\vartheta,\Xi}([a,b],E)=\{u:[a,b]\to E:(\Xi(t)-\Xi(a))^{\vartheta}u(t)\in BC([a,b],E)\}.$$

Let $E$ be the space of fuzzy numbers, i.e., normal, fuzzy convex, upper semicontinuous, and compactly supported fuzzy subsets of the real numbers. For $l\in[0,1]$, denote $[w]^l=\{\bar{w}\in\mathbb{R}|w(\bar{w})\ge l\}$ and $[w]^0=\overline{\{\bar{w}\in\mathbb{R}|w(\bar{w})>0\}}$. Then it is well-known that the $l$-level set of $w$, $[w]^l=[\underline{w}(l),\overline{w}(l)]$, is a bounded closed interval, for any $l\in[0,1]$. For $w\in E$, we define the diameter of the $l$-level set of $w$ as $d([w]^l)=\overline{w}(l)-\underline{w}(l)$.

**Definition 11.1.1.** *Let $w,u\in E$. If there exists $\bar{w}\in E$ such that $w=u+\bar{w}$, then $\bar{w}$ is called the Hukuhara difference of $w$ and $u$, and it is denoted by $w\ominus u$.*

**Definition 11.1.2.** *Let $w,u\in E$. The Hausdorff distance between $w$ and $u$ is defined by:*

$$D_0[w,u]=\sup_{l\in[0,1]}\max\{|\underline{w}(l)-\underline{u}(l)|,|\overline{w}(l)-\overline{u}(l)|\}.$$

**Definition 11.1.3.** *[7] Let $w:(a,b)\to E$ and $t\in(a,b)$. The fuzzy function $w$ is said to be generalized Hukuhara differentiability (gH-differentiability)* *at t if there exists an element $w^{'}(t) \in E$ such that*

$$w^{'}(t) = \lim_{h \to 0} \frac{w(t+h) \ominus_{gH} w(t)}{h},$$

*where $\ominus_{gH}$ represents the generalized Hukuhara difference of two fuzzy numbers $w, u \in E$ (gH-differences for short) defined as follows:*

$$w \ominus_{gH} u = \bar{w} \iff \begin{cases} (i) & w = u + \bar{w}, \quad or \\ (ii) & u = w + (-1)\bar{w}. \end{cases}$$

**Definition 11.1.4.** *[7] Let $w : [a,b] \to E$. We say that $w$ is d-increasing (d-decreasing) on $[a,b]$ if for every $l \in [0,1]$ the function $t \longmapsto d([w(t)]^l)$ is nondecreasing (nonincreasing) on $[a,b]$. In addition, if $w$ is d-increasing or d-decreasing on $[a,b]$, then we say that $w$ is d-monotone on $[a,b]$.*

**Definition 11.1.5.** *The left-sided tempered $\Xi$-RL fractional integral of order $\zeta_1 > 0$ of the fuzzy function $w$ is defined as:*

$$_{a^+}^{T}\mathfrak{I}_{\Xi(t)}^{\zeta_1,\mu} w(t) = \frac{1}{\Gamma(\zeta_1)} \int_a^t \Xi^{'}(\xi)(\Xi(t) - \Xi(\xi))^{\zeta_1 - 1} e^{-\mu(\Xi(t) - \Xi(\xi))} w(\xi) d\xi,$$

*provided the integral exists.*

**Definition 11.1.6.** *The left-sided fuzzy tempered generalized Hukuhara fractional derivative in a sense $\Xi$-type RL of order $\zeta_1$ of $w$ is defined as follows:*

$$\begin{aligned} _{a^+}^{TRL}\mathfrak{D}_{\Xi(t)}^{\zeta_1,\mu} w(t) =& \left( \frac{1}{\Xi^{'}(t)} \frac{d}{dt} + \mu \right) {}_{a^+}^{T}\mathfrak{I}_{\Xi(t)}^{1-\zeta_1,\mu} w(t) \\ =& \frac{1}{\Gamma(1-\zeta_1)} \left( \frac{1}{\Xi^{'}(t)} \frac{d}{dt} + \mu \right) \int_a^t \Xi^{'}(\xi)(\Xi(t) - \Xi(\xi))^{-\zeta_1} \\ & \times e^{-\mu(\Xi(t) - \Xi(\xi))} w(\xi) d\xi, \end{aligned}$$

if the gH-derivative $w^{'}_{1-\zeta_1,\Xi}(t)$ exists for $t \in (a,b)$, where,

$$w_{1-\zeta_1,\Xi}(t) = {}_{a^+}^{T}\mathfrak{I}_{\Xi(t)}^{1-\zeta_1,\mu} w(t)$$

$$=\frac{1}{\Gamma(1-\zeta_1)}\int_a^t \Xi^{'}(\xi)(\Xi(t)-\Xi(\xi))^{-\zeta_1}e^{-\mu(\Xi(t)-\Xi(\xi))}w(\xi)d\xi.$$

**Definition 11.1.7.** *The tempered Ξ-Hilfer-fuzzy-type gH-fractional derivative of order $\zeta_1$ and type $\zeta_2$ and index $\mu$ of a function $w$ is defined by:*

$${}^{TH}_{a^+}\mathfrak{D}^{\zeta_1,\zeta_2,\mu}_{\Xi(t)}w(t) = {}^{T}_{a^+}\mathfrak{I}^{\zeta_2(1-\zeta_1)}_{\Xi(t)}\left(\frac{1}{\Xi'(t)}\cdot\frac{d}{dt}+\mu\right){}^{T}_{a^+}\mathfrak{I}^{(1-\zeta_2)(1-\zeta_1)}_{\Xi(t)}w(t).$$

*In other way,*

$${}^{TH}_{a^+}\mathfrak{D}^{\zeta_1,\zeta_2,\mu}_{\Xi(t)}w(t) = {}^{T}_{a^+}\mathfrak{I}^{\vartheta-\zeta_1,\mu}_{\Xi(t)}\,{}^{TR}_{a^+}\mathfrak{D}^{\vartheta,\mu}_{\Xi(t)}w(t).$$

**Lemma 11.1.1.** *[29] Let $\zeta_1 > 0$, $0 \le \vartheta < 1$ and $w \in L^1([a,b],E)$. Then,*

$${}^{T}_{a^+}\mathfrak{I}^{\zeta_1,\mu}_{\Xi(t)}\,{}^{T}_{a^+}\mathfrak{I}^{\zeta_2,\mu}_{\Xi(t)}w(t) = {}^{T}_{a^+}\mathfrak{I}^{\zeta_1+\zeta_2,\mu}_{\Xi(t)}w(t), \quad a.e. \quad t\in[a,b].$$

*In particular,*

*(i) if $w \in C_{\vartheta,\Xi}([a,b],E)$. Then, ${}^{T}_{a^+}\mathfrak{I}^{\zeta_1,\mu}_{\Xi(t)}\,{}^{T}_{a^+}\mathfrak{I}^{\zeta_2,\mu}_{\Xi(t)}w(t) = {}^{T}_{a^+}\mathfrak{I}^{\zeta_1+\zeta_2,\mu}_{\Xi(t)}w(t)$, $t \in (a,b]$.*

*(ii) If $w \in C([a,b],E)$. Then, ${}^{T}_{a^+}\mathfrak{I}^{\zeta_1,\mu}_{\Xi(t)}\,{}^{T}_{a^+}\mathfrak{I}^{\zeta_2,\mu}_{\Xi(t)}w(t) = {}^{T}_{a^+}\mathfrak{I}^{\zeta_1+\zeta_2,\mu}_{\Xi(t)}w(t)$, $t \in [a,b]$.*

**Lemma 11.1.2.** *[29] Let $\zeta_1 > 0$, $0 \le \zeta_2 \le 1$ and $0 \le \vartheta < 1$.*
*If $w \in C_{\vartheta,\Xi}([a,b],E)$, then,*

$${}^{TH}_{a^+}\mathfrak{D}^{\zeta_1,\zeta_2,\mu}_{\Xi(t)}\,{}^{T}_{a^+}\mathfrak{I}^{\zeta_1,\mu}_{\Xi(t)}w(t) = w(t), \quad t\in(a,b].$$

*If $w \in C^1([a,b],E)$. Then,*

$${}^{TH}_{a^+}\mathfrak{D}^{\zeta_1,\zeta_2,\mu}_{\Xi(t)}\,{}^{T}_{a^+}\mathfrak{I}^{\zeta_1,\mu}_{\Xi(t)}w(t) = w(t), \quad t\in[a,b].$$

**Lemma 11.1.3.** *[22] Let $t > a$, $\zeta_1 \geq 0$, and $\nu > 0$. Then,*

$$
{}_{a^+}^{T}\mathfrak{I}_{\Xi(t)}^{\zeta_1,\mu}(\Xi(t)-\Xi(a))^{\nu-1} = \frac{\Gamma(\nu)}{\Gamma(\nu+\zeta_1)}(\Xi(t)-\Xi(a))^{\zeta_1+\nu-1} \times e^{-\mu(\Xi(t)-\Xi(a))},
$$

*and if $0 < \zeta_1 < 1$, we have*

$$
{}_{a^+}^{TRL}\mathfrak{D}_{\Xi(t)}^{\zeta_1,\mu}(\Xi(t)-\Xi(a))^{\zeta_1-1} = 0.
$$

**Lemma 11.1.4.** *[29] Let $0 < \zeta_1 < 1$, $0 \leq \vartheta < 1$.*
*If $w \in C_{1-\vartheta,\Xi}([a,b],E)$ and ${}_{a^+}^{T}\mathfrak{I}_{\Xi(t)}^{1-\vartheta,\mu} \in C^1_{1-\vartheta,\Xi}([a,b],E)$. Then,*

$$
{}_{a^+}^{T}\mathfrak{I}_{\Xi(t)}^{\zeta_1,\mu}{}_{a^+}^{TH}\mathfrak{D}_{\Xi(t)}^{\zeta_1,\zeta_2,\mu}w(t) = w(t) \ominus \frac{{}_{a^+}^{T}\mathfrak{I}_{\Xi(t)}^{1-\vartheta,\mu}w(a)}{\Gamma(\vartheta)}(\Xi(t)-\Xi(a))^{\vartheta-1} \times e^{-\mu(\Xi(t)-\Xi(a))}.
$$

**Theorem 11.1.1.** *[8] let $\mathbb{K}$ be a nonempty closed subset of a Banach space $Y$, and let $k_n \leq 0$, $n \in \mathbb{N} \cup \{0\}$, be a sequence such that $\sum_{n=0}^{\infty} k_n$ converges. Moreover, let the mapping $\mathbb{P} : \mathbb{K} \to \mathbb{K}$ satisfy the inequality*

$$
\|\mathbb{P}^n x - \mathbb{P}^n y\| \leq k_n \|x - y\|,
$$

*for every $n \in \mathbb{N} \cup \{0\}$ and every $x, y \in \mathbb{K}$. Then, $\mathbb{P}$ has a uniquely defined fixed point $x^*$. In addition, the sequence $\{\mathbb{P}^n x_0\}_{n=1}^{\infty}$ converges to this fixed point $x^*$ for every $x_0 \in \mathbb{P}$.*

**Theorem 11.1.2.** *[18] Let $(Y, d)$ be a nonempty complete metric space, let $k \in [0, 1)$, and let $S : Y \to Y$ be the map such that, for every $x_1, y_1 \in Y$, the relation*

$$
d(Sx_1, Sy_1) \leq kd(x_1, y_1),
$$

*holds. Then, the operator $S$ has a unique fixed point $x_1^* \in Y$. In addition, if $S^k$, $k \in \mathbb{N}$ is the sequence of operators defined by:*

$$S^k = S, \quad and \quad S^k = SS^{k-1}, \quad k \in \mathbb{N}\backslash\{1\},$$

*then, for any $z \in Y$, the sequence $\{S^k z\}_{k=1}^{\infty}$ converges to the fixed point $x_1^*$.*

**Lemma 11.1.5.** *(**Gronwall's Lemma**)[22] Let $w : [a,b] \to \mathbb{R}^+$ be a real-valued function and $Q(.)$ is a nonnegative, locally integrable function on $[a,b]$. Assume that there is a positive constant $H$ such that for $\zeta_1 \in (0,1)$*

$$w(t) \leq Q(t) + H(t)\int_a^t \Xi^{'}(\xi)(\Xi(t)-\Xi(\xi))^{\zeta_1-1}e^{-\mu(\Xi(t)-\Xi(\xi))}w(\xi)d\xi,$$

*then, for all $t \in [a,b]$, we have,*

$$\begin{aligned} w(t) \leq & Q(t) + \int_a^t \sum_{k=1}^{\infty} \frac{[H(t)\Gamma(\zeta_1)]^k}{\Gamma(\zeta_1 k)} \Xi^{'}(\xi)(\Xi(t)-\Xi(\xi))^{\zeta_1 k-1} \\ & \times e^{-\mu(\Xi(t)-\Xi(\xi))}Q(\xi)d\xi. \end{aligned}$$

*Further, if $Q$ is a non-decreasing function on $[a,b]$, then,*

$$w(t) \leq Q(t)\mathbb{E}_{\zeta_1}(H(t)\Gamma(\zeta_1)(\Xi(t)-\Xi(a))^{\zeta_1}).$$

---

## 11.2 Main Results

Suppose that the function $g : (a,b] \times E \times E \to E$ is continuous and satisfies the following assumptions:

($\mathbb{A}1$) The function $w :\longmapsto g(t,w,u)$ and $u :\longmapsto g(t,w,u)$ are continuous on $BC_{1-\vartheta,\Xi}([a,b],E)$ for $t \in [a,b]$.

($\mathbb{A}2$) There exists a continuous real-valued function $m : [a,b] \to \mathbb{R}^+$ and a constant $L_g \in (0,1)$ such that for each $t \in [a,b]$, and all

$w_1, w_2, u_1, u_2 \in E$, we have,

$$D_{C_{1-\vartheta,\Xi}[a,b]}[g(t,w_1,w_2), g(t,u_1,u_2)] \leq m(t) D_{C_{1-\vartheta,\Xi}[a,b]}[w_1,u_1] + L_g D_{C_{1-\vartheta,\Xi}[a,b]}[w_2,u_2],$$

where $M = \max\{m(t) : t \in [a,b]\}$.

**Lemma 11.2.1.** *Assume that $g : (a,b] \times E \times E \to E$ is a continuous function. A fuzzy function $w$ is a solution of* (11.0.1) *if and only if $w$ satisfies the following fuzzy integral equation:*

$$\begin{cases} w(t) \ominus_{gH} \dfrac{(\Xi(t)-\Xi(a))^{\vartheta-1} e^{-\mu(\Xi(t)-\Xi(a))}}{\Gamma(\vartheta)} w_0 \\ \quad = \dfrac{1}{\Gamma(\zeta_1)} \displaystyle\int_a^t \Xi'(\xi)(\Xi(t)-\Xi(\xi))^{\zeta_1-1} e^{-\mu(\Xi(t)-\Xi(\xi))} \\ \quad \times g\big(\xi, w(\xi), {}^{TH}_{a^+}\mathfrak{D}^{\zeta_1,\zeta_2,\mu}_{\Xi(t)} w(\xi)\big) d\xi, \end{cases} \tag{11.2.2}$$

*provided that $d([w(t)]^l) - d([\rho(t)]^l)$ has a constant sign on $(a,b]$.*

**Theorem 11.2.1.** *Assume that* ($\mathbb{A}1$) *and* ($\mathbb{A}2$) *are fulfilled. If*

$$T = \min\left\{b, \Xi^{-1}\left[\left(\frac{(1-L_g)\Gamma(\zeta_1+\vartheta)}{Me^{-\mu(\Xi(t)-\Xi(a))}\Gamma(\zeta_1)}\right)^{\frac{1}{\zeta_1}} + \Xi(a)\right]\right\},$$

*then the problem* (11.0.1) *has a unique solution.*

*Proof.* Consider the operator $S : C_{1-\vartheta,\Xi}([a,b],E) \to C_{1-\vartheta,\Xi}([a,b],E)$

$$\begin{cases} (Sw(t)) \ominus_{gH} \rho(t) \\ \quad = \dfrac{1}{\Gamma(\zeta_1)} \displaystyle\int_a^t \Xi'(\xi)(\Xi(t)-\Xi(\xi))^{\zeta_1-1} e^{-\mu(\Xi(t)-\Xi(\xi))} \\ \quad \times g\big(\xi, w(\xi), {}^{TH}_{a^+}\mathfrak{D}^{\zeta_1,\zeta_2,\mu}_{\Xi(t)} w(\xi)\big) d\xi, \end{cases} \tag{11.2.3}$$

where,

$$\rho(t) = \frac{(\Xi(t)-\Xi(a))^{\vartheta-1} e^{-\mu(\Xi(t)-\Xi(a))}}{\Gamma(\vartheta)} w_0.$$

Let $a = t_0 < t_1 < \cdots < t_n = b$. Let us consider the complete metric space $C_{1-\vartheta,\Xi}([a,t_1],E)$ with respect to the metric

$$D_{C_{1-\vartheta,\Xi}[a,t_1]}[w,u] = \sup_{t\in[a,t_1]} (\Xi(t)-\Xi(a))^{1-\vartheta} D_0[w(t),u(t)].$$

Now, we show that the operator $S$ is a contraction. Choose $t_1 \in (a,b]$ such that

$$\frac{\Gamma(\vartheta)(\Xi(t_1)-\Xi(a))^{\zeta_1} e^{-\mu(\Xi(t_1)-\Xi(a))} M}{\Gamma(\vartheta+\zeta_1)(1-L_g)} < 1, \tag{11.2.4}$$

holds, where $M$ and $L_g$ are defined as in an assumption ($\mathbb{A}2$). Let $w,u \in C_{1-\vartheta,\Xi}([a,t_1],E)$, then by ($\mathbb{A}2$) one has

$$\begin{cases} D_{C_{1-\vartheta,\Xi}[a,t_1]}[{}^{TH}_{a+}\mathfrak{D}^{\zeta_1,\zeta_2,\mu}_{\Xi(t)} w, {}^{TH}_{a+}\mathfrak{D}^{\zeta_1,\zeta_2,\mu}_{\Xi(t)} u] \leq m(t) D_{C_{1-\gamma,\Xi}[a,t_1]}[w,u] \\ +L_g D_{C_{1-\gamma,\Xi}[a,t_1]}[{}^{TH}_{a+}\mathfrak{D}^{\zeta_1,\zeta_2,\mu}_{\Xi(t)} w, {}^{TH}_{a+}\mathfrak{D}^{\zeta_1,\zeta_2,\mu}_{\Xi(t)} u], \end{cases}$$

and so we get,

$$D_{C_{1-\vartheta,\Xi}[a,t_1]}[{}^{TH}_{a+}\mathfrak{D}^{\zeta_1,\zeta_2,\mu}_{\Xi(t)} w, {}^{TH}_{a+}\mathfrak{D}^{\zeta_1,\zeta_2,\mu}_{\Xi(t)} u] \leq \frac{m(t)}{1-L_g} D_{C_{1-\gamma,\Xi}[a,t_1]}[w,u]. \tag{11.2.5}$$

Setting $M = \max_{t\in[a,b]} m(t)$. Then, the (11.2.5) yields that

$$\begin{cases} D_{C_{1-\vartheta,\Xi}[a,t_1]}[(\Xi(t)-\Xi(a))^{1-\vartheta}(Sw), (\Xi(t)-\Xi(a))^{1-\vartheta}(Su)] \\ = D_{C_{1-\vartheta,\Xi}[a,t_1]}[(\Xi(t)-\Xi(a))^{1-\vartheta}(Sw \ominus_{gH} \rho(t)), \\ \quad (\Xi(t)-\Xi(a))^{1-\vartheta}(Su \ominus_{gH} \rho(t))] \end{cases}$$

$$\leq \sup_{t\in[a,t_1]} \frac{(\Xi(t)-\Xi(a))^{1-\vartheta}}{\Gamma(\zeta_1)} \int_a^t \Xi'(\xi)(\Xi(t)-\Xi(\xi))^{\zeta_1-1} e^{-\mu(\Xi(t)-\Xi(\xi))}$$

$$\times D_0\left[g(\xi, w(\xi), {}^{TH}_{a+}\mathfrak{D}^{\zeta_1,\zeta_2,\mu}_{\Xi(t)} w(\xi)), g(\xi, u(\xi), {}^{TH}_{a+}\mathfrak{D}^{\zeta_1,\zeta_2,\mu}_{\Xi(t)} u(\xi))\right] d\xi$$

$$\leq \sup_{t\in[a,t_1]}\left(\frac{m(t)}{1-L_g}D_{C_{1-\gamma,\Xi}[a,t_1]}[w,u]\right)(\Xi(t)-\Xi(a))^{1-\vartheta}$$
$$\times\, {}^{TH}_{a^+}\mathfrak{I}^{\zeta_1,\mu}_{\Xi(t)}[(\Xi(t)-\Xi(a))^{\vartheta-1}]$$
$$\leq \frac{\Gamma(\vartheta)[(\Xi(t_1)-\Xi(a))]^{\zeta_1}}{\Gamma(\vartheta+\zeta_1)}\frac{[(\Xi(t_1)-\Xi(a))]^{\vartheta}}{(\Xi(t_1)-\Xi(a))}e^{-\mu(\Xi(t_1)-\Xi(a))}$$
$$\times\left(\frac{M}{1-L_g}D_{C_{1-\gamma,\Xi}[a,t_1]}[w,u]\right).$$

Since $\frac{[(\Xi(t_1)-\Xi(a))]^{\vartheta}}{(\Xi(t_1)-\Xi(a))}<1$. By (11.2.4), the operator $S$ is a contraction. Hence, by Theorem 11.1.2, $S$ has a unique fixed point which is a unique solution of $w^*\in C_{1-\vartheta,\Xi}([a,t_1],E)$. This solution $w^*$ is obtained as a limit of a convergent sequence $S^k v$ :

$$\lim_{k\to\infty} D_{C_{1-\gamma,\Xi}[a,t_1]}[S^k v,w^*]=0.$$

For any function $v\in C_{1-\vartheta,\Xi}([a,t_1],E)$ and for $k\in\mathbb{N}$

$$(S^k v)(t)\ominus_{gH}\rho(t)=(SS^{k-1}v)(t)\ominus_{gH}\rho(t)$$
$$=\frac{1}{\Gamma(\zeta_1)}\int_a^t \Xi'(\xi)(\Xi(t)-\Xi(\xi))^{\zeta_1-1}e^{-\mu(\Xi(t)-\Xi(\xi))}$$
$$\times\, g\big(\xi,(S^{k-1}v)(\xi),{}^{TH}_{a^+}\mathfrak{D}^{\zeta_1,\zeta_2,\mu}_{\Xi(t)}(S^{k-1}v)(\xi)\big)d\xi.$$

Put $v(t)=\rho(t)$, where

$$\rho(t)=\frac{(\Xi(t)-\Xi(a))^{\vartheta-1}}{\Gamma(\vartheta)}e^{-\mu(\Xi(t)-\Xi(a))}w_0,$$

and denote, for $k\in\mathbb{N}$, $w_k(t)=(S^k v)(t)$. Then, we get for $t\in[a,t_1]$

$$\begin{cases} w_k(t)\ominus_{gH}\rho(t) \\ \quad =\dfrac{1}{\Gamma(\zeta_1)}\displaystyle\int_a^t \Xi'(\xi)(\Xi(t)-\Xi(\xi))^{\zeta_1-1}e^{-\mu(\Xi(t)-\Xi(\xi))} \\ \quad \times g\big(\xi,w_{k-1}(\xi),{}^{TH}_{a^+}\mathfrak{D}^{\zeta_1,\zeta_2,\mu}_{\Xi(t)}w_{k-1}(\xi)\big)d\xi, \end{cases} \tag{11.2.6}$$

and $\lim_{k\to\infty} D_{C_{1-\vartheta,\Xi}[a,t_1]}[w_k, w^*] = 0$.

Now, we define the interval $[t_1, t_2]$, where $t_2 = t_1 + h_1$, $h_1 > 0$ and $t_2 < b$. From (11.2.2) becomes on $[t_1, t_2]$ as follows:

$$\begin{cases} w(t) \ominus_{gH} \rho(t) \\ \quad = \dfrac{1}{\Gamma(\zeta_1)} \displaystyle\int_{t_1}^{t} \Xi^{'}(\xi)(\Xi(t) - \Xi(\xi))^{\zeta_1 - 1} e^{-\mu(\Xi(t)-\Xi(\xi))} \\ \quad \times g\big(\xi, w(\xi), {}_{a^+}^{TH}\mathfrak{D}_{\Xi(t)}^{\zeta_1,\zeta_2,\mu} w(\xi)\big) d\xi, \end{cases}$$

where $\rho_1(t)$ is defined by:

$$\rho_1(t) = \begin{cases} \rho(t) + \dfrac{1}{\Gamma(\zeta_1)} \displaystyle\int_{a}^{t} \Xi^{'}(\xi)(\Xi(t) - \Xi(\xi))^{\zeta_1 - 1} e^{-\mu(\Xi(t)-\Xi(\xi))} \\ \times g\big(\xi, w(\xi), {}_{a^+}^{TH}\mathfrak{D}_{\Xi(t)}^{\zeta_1,\zeta_2,\mu} w(\xi)\big) d\xi, \\ \text{if } w \text{ is } d\text{-increasing on } [a, t_2], \\ \rho(t) \ominus \dfrac{(-1)}{\Gamma(\zeta_1)} \displaystyle\int_{a}^{t} \Xi^{'}(\xi)(\Xi(t) - \Xi(\xi))^{\zeta_1 - 1} e^{-\mu(\Xi(t)-\Xi(\xi))} \\ \times g\big(\xi, w(\xi), {}_{a^+}^{TH}\mathfrak{D}_{\Xi(t)}^{\zeta_1,\zeta_2,\mu} w(\xi)\big) d\xi, \\ \text{if } w \text{ is } d\text{-decreasing on } [a, t_2]. \end{cases}$$

By $(\mathbb{A}2)$ and using the same argument as above, we observe that $\rho_1(t) \in C_{1-\vartheta,\Xi}([t_1, t_2], E)$, and it follows that there exists a unique solution $w^* \in C_{1-\vartheta,\Xi}([t_1, t_2], E)$ for the integral (11.2.2) on $[t_1, t_2]$. By repeating this process for the next intervals, we can conclude that there exists a unique solution $w^* \in C_{1-\vartheta,\Xi}([a, b], E)$ for the integral (11.2.2) on $[a, b]$ provided that

$$\frac{\Gamma(\vartheta)(\Xi(b) - \Xi(a))^{\zeta_1} e^{-\mu(\Xi(b)-\Xi(a))} M}{\Gamma(\vartheta + \zeta_1)(1 - L_g)} < 1.$$

Finally, we have to show that the obtained $w^* \in C_{1-\vartheta,\Xi}([a, b], E)$ of problem (11.2.2) holds problem (11.0.1). So, ${}_{a^+}^{TH}\mathfrak{D}_{\Xi(t)}^{\zeta_1,\zeta_2,\mu} w^*(t) \in C_{1-\vartheta,\Xi}([a, b], E)$ and $w^*$ satisfies the problem (11.0.1) provided that $d([w^*(t)]^l) - d([\rho(t)]^l)$ has a constant sign on $(a, b]$. We know that $w^*$ is the limit of the sequence $w_k \in C_{1-\vartheta,\Xi}([a,b],E)$, $k \in \mathbb{N}$, i.e,

$$D_{C_{1-\vartheta,\Xi}[a,b]}[w_k, w^*] \to 0, \quad \text{as} \quad k \to \infty. \tag{11.2.7}$$

Then, with the help of (11.2.6), the Lipscitz condition ($\mathbb{A}2$), and by Lemma 11.2.1 one has

$$\begin{cases} D_{C_{1-\vartheta,\Xi}[a,b]}\big[{}^{TH}_{a^+}\mathfrak{D}^{\zeta_1,\zeta_2,\mu}_{\Xi(t)}w_k, {}^{TH}_{a^+}\mathfrak{D}^{\zeta_1,\zeta_2,\mu}_{\Xi(t)}w^*\big] \\ = D_{C_{1-\vartheta,\Xi}[a,b]}\big[g\big(\xi, w_{k-1}(\xi), {}^{TH}_{a^+}\mathfrak{D}^{\zeta_1,\zeta_2,\mu}_{\Xi(t)}w_{k-1}(\xi)\big), \\ \quad g\big(\xi, w^*(\xi), {}^{TH}_{a^+}\mathfrak{D}^{\zeta_1,\zeta_2,\mu}_{\Xi(t)}w^*(\xi)\big)\big] \\ \leq m(t) D_{C_{1-\gamma,\Xi}[a,b]}[w_{k-1}, w^*] \\ \quad + L_g D_{C_{1-\gamma,\Xi}[a,b]}\big[{}^{TH}_{a^+}\mathfrak{D}^{\zeta_1,\zeta_2,\mu}_{\Xi(t)}w_{k-1}, {}^{TH}_{a^+}\mathfrak{D}^{\zeta_1,\zeta_2,\mu}_{\Xi(t)}w^*\big], \end{cases} \tag{11.2.8}$$

and so,

$$\begin{aligned} &D_{C_{1-\vartheta,\Xi}[a,b]}\big[{}^{TH}_{a^+}\mathfrak{D}^{\zeta_1,\zeta_2,\mu}_{\Xi(t)}w_k, {}^{TH}_{a^+}\mathfrak{D}^{\zeta_1,\zeta_2,\mu}_{\Xi(t)}w^*\big] \\ &\qquad \leq \frac{M}{1-L_g} D_{C_{1-\vartheta,\Xi}[a,b]}[w_{k-1}, w^*]. \end{aligned}$$

By (11.2.7), we get,

$$\lim_{k\to\infty} D_{C_{1-\vartheta,\Xi}[a,b]}\big[{}^{TH}_{a^+}\mathfrak{D}^{\zeta_1,\zeta_2,\mu}_{\Xi(t)}w_k, {}^{TH}_{a^+}\mathfrak{D}^{\zeta_1,\zeta_2,\mu}_{\Xi(t)}w_k^*\big] = 0. \tag{11.2.9}$$

Then, because

$${}^{TH}_{a^+}\mathfrak{D}^{\zeta_1,\zeta_2,\mu}_{\Xi(t)}w_k(t) = g(t, w_{k-1}(t), {}^{TH}_{a^+}\mathfrak{D}^{\zeta_1,\zeta_2,\mu}_{\Xi(t)}w_{k-1}(t)).$$

From (11.2.8)-(11.2.9), we have

$$g(., w_{k-1}(.), {}^{TH}_{a^+}\mathfrak{D}^{\zeta_1,\zeta_2,\mu}_{\Xi(t)}w_{k-1}(.)) \to g(., w^*(.), {}^{TH}_{a^+}\mathfrak{D}^{\zeta_1,\zeta_2,\mu}_{\Xi(t)}w^*(.))$$

as $k \to \infty$. Hence, (11.2.9) we conclude that ${}^{TH}_{a^+}\mathfrak{D}^{\zeta_1,\zeta_2,\mu}_{\Xi(t)}w^*(t)$ belongs to $C_{1-\vartheta,\Xi}([a,b],E)$. Therefore, $w^*$ satisfies the problem (11.0.1).

## 11.3 Continuous dependence of solution and $\delta$-approximate solution

Consider the system of tempered Ξ-Hilfer-type implicit FFDEs

$$\begin{cases} {}^{TH}_{a^+}\mathfrak{D}^{\zeta_1-\eta,\zeta_2,\mu}_{\Xi(t)}w(t) = g_1(t, w(t), {}^{TH}_{a^+}\mathfrak{D}^{\zeta_1-\eta,\zeta_2,\mu}_{\Xi(t)}w(t), \chi_1), \\ {}^{T}_{a^+}\mathfrak{I}^{1-\vartheta-\eta(\zeta_2-1),\mu}_{\Xi(t)}w(a) = w_0, \quad \zeta_1 \leq \vartheta = \zeta_1 + \zeta_2 - \zeta_1\zeta_2 < 1, \end{cases} \tag{11.3.10}$$

and

$$\begin{cases} {}^{TH}_{a^+}\mathfrak{D}^{\zeta_1,\zeta_2,\mu}_{\Xi(t)}u(t) = g_2(t, u(t), {}^{TH}_{a^+}\mathfrak{D}^{\zeta_1,\zeta_2,\mu}_{\Xi(t)}u(t), \chi_2), \\ {}^{T}_{a^+}\mathfrak{I}^{1-\vartheta}_{\Xi(t)}u(a) = u_0, \end{cases} \tag{11.3.11}$$

where $\chi_1, \chi_2$ are fuzzy parameters, $\eta > 0$ satisfies $0 < \zeta_1 - \eta < \zeta_1 \leq 1$.

**Theorem 11.3.1.** *Assume that $g_1, g_2 : (a, b] \times E \times E \times E \to E$ satisfy the following assumptions:*

*(i) There exist $\epsilon, \epsilon_1 > 0$ such that for each $\bar{w} \in E$ for $t \in [a, b]$ and $\chi_1 \in E$*

$$\begin{aligned} D_0[&g_1(t, \bar{w}(t), {}^{TH}_{a^+}\mathfrak{D}^{\zeta_1-\eta,\zeta_2,\mu}_{\Xi(t)}\bar{w}(t), \chi_1), \\ &g_2(t, \bar{w}(t), {}^{TH}_{a^+}\mathfrak{D}^{\zeta_1-\eta,\zeta_2,\mu}_{\Xi(t)}\bar{w}(t), \chi_1)] \leq \epsilon_1, \\ D_0[&g_2(t, \bar{w}(t), {}^{TH}_{a^+}\mathfrak{D}^{\zeta_1-\eta,\zeta_2,\mu}_{\Xi(t)}\bar{w}(t), \chi_1), \\ &g_2(t, \bar{w}(t), {}^{TH}_{a^+}\mathfrak{D}^{\zeta_1-\eta,\zeta_2,\mu}_{\Xi(t)}\bar{w}(t), \chi_1)] \leq \epsilon_2. \end{aligned}$$

*(ii) There exists a continuous real-valued function $m : [a, b] \to \mathbb{R}^+$ and a constant $L_g \in (0, 1)$ such that for each $t \in [a, b]$, and all $\bar{w}, v \in E$,*

*we have*

$$\begin{cases} D_0[g_*(t, \bar{w}(t), {}^{TH}_{a^+}\mathfrak{D}^{\zeta_1-\eta,\zeta_2,\mu}_{\Xi(t)}\bar{w}(t), \chi_1), \\ \quad g_*(t, v(t), {}^{TH}_{a^+}\mathfrak{D}^{\zeta_1-\eta,\zeta_2,\mu}_{\Xi(t)}v(t), \chi_1)] \\ \leq m(t)D_0[\bar{w}, v] + L_g D_0[{}^{TH}_{a^+}\mathfrak{D}^{\zeta_1-\eta,\zeta_2,\mu}_{\Xi(t)}\bar{w}, {}^{TH}_{a^+}\mathfrak{D}^{\zeta_1-\eta,\zeta_2,\mu}_{\Xi(t)}v], \end{cases}$$

*where* $g_*$ *denotes the functions* $g_1, g_2$.

(*iii*) *There exists a continuous real-valued function* $p : [a, b] \to \mathbb{R}^+$ *such that for each* $\bar{w} \in E$ *and for* $t \in [a, b]$

$$\begin{cases} D_0[g_*(t, \bar{w}(t), {}^{TH}_{a^+}\mathfrak{D}^{\zeta_1-\eta,\zeta_2,\mu}_{\Xi(t)}\bar{w}(t), \chi_1), \\ \qquad g_*(t, \bar{w}(t), {}^{TH}_{a^+}\mathfrak{D}^{\zeta_1-\eta,\zeta_2,\mu}_{\Xi(t)}\bar{w}(t), \chi_2)] \leq p(t)D_0[\chi_1, \chi_2], \end{cases}$$

*where* $g_*$ *denotes the functions* $g_1, g_2$.

(*iv*) *There exist* $K > 0$ *such that for each* $\bar{w}, v \in E$ *and for* $t \in [a, b]$

$$\max\Big\{ D_0[g_1(t, \bar{w}(t), {}^{TH}_{a^+}\mathfrak{D}^{\zeta_1-\eta,\zeta_2,\mu}_{\Xi(t)}\bar{w}(t), \chi_1), \widehat{0}], \\ D_0[g_2(t, v(t), {}^{TH}_{a^+}\mathfrak{D}^{\zeta_1-\eta,\zeta_2,\mu}_{\Xi(t)}v(t), \chi_2), \widehat{0}] \Big\} \leq K.$$

*Then, if* $w(t), u(t)$ *are the solutions of* (11.3.10) *and* (11.3.11), *transform, we get the following estimate:*
$D_0[w(t), u(t)]$

$$\begin{cases} \leq Q(t) + \int_a^t \Bigg[ \sum_{k=1}^{\infty} \left( \frac{M\Gamma(\zeta_1 - \eta)}{(1 - L_g)\Gamma(\zeta_1)} \right)^k \\ \quad \times \frac{\Xi'(\xi)(\Xi(t) - \Xi(\xi))^{k(\zeta_1-\eta)-1} e^{-\mu(\Xi(t)-\Xi(a))}}{\Gamma(k(\zeta_1 - \eta))} Q(\xi) \Bigg] d\xi, \end{cases} \tag{11.3.12}$$

*where*

$$\begin{cases} Q(t) = D_0\Bigg[\dfrac{(\Xi(t)-\Xi(a))^{\vartheta+\eta(\zeta_2-1)-1}e^{-\mu(\Xi(t)-\Xi(a))}}{\Gamma(\vartheta+\eta(\zeta_2-1))}w_0, \\ \dfrac{(\Xi(t)-\Xi(a))^{\vartheta-1}e^{-\mu(\Xi(t)-\Xi(a))}}{\Gamma(\vartheta)}u_0\Bigg] \\ +K\Bigg|\dfrac{(\Xi(t)-\Xi(a))^{\zeta_1-\eta}e^{-\mu(\Xi(t)-\Xi(a))}}{\Gamma(\zeta_1-\eta+1))} \\ -\dfrac{(\Xi(t)-\Xi(a))^{\zeta_1-\eta}e^{-\mu(\Xi(t)-\Xi(a))}}{\Gamma(\zeta_1-\eta))\Gamma(\zeta_1)}\Bigg| \\ +K\Bigg|\dfrac{(\Xi(t)-\Xi(a))^{\zeta_1-\eta}e^{-\mu(\Xi(t)-\Xi(a))}}{\Gamma(\zeta_1-\eta))\Gamma(\zeta_1)} \\ -\dfrac{(\Xi(t)-\Xi(a))^{\zeta_1}e^{-\mu(\Xi(t)-\Xi(a))}}{\Gamma(\zeta_1+1))}\Bigg| \\ +\epsilon_2\dfrac{(\Xi(t)-\Xi(a))^{\zeta_1-\eta}e^{-\mu(\Xi(t)-\Xi(a))}}{\Gamma(\zeta_1-\eta))\Gamma(\zeta_1)} \\ +\dfrac{1}{(1-L_g)\Gamma(\zeta_1)}\Bigg(\dfrac{(\Xi(t)-\Xi(a))^{\zeta_1-\eta}e^{-\mu(\Xi(t)-\Xi(a))}}{(\zeta_1-\eta)} \\ +\dfrac{PD_0[\chi_1,\chi_2](\Xi(t)-\Xi(a))^{\zeta_1-\eta}e^{-\mu(\Xi(t)-\Xi(a))}}{(\zeta_1-\eta)}\Bigg). \end{cases} \tag{11.3.13}$$

*Proof.* The problems (11.0.1) and (11.3.10)-(11.3.11) are given by

$$\begin{cases} w(t)\ominus_{gH}\dfrac{(\Xi(t)-\Xi(a))^{\vartheta+\eta(\zeta_2-1)-1}e^{-\mu(\Xi(t)-\Xi(a))}}{\Gamma(\vartheta+\eta(\zeta_2-1))}w_0 \\ =\dfrac{1}{\Gamma(\zeta_1-\eta)}\displaystyle\int_a^t \Xi^{'}(\xi)(\Xi(t)-\Xi(\xi))^{\zeta_1-\eta-1}e^{-\mu(\Xi(t)-\Xi(\xi))} \\ \times g_1\big(\xi, w(\xi), {}_{a^+}^{TH}\mathfrak{D}_{\Xi(t)}^{\zeta_1-\eta,\zeta_2,\mu}w(\xi),\chi_1\big)d\xi, \quad t>a, \end{cases} \tag{11.3.14}$$

and

$$\begin{cases} u(t)\ominus_{gH}\dfrac{(\Xi(t)-\Xi(a))^{\vartheta-1}e^{-\mu(\Xi(t)-\Xi(a))}}{\Gamma(\vartheta)}u_0 \\ =\dfrac{1}{\Gamma(\zeta_1)}\displaystyle\int_a^t \Xi^{'}(\xi)(\Xi(t)-\Xi(\xi))^{\zeta_1-1}e^{-\mu(\Xi(t)-\Xi(\xi))} \\ \times g_2\big(\xi, u(\xi), {}_{a^+}^{TH}\mathfrak{D}_{\Xi(t)}^{\zeta_1,\zeta_2,\mu}u(\xi),\chi_2\big)d\xi, \quad t>a. \end{cases} \tag{11.3.15}$$

Furthermore, by $(i)$-$(iii)$ one has

$$
\begin{aligned}
&D_0[{}^{TH}_{a^+}\mathfrak{D}^{\zeta_1-\eta,\zeta_2,\mu}_{\Xi(t)}w(t), {}^{TH}_{a^+}\mathfrak{D}^{\zeta_1-\eta,\zeta_2,\mu}_{\Xi(t)}u(t)]\\
&\quad = D_0[g_1\big(t, w(t), {}^{TH}_{a^+}\mathfrak{D}^{\zeta_1-\eta,\zeta_2,\mu}_{\Xi(t)}w(t), \chi_1\big),\\
&\quad\quad g_2\big(t, u(t), {}^{TH}_{a^+}\mathfrak{D}^{\zeta_1-\eta,\zeta_2,\mu}_{\Xi(t)}u(t), \chi_2\big)]\\
&\quad \leq D_0[g_1\big(t, w(t), {}^{TH}_{a^+}\mathfrak{D}^{\zeta_1-\eta,\zeta_2,\mu}_{\Xi(t)}w(t), \chi_1\big),\\
&\quad\quad g_2\big(t, w(t), {}^{TH}_{a^+}\mathfrak{D}^{\zeta_1-\eta,\zeta_2,\mu}_{\Xi(t)}w(t), \chi_1\big)]\\
&\quad + D_0[g_2\big(t, w(t), {}^{TH}_{a^+}\mathfrak{D}^{\zeta_1-\eta,\zeta_2,\mu}_{\Xi(t)}w(t), \chi_1\big),\\
&\quad\quad g_2\big(t, u(t), {}^{TH}_{a^+}\mathfrak{D}^{\zeta_1-\eta,\zeta_2,\mu}_{\Xi(t)}u(t), \chi_2\big)]\\
&\quad \leq \epsilon_1 + m(t)D_0[w, u] + L_g D_0[{}^{TH}_{a^+}\mathfrak{D}^{\zeta_1-\eta,\zeta_2,\mu}_{\Xi(t)}w, {}^{TH}_{a^+}\mathfrak{D}^{\zeta_1-\eta,\zeta_2,\mu}_{\Xi(t)}u]\\
&\quad\quad + p(t)D_0[\chi_1, \chi_2].
\end{aligned}
$$

It follows that
$D_0[{}^{TH}_{a^+}\mathfrak{D}^{\zeta_1-\eta,\zeta_2,\mu}_{\Xi(t)}w(t), {}^{TH}_{a^+}\mathfrak{D}^{\zeta_1-\eta,\zeta_2,\mu}_{\Xi(t)}u(t)]$

$$
\leq \frac{1}{1-L_g}(\epsilon_1 + p(t)D_0[\chi_1, \chi_2] + m(t)D_0[w, u]). \tag{11.3.16}
$$

With the help of (11.3.14)-(11.3.15) and by $(i)$-$(iv)$, we get $D_0[w(t), u(t)]$

$$
\begin{aligned}
&\leq D_0\Bigg[\frac{(\Xi(t)-\Xi(a))^{\vartheta+\eta(\zeta_2-1)-1}}{\Gamma(\vartheta+\eta(\zeta_2-1))}e^{-\mu(\Xi(t)-\Xi(a))}w_0,\\
&\quad \frac{(\Xi(t)-\Xi(a))^{\vartheta-1}}{\Gamma(\vartheta)}e^{-\mu(\Xi(t)-\Xi(a))}u_0\Bigg]\\
&\quad + D_0\Bigg[\frac{1}{\Gamma(\zeta_1-\eta)}\int_a^t \Xi'(\xi)(\Xi(t)-\Xi(\xi))^{\zeta_1-\eta-1}e^{-\mu(\Xi(t)-\Xi(\xi))}\\
&\quad \times g_1\big(\xi, w(\xi), {}^{TH}_{a^+}\mathfrak{D}^{\zeta_1-\eta,\zeta_2,\mu}_{\Xi(t)}w(\xi), \chi_1\big)d\xi,\\
&\quad \frac{1}{\Gamma(\zeta_1)}\int_a^t \Xi'(\xi)(\Xi(t)-\Xi(\xi))^{\zeta_1-1}e^{-\mu(\Xi(t)-\Xi(\xi))}\\
&\quad \times g_2\big(\xi, u(\xi), {}^{TH}_{a^+}\mathfrak{D}^{\zeta_1,\zeta_2,\mu}_{\Xi(t)}u(\xi), \chi_2\big)d\xi\Bigg].
\end{aligned}
$$

Therefore, we have

$$
\begin{aligned}
&D_0[w(t), u(t)]\\
&\leq D_0\Bigg[\frac{(\Xi(t)-\Xi(a))^{\vartheta+\eta(\zeta_2-1)-1}e^{-\mu(\Xi(t)-\Xi(a))}}{\Gamma(\vartheta+\eta(\zeta_2-1))}w_0,\\
&\quad \frac{(\Xi(t)-\Xi(a))^{\vartheta-1}e^{-\mu(\Xi(t)-\Xi(a))}}{\Gamma(\vartheta)}u_0\Bigg]\\
&\quad + D_0\Bigg[\frac{1}{\Gamma(\zeta_1-\eta)}\int_a^t \Xi^{'}(\xi)(\Xi(t)-\Xi(\xi))^{\zeta_1-\eta-1}e^{-\mu(\Xi(t)-\Xi(\xi))}\\
&\quad \times g_1\big(\xi, w(\xi), {}_{a^+}^{TH}\mathfrak{D}_{\Xi(t)}^{\zeta_1-\eta,\zeta_2,\mu}w(\xi), \chi_1\big)d\xi,\\
&\quad \frac{1}{\Gamma(\zeta_1)}\int_a^t \Xi^{'}(\xi)(\Xi(t)-\Xi(\xi))^{\zeta_1-\eta-1}e^{-\mu(\Xi(t)-\Xi(\xi))}\\
&\quad \times g_1\big(\xi, w(\xi), {}_{a^+}^{TH}\mathfrak{D}_{\Xi(t)}^{\zeta_1-\eta,\zeta_2,\mu}w(\xi), \chi_1\big)d\xi\Bigg]\\
&\quad + D_0\Bigg[\frac{1}{\Gamma(\zeta_1)}\int_a^t \Xi^{'}(\xi)(\Xi(t)-\Xi(\xi))^{\zeta_1-\eta-1}e^{-\mu(\Xi(t)-\Xi(\xi))}\\
&\quad \times g_1\big(\xi, w(\xi), {}_{a^+}^{TH}\mathfrak{D}_{\Xi(t)}^{\zeta_1-\eta,\zeta_2,\mu}w(\xi), \chi_1\big)d\xi,\\
&\quad \frac{1}{\Gamma(\zeta_1)}\int_a^t \Xi^{'}(\xi)(\Xi(t)-\Xi(\xi))^{\zeta_1-\eta-1}e^{-\mu(\Xi(t)-\Xi(\xi))}\\
&\quad \times g_2\big(\xi, u(\xi), {}_{a^+}^{TH}\mathfrak{D}_{\Xi(t)}^{\zeta_1-\eta,\zeta_2,\mu}u(\xi), \chi_2\big)d\xi\Bigg]\\
&\quad + D_0\Bigg[\frac{1}{\Gamma(\zeta_1)}\int_a^t \Xi^{'}(\xi)(\Xi(t)-\Xi(\xi))^{\zeta_1-\eta-1}e^{-\mu(\Xi(t)-\Xi(\xi))}\\
&\quad \times g_2\big(\xi, u(\xi), {}_{a^+}^{TH}\mathfrak{D}_{\Xi(t)}^{\zeta_1-\eta,\zeta_2,\mu}u(\xi), \chi_2\big)d\xi,\\
&\quad \frac{1}{\Gamma(\zeta_1)}\int_a^t \Xi^{'}(\xi)(\Xi(t)-\Xi(\xi))^{\zeta_1-\eta-1}e^{-\mu(\Xi(t)-\Xi(\xi))}\\
&\quad \times g_2\big(\xi, u(\xi), {}_{a^+}^{TH}\mathfrak{D}_{\Xi(t)}^{\zeta_1,\zeta_2,\mu}u(\xi), \chi_2\big)d\xi\Bigg].
\end{aligned}
$$

This implies

$$
\begin{cases}
D_0[w(t),u(t)] \\
\quad \leq D_0\left[\dfrac{(\Xi(t)-\Xi(a))^{\vartheta+\eta(\zeta_2-1)-1}e^{-\mu(\Xi(t)-\Xi(a))}}{\Gamma(\vartheta+\eta(\zeta_2-1))}w_0,\right. \\
\quad \left.\dfrac{(\Xi(t)-\Xi(a))^{\vartheta-1}e^{-\mu(\Xi(t)-\Xi(a))}}{\Gamma(\vartheta)}u_0\right] \\
\quad +K\left|\dfrac{(\Xi(t)-\Xi(a))^{\zeta_1-\eta}e^{-\mu(\Xi(t)-\Xi(a))}}{\Gamma(\zeta_1-\eta+1)}\right. \\
\quad \left.-\dfrac{(\Xi(t)-\Xi(a))^{\zeta_1-\eta}e^{-\mu(\Xi(t)-\Xi(a))}}{\Gamma(\zeta_1)\Gamma(\zeta_1-\eta)}\right| \\
\quad +\dfrac{1}{(1-L_g)\Gamma(\zeta_1)}\displaystyle\int_a^t \Xi^{'}(\xi)(\Xi(t)-\Xi(\xi))^{\zeta_1-\eta-1}e^{-\mu(\Xi(t)-\Xi(\xi))} \\
\quad \times(\epsilon_1+m(\xi)D_0[w(\xi),u(\xi)]+p(\xi)D_0[\chi_1,\chi_2])d\xi \\
\quad +K\left|\dfrac{(\Xi(t)-\Xi(a))^{\zeta_1-\eta}e^{-\mu(\Xi(t)-\Xi(a))}}{\Gamma(\zeta_1)\Gamma(\zeta_1-\eta)}\right. \\
\quad \left.-\dfrac{(\Xi(t)-\Xi(a))^{\zeta_1}e^{-\mu(\Xi(t)-\Xi(a))}}{\Gamma(\zeta_1+1)}\right| \\
\quad +\epsilon_2\dfrac{(\Xi(t)-\Xi(a))^{\zeta_1-\eta}}{\Gamma(\zeta_1-\eta)\Gamma(\zeta_1)}e^{-\mu(\Xi(t)-\Xi(a))}.
\end{cases}
\tag{11.3.17}
$$

Set $M=\max_{t\in[a,b]} m(t)$ and $P=\max_{t\in[a,b]} p(t)$, one has

$$
\begin{aligned}
&D_0[w(t),u(t)] \\
&\quad \leq Q(t)+\frac{M}{(1-L_g)\Gamma(\zeta_1)}\int_a^t \Xi^{'}(\xi)(\Xi(t)-\Xi(\xi))^{\zeta_1-\eta-1}e^{-\mu(\Xi(t)-\Xi(\xi))} \\
&\quad \times D_0[w(\xi),u(\xi)]d\xi,
\end{aligned}
$$

where $Q(t)$ is defined as in (11.3.13). By applying generalized Gronwall's inequality

$$
\begin{aligned}
D_0[w(t),u(t)] \leq & Q(t)+\int_a^t\left[\sum_{k=1}^{\infty}\left(\frac{M\Gamma(\zeta_1-\eta)}{(1-L_g)\Gamma(\zeta_1)}\right)^k\right. \\
& \left.\times\frac{\Xi^{'}(\xi)(\Xi(t)-\Xi(\xi))^{k(\zeta_1-\eta)-1}e^{-\mu(\Xi(t)-\Xi(a))}}{\Gamma(k(\zeta_1-\eta))}Q(\xi)\right]d\xi.
\end{aligned}
$$

Hence, we obtain (11.3.12).

**Definition 11.3.1.** *A function $w \in C_{1-\vartheta,\Xi}([a,b],E)$ satisfying the tempered Ξ-Hilfer implicit fuzzy fractional differential inequality*

$$D_0[{}^{TH}_{a^+}\mathfrak{D}^{\zeta_1,\zeta_2,\mu}_{\Xi(t)}w(t), g_1(t,w,{}^{TH}_{a^+}\mathfrak{D}^{\zeta_1,\zeta_2,\mu}_{\Xi(t)}w)] \leq \delta, \quad t \in [a,b],$$

*and ${}^{T}_{a^+}\mathfrak{J}^{1-\vartheta,\mu}_{\Xi(t)}w(a) = w_0$, is called $\delta$-approximate solutions of the problem (11.0.1).*

**Theorem 11.3.2.** *Let $g : (a,b] \times E \times E \to E$ satisfy*

$$\begin{cases} D_0[g(t,w,{}^{TH}_{a^+}\mathfrak{D}^{\zeta_1,\zeta_2,\mu}_{\Xi(t)}w), g(t,u,{}^{TH}_{a^+}\mathfrak{D}^{\zeta_1,\zeta_2,\mu}_{\Xi(t)}u)] \\ \quad \leq m(t)D_0[w,u] + L_g D_0[{}^{TH}_{a^+}\mathfrak{D}^{\zeta_1,\zeta_2,\mu}_{\Xi(t)}w, {}^{TH}_{a^+}\mathfrak{D}^{\zeta_1,\zeta_2,\mu}_{\Xi(t)}u], \end{cases}$$

*where $L_g \in (0,1)$ is a constant and $w,u \in E$, $m : [a,b] \to \mathbb{R}^+$ is a continuous function. Let $w,u \in C_{1-\vartheta,\Xi}([a,b],E)$ be a $\delta_w,\delta_u$-approximate solutions of the problem (11.0.1) corresponding to the initial conditions $w_0,u_0$, respectively. Then,*

$$\begin{cases} D_{C_{1-\vartheta,\Xi}[a,b]}[w,u] \leq & D_0[w_0,u_0]\mathbb{E}_{\zeta_1,\vartheta}\left(\dfrac{M(\Xi(b)-\Xi(a))^{\zeta_1}}{1-L_g}\right) \\ & +(\Xi(b)-\Xi(a))^{\zeta_1}\dfrac{(\delta_w+\delta_u)}{(1-L_g)}\mathbb{E}_{\zeta_1,\zeta_1+1} \\ & \times\left(\dfrac{M}{1-L_g}(\Xi(b)-\Xi(a))^{\zeta_1}\right). \end{cases} \tag{11.3.18}$$

*Proof.* Let $w,u \in C_{1-\vartheta}((a,b],E)$ be a $\delta_w$-approximate and $\delta_u$-approximate solutions of the problem (11.0.1). Then, we have

$$\begin{aligned} &D_0[{}^{TH}_{a^+}\mathfrak{D}^{\zeta_1,\zeta_2,\mu}_{\Xi(t)}w(t), g(t,w,{}^{TH}_{a^+}\mathfrak{D}^{\zeta_1,\zeta_2,\mu}_{\Xi(t)}w)] \leq \delta_w, \\ &{}^{T}_{a^+}\mathfrak{J}^{1-\vartheta,\mu}_{\Xi(t)}w(a) = w_0, \end{aligned}$$

and

$$D_0[{}^{TH}_{a^+}\mathfrak{D}^{\zeta_1,\zeta_2,\mu}_{\Xi(t)}u(t), g(t,u,{}^{TH}_{a^+}\mathfrak{D}^{\zeta_1,\zeta_2,\mu}_{\Xi(t)}u)] \leq \delta_u, {}^{T}_{a^+}\mathfrak{J}^{1-\vartheta,\mu}_{\Xi(t)}u(a) = u_0.$$

This yields that

$$\begin{cases} \dfrac{1}{\Gamma(\zeta_1)}\displaystyle\int_a^t \Xi'(\xi)(\Xi(t)-\Xi(\xi))^{\zeta_1-1}e^{-\mu(\Xi(t)-\Xi(\xi))}\delta_w d\xi \\ \geq \dfrac{1}{\Gamma(\zeta_1)}\displaystyle\int_a^t \Xi'(\xi)(\Xi(t)-\Xi(\xi))^{\zeta_1-1}e^{-\mu(\Xi(t)-\Xi(\xi))} \\ \quad \times D_0[{}_{a^+}^{TH}\mathfrak{D}_{\Xi(t)}^{\zeta_1,\zeta_2,\mu}w(t), g(t,w,{}_{a^+}^{TH}\mathfrak{D}_{\Xi(t)}^{\zeta_1,\zeta_2,\mu}w)]d\xi \\ \geq D_0\Bigg[w(t)\ominus_{gH}\dfrac{(\Xi(t)-\Xi(a))^{\vartheta-1}}{\Gamma(\vartheta)}e^{-\mu(\Xi(t)-\Xi(\xi))}w_0, \\ \quad {}_{a^+}^{T}\mathfrak{I}_{\Xi(t)}^{\zeta_1,\mu}g(t,w,{}_{a^+}^{TH}\mathfrak{D}_{\Xi(t)}^{\zeta_1,\zeta_2,\mu}w)\Bigg] \end{cases}$$

and so one has

$$\begin{cases} \dfrac{\delta_w(\Xi(t)-\Xi(a))^{\zeta_1}e^{-\mu(\Xi(t)-\Xi(a))}}{\Gamma(\zeta_1+1)} \\ \geq D_0\Bigg[w(t)\ominus_{gH}\dfrac{(\Xi(t)-\Xi(a))^{\vartheta-1}}{\Gamma(\vartheta)}e^{-\mu(\Xi(t)-\Xi(\xi))}w_0, \\ \quad {}_{a^+}^{T}\mathfrak{I}_{\Xi(t)}^{\zeta_1,\mu}g(t,w,{}_{a^+}^{TH}\mathfrak{D}_{\Xi(t)}^{\zeta_1,\zeta_2,\mu}w)\Bigg]. \end{cases}$$

Using the fact $D_0[w,\widehat{0}] - D_0[u,\widehat{0}] \leq D_0[w,u] \leq D_0[w,\widehat{0}] + D_0[u,\widehat{0}]$ in the above inequality, we have

$$\begin{cases} \dfrac{(\delta_w+\delta_u)(\Xi(t)-\Xi(a))^{\zeta_1}e^{-\mu(\Xi(t)-\Xi(a))}}{\Gamma(\zeta_1+1)} \\ \geq D_0\Bigg[w(t)\ominus_{gH}\dfrac{(\Xi(t)-\Xi(a))^{\vartheta-1}}{\Gamma(\vartheta)}e^{-\mu(\Xi(t)-\Xi(a))}w_0, \\ \quad {}_{a^+}^{T}\mathfrak{I}_{\Xi(t)}^{\zeta_1,\mu}g(t,w,{}_{a^+}^{TH}\mathfrak{D}_{\Xi(t)}^{\zeta_1,\zeta_2,\mu}w)\Bigg] \\ \quad +D_0\Bigg[u(t)\ominus_{gH}\dfrac{(\Xi(t)-\Xi(a))^{\vartheta-1}}{\Gamma(\vartheta)}e^{-\mu(\Xi(t)-\Xi(a))}u_0, \\ \quad {}_{a^+}^{T}\mathfrak{I}_{\Xi(t)}^{\zeta_1,\mu}g(t,u,{}_{a^+}^{TH}\mathfrak{D}_{\Xi(t)}^{\zeta_1,\zeta_2,\mu}u)\Bigg] \end{cases}$$

$$\begin{cases} \geq D_0\Big[w(t)\ominus_{gH}\dfrac{(\Xi(t)-\Xi(a))^{\vartheta-1}}{\Gamma(\vartheta)}e^{-\mu(\Xi(t)-\Xi(a))}w_0, \\ \quad u(t)\ominus_{gH}\dfrac{(\Xi(t)-\Xi(a))^{\vartheta-1}}{\Gamma(\vartheta)}e^{-\mu(\Xi(t)-\Xi(a))}u_0\Big] \\ \quad -D_0\Big[{}^{T}_{a+}\mathfrak{I}^{\zeta_1,\mu}_{\Xi(t)}g(t,w,{}^{TH}_{a+}\mathfrak{D}^{\zeta_1,\zeta_2,\mu}_{\Xi(t)}w),{}^{T}_{a+}\mathfrak{I}^{\zeta_1,\mu}_{\Xi(t)}g(t,u,{}^{TH}_{a+}\mathfrak{D}^{\zeta_1,\zeta_2,\mu}_{\Xi(t)}u)\Big]. \end{cases}$$

By ($\mathbb{A}2$), we have

$$\begin{cases} D_0[w(t),u(t)] \\ \leq \dfrac{(\Xi(t)-\Xi(a))^{\vartheta-1}e^{-\mu(\Xi(t)-\Xi(a))}}{\Gamma(\vartheta)}D_0[w_0,u_0] \\ \quad +\dfrac{(\delta_w+\delta_u)(\Xi(t)-\Xi(a))^{\zeta_1}e^{-\mu(\Xi(t)-\Xi(a))}}{\Gamma(\zeta_1+1)} \\ \quad +\dfrac{1}{\Gamma(\zeta_1)}\displaystyle\int_a^t \Xi'(\xi)(\Xi(t)-\Xi(\xi))^{\zeta_1-1}e^{-\mu(\Xi(t)-\Xi(\xi))} \\ \quad \times(m(\xi)D_0[w(\xi),u(\xi)] \\ \quad +L_gD_0[{}^{TH}_{a+}\mathfrak{D}^{\zeta_1,\zeta_2,\mu}_{\Xi(t)}w(\xi),{}^{TH}_{a+}\mathfrak{D}^{\zeta_1,\zeta_2,\mu}_{\Xi(t)}u(\xi)])d\xi. \end{cases} \tag{11.3.19}$$

On the other hand, one also has

$$\begin{aligned} \delta_w+\delta_u &\geq D_0\big[{}^{TH}_{a+}\mathfrak{D}^{\zeta_1,\zeta_2,\mu}_{\Xi(t)}w(t)\ominus_{gH}g(t,w,{}^{TH}_{a+}\mathfrak{D}^{\zeta_1,\zeta_2,\mu}_{\Xi(t)}w),\widehat{0}\big] \\ &\quad + D_0\big[{}^{TH}_{a+}\mathfrak{D}^{\zeta_1,\zeta_2,\mu}_{\Xi(t)}u(t)\ominus_{gH}g(t,u,{}^{TH}_{a+}\mathfrak{D}^{\zeta_1,\zeta_2,\mu}_{\Xi(t)}u),\widehat{0}\big] \\ &\geq D_0\big[{}^{TH}_{a+}\mathfrak{D}^{\zeta_1,\zeta_2,\mu}_{\Xi(t)}w(t)\ominus_{gH}{}^{TH}_{a+}\mathfrak{D}^{\zeta_1,\zeta_2,\mu}_{\Xi(t)}u(t), \\ &\qquad g(t,w,{}^{TH}_{a+}\mathfrak{D}^{\zeta_1,\zeta_2,\mu}_{\Xi(t)}w)\ominus_{gH}g(t,u,{}^{TH}_{a+}\mathfrak{D}^{\zeta_1,\zeta_2,\mu}_{\Xi(t)}u)\big] \\ &\geq D_0\big[{}^{TH}_{a+}\mathfrak{D}^{\zeta_1,\zeta_2,\mu}_{\Xi(t)}w(t)\ominus_{gH}{}^{TH}_{a+}\mathfrak{D}^{\zeta_1,\zeta_2,\mu}_{\Xi(t)}u(t),\widehat{0}\big] \\ &\qquad - D_0\big[g(t,w,{}^{TH}_{a+}\mathfrak{D}^{\zeta_1,\zeta_2,\mu}_{\Xi(t)}w)\ominus_{gH}g(t,u,{}^{TH}_{a+}\mathfrak{D}^{\zeta_1,\zeta_2,\mu}_{\Xi(t)}u),\widehat{0}\big], \end{aligned}$$

and so it yields that

$$\begin{cases} D_0\left[{}^{TH}_{a+}\mathfrak{D}^{\zeta_1,\zeta_2,\mu}_{\Xi(t)}w(t), {}^{TH}_{a+}\mathfrak{D}^{\zeta_1,\zeta_2,\mu}_{\Xi(t)}u(t)\right] \\ \leq \dfrac{(\delta_w+\delta_u)}{1-L_g} + \dfrac{M}{1-L_g}D_0[w(t),u(t)]. \end{cases} \tag{11.3.20}$$

From Eqns. (11.3.19)-(11.3.20), we get

$$\begin{aligned} & D_0[w(t),u(t)] \\ & \leq \frac{(\Xi(t)-\Xi(a))^{\vartheta-1}e^{-\mu(\Xi(t)-\Xi(a))}}{\Gamma(\vartheta)}D_0[w_0,u_0] \\ & \quad + \frac{(\delta_w+\delta_u)(\Xi(t)-\Xi(a))^{\zeta_1}e^{-\mu(\Xi(t)-\Xi(a))}}{\Gamma(\zeta_1+1)(1-L_g)} \\ & \quad + \frac{M}{(1-L_g)\Gamma(\zeta_1)}\int_a^t \Xi'(\xi)(\Xi(t)-\Xi(\xi))^{\zeta_1-1} \\ & \quad \times e^{-\mu(\Xi(t)-\Xi(\xi))}D_0[w(\xi),u(\xi)]d\xi. \end{aligned}$$

Therefore, by the generalized Gronwall's inequality, we have

$$\begin{cases} D_0[w(t),v(t)] \\ \leq Q(t) + \int_a^t \sum_{k=1}^{\infty}\left(\dfrac{M}{1-L_g}\right)^k \dfrac{\Xi'(\xi)(\Xi(t)-\Xi(\xi))^{k\zeta_1-1}}{\Gamma(k\zeta_1)} \\ \times e^{-\mu(\Xi(t)-\Xi(\xi))}Q(\xi)d\xi, \end{cases}$$

where

$$\begin{aligned} Q(t) := & \frac{(\Xi(t)-\Xi(a))^{\vartheta-1}e^{-\mu(\Xi(t)-\Xi(a))}}{\Gamma(\vartheta)}D_0[w_0,u_0] \\ & + \frac{(\delta_w+\delta_u)(\Xi(t)-\Xi(a))^{\zeta_1}e^{-\mu(\Xi(t)-\Xi(a))}}{(1-L_g)\Gamma(\zeta_1+1)}, \end{aligned}$$

which is the same as

$$D_0[w(t),u(t)] \leq (\Xi(t)-\Xi(a))^{\vartheta-1}D_0[w_0,u_0]\mathbb{E}_{\zeta_1,\vartheta}\left(\frac{M(\Xi(t)-\Xi(a))^{\zeta_1}}{1-L_g}\right)$$

$$+ (\Xi(t) - \Xi(a))^{\zeta_1} \frac{(\delta_w + \delta_u)}{(1 - L_g)}$$
$$\times \mathbb{E}_{\zeta_1, \zeta_1 + 1}\left(\frac{M}{1 - L_g}(\Xi(t) - \Xi(a))^{\zeta_1}\right).$$

Then

$$D_{C_{1-\vartheta,\Xi}[a,b]}[w, u] \leq D_0[w_0, u_0]\mathbb{E}_{\zeta_1,\vartheta}\left(\frac{M(\Xi(b) - \Xi(a))^{\zeta_1}}{1 - L_g}\right)$$
$$+ (\Xi(b) - \Xi(a))^{\zeta_1} \frac{(\delta_w + \delta_u)}{(1 - L_g)} \times \mathbb{E}_{\zeta_1, \zeta_1 + 1}\left(\frac{M}{1 - L_g}(\Xi(b) - \Xi(a))^{\zeta_1}\right),$$

which gives the stated result.

---

## 11.4 An Example

**Example 11.4.1.** *Consider the following IVP*

$$\begin{cases} {}^{TH}_{0^+}\mathfrak{D}^{\frac{1}{2},\frac{1}{5},1}_{\Xi(t)} w(t) = g(t, w(t), {}^{TH}_{0^+}\mathfrak{D}^{\frac{1}{2},\frac{1}{5},1}_{\Xi(t)} w(t)), \quad t \in (0, 1], \\ {}^{T}_{a^+}\mathfrak{I}^{1-\frac{3}{5}}_{\Xi(t)} w(0) = w_0, \end{cases} \tag{11.4.21}$$

where $\zeta_1 = \frac{1}{2}$, $\zeta_2 = \frac{1}{5}$, $\mu = 1$, $\vartheta = \frac{3}{5}$, and $\Xi(t) = \sqrt{t} - 1$. Set

$$g(t, w(t), {}^{TH}_{0^+}\mathfrak{D}^{\frac{1}{2},\frac{1}{5},1}_{\Xi(t)} w(t)) = \frac{1}{9e^{-t+5}}\left(1 + |w(t)| + |{}^{TH}_{0^+}\mathfrak{D}^{\frac{1}{2},\frac{1}{5},1}_{\Xi(t)} w(t)|\right)$$
$$+ \frac{log(\sqrt{t} + 1)}{\sqrt{t} - 1}, \quad t \in (0, 1].$$

Clearly, the function $g \in C_{1-\frac{3}{5},\Xi}([0, 1], E)$. Hence the assumption ($\mathbb{A}1$) is satisfied. For each $w_1, w_2, u_1, u_2 \in E$ and $t \in (0, 1]$

$$
\begin{aligned}
&D_{C_{1-\frac{3}{5},\Xi}([0,1],E)}[g(t,w_1,w_2),g(t,u_1,u_2)] \\
&\quad = \max_{t\in[0,1]} (\Xi(t)-\Xi(0))^{\frac{2}{5}} D_0[g(t,w_1,w_2),g(t,u_1,u_2)] \\
&\quad \leq \frac{1}{9e} \max_{t\in[0,1]} (\Xi(t)-\Xi(0))^{\frac{2}{5}} (D_0[w_1,u_1]+D_0[w_2,u_2]) \\
&\quad \leq \frac{1}{9e} D_{C_{1-\frac{3}{5},\Xi}([0,1],E)}[w_1,u_1] + \frac{1}{9e} D_{C_{1-\frac{3}{5},\Xi}([0,1],E)}[w_2,u_2].
\end{aligned}
$$

Therefore, $(\mathbb{A}2)$ is verified with $M = L_g = \frac{1}{9e}$. Thus, from (11.2.4) we have

$$
\begin{aligned}
\left(\frac{\Gamma(\vartheta)(\Xi(t)-\Xi(a))^{\zeta_1} e^{-\mu(\Xi(t)-\Xi(a))} M}{\Gamma(\vartheta+\zeta_1)(1-L_g)}\right) &= \left(\frac{\Gamma(\frac{3}{5})(\sqrt{t})^{\frac{1}{2}} e^{-1}}{\Gamma(\frac{3}{5}+\frac{1}{2})} \frac{\frac{1}{9e}}{(1-\frac{1}{9e})}\right) \\
&\approx 0.02448 < 1,
\end{aligned}
$$

is satisfied with $b = 1$, $a = 0$. Now we can apply Theorem 11.2.1 and conclude that the problem (11.4.21) has a unique solution in $C_{1-\frac{3}{5},\Xi}([0,1],E)$.

---

## 11.5 Conclusion

In this chapter, the existence and uniqueness of solutions for implicit fuzzy differential equations with tempered $\Xi$-Hilfer type fractional derivatives were discussed. Furthermore, the generalized Gronwall's inequality is employed to get uniqueness and continuous dependence of the $\delta$-approximate solution. The acquired results which are obtained in the current chapter are more general and cover many of the parallel problems that contain special cases of function $\Xi$ because our proposed problem contains a general fractional derivative that integrates many classical fractional derivatives.

# References

[1] Allahviranloo, T., Salahshour, S., & Abbasbandy, S., (2012). Explicit solutions of fractional differential equations with uncertainty. *Soft Comput., 16*, 297–302.

[2] Almalahi, M. A., Abdo, M. S., & Panchal, S. K., (2020) On the theory of fractional terminal value problem with $\psi$-Hilfer fractional derivative. *AIMS Math., 5*(5), 4889–4908.

[3] Almalahi, M. A., & Panchal, S. K., (2020). Some existence and stability results for $\psi$-Hilfer fractional implicit differential equation with periodic conditions: $\psi$-Hilfer fractional implicit differential equation. *J. Math. Model. Anal., 1*(1), 1–19.

[4] An, T. V., & Hoa, N. V., (2020). Fuzzy fractional integral equations involving the kernel $\psi$-functions. *J. Int. Fuzzy Syst., 38*(4), 5127–5141.

[5] Bede, B., Rudas, I. J., & Attila, L., (2007). First-order linear fuzzy differential equations under generalized differentiability. *Inf. Sci., 177*, 3627–3635.

[6] Bede, B., & Gal, S. G., (2005). Generalizations of the differentiability of fuzzy-number-valued functions with applications to fuzzy differential equations. *Fuzzy Sets Syst., 151*, 581–599.

[7] Bede, B., & Stefanini, L., (2013). Generalized differentiability of fuzzy valued functions. *Fuzzy Sets Syst., 230*, 119–141.

[8] Diethelm, K., & Ford, N. J., (2002). Analysis of fractional differential equations, *J. Math. Anal. Appl., 265*, 229–248.

[9] Fahad, H. M., Fernandez, A., Rehman, M. U., & Siddiqi, M., (2021). Tempered and Hadamard-type fractional calculus with respect to functions. *Mediterr. J. Math., 18*(4), 143.

[10] Fernandez, A., & Ustaoglu, C., (2020). On some analytic properties of tempered fractional calculus. *J. Comput. Appl. Math., 366*, 112400.

[11] Hilfer, R., (2000). *Applications of Fractional Calculus in Physics*, World Scientific, Singapore.

[12] Hoa, N. V., (2021). On the stability for implicit uncertain fractional integral equations with fuzzy concept. *Iran. J. Fuzzy Syst., 18*(1), 185–201.

[13] Hoa, N. V., Lupulescu, V., & O'Regan, D., (2017). Solving interval-valued fractional initial value problems under Caputo gH-fractional differentiability. *Fuzzy Sets Syst., 309*, 1–34.

[14] Hoa, N. V., Lupulescu, V., & O'Regan, D., (2018). A note on initial value problems for fractional fuzzy differential equations. *Fuzzy Sets Syst., 347*, 54–69.

[15] Hoa, N. V., & Vu, H., (2020). A survey on the initial value problems of fuzzy implicit fractional differential equations. *Fuzzy Sets Syst., 400*, 90–133.

[16] Kanagarajan, K., Vivek, R., Vivek, D., & Elsayed, E. M., (2022). Existence Results for Fuzzy Differential Equations with $\psi$-Hilfer Fractional Derivative. *Ann. Commun. Math., 5*(1), 38–54.

[17] Kharade, J. P., & Kucche, K. D., (2020). On the impulsive implicit $\psi$-Hilfer fractional differential equations with delay. *Math. Methods. Appl. Sci., 43*(4), 1938–1952.

[18] Kilbas, A. A., Srivastava, H. M., & Trujillo, J. J., (2006). *Theory and Applications of fractional differential equations*, Elsevier, Amsterdam.

[19] Kucche, K. D., Mali, A. D., Fernandez, A., & Fahad, H. M., (2022). On tempered Hilfer fractional derivatives with respect to functions and the associated fractional differential equations. *Chaos, Solitons. Fractals., 163,* 112547.

[20] Li, C., Deng, W., & Zhao, L., (2017). Well-posedness and numerical approximation of tempered fractional ordinary differential equations. *Discrete. Contin. Dyn. Syst. Ser, 20*(5), 1239–1262.

[21] Long, H. V., Son, N. T. K., & Hoa, N. V., (2017). Fuzzy fractional partial differential equations in partial ordered metric spaces. *Iran. J. Fuzzy Syst., 14*, 107–126.

[22] Mali, A. D., Kucche, K. D., Fernandez, A., & Fahad, H. M., (2022). On tempered fractional calculus with respect to functions and the associated fractional differential equations. *Math. Methods Appl. Sci., 45* 11134–11157.

[23] Mazandarani, M., & Kamyad, A. V., (2013). Modified fractional Euler method for solving fuzzy fractional initial value problem. *Commun. Nonlinear Sci. Numer., 18*, 12–21.

[24] Medved, M., & Brestovanska, E., (2021). Differential equations with tempered $\psi$-Caputo fractional derivative. *Math. Model. Anal., 26*, 631–650.

[25] Obeidat, N. A., & Bentil, D. E., (2021). New theories and applications of tempered fractional differential equations. *Nonlinear Dyn., 105*(2), 1689–1702.

[26] Podlubny, I., (1999). *Fractional differential equation*, Academic Press, San Diego.

[27] Rashid, S., Jarad, F., & Abualnaja, K. M., (2021). On fuzzy Volterra-Fredholm integrodifferential equation associated with Hilfer-generalized proportional fractional derivative. *AIMS Math., 6*, 10920–10946.

[28] Sabzikar, F., Meerschaert, M. M., & Chen, J., (2015). Tempered fractional calculus. *J. Comput. Phys., 293*, 14–28.

[29] Sousa, J. V. C., & Capelas de Oliveira, E., (2018). On the $\psi$-Hilfer fractional derivative. *Commun. Nonlinear Sci. Numer. Simul., 60*, 72–91.

[30] Vivek, D., Kanagarajan, K., & Elsayed, E. M., (2018). Some existence and stability results for Hilfer fractional implicit differential equations with nonlocal conditions. *Mediterr. J. Math., 15*(1), 15.

[31] Vivek, R., Vivek, D., Kanagarajan, K., & Elsayed, E. M., (2022). Some results on the study of Ξ-Hilfer type fuzzy fractional differential equations with time delay. *Proc. Int. Math. Sci., 4*(2), 65–76.

[32] Vivek, R., Elsayed, E. M., Kanagarajan, K., & Vivek, D., (2022). Qualitative analysis of quaternion fuzzy fractional differential equations with Ξ-Hilfer fractional derivative. *Pure Appl. Anal.*, 2022, 6–6.

[33] Vivek, R., Vivek, D., Kanagarajan, K., & Elsayed, E. M., (2022). Existence and finite-time stability of fuzzy delay differential equations with tempered-regularized Hilfer fractional derivative. *Math. Sci. Lett.*, *11*(3), 95–102.

# Chapter 12

## Computational Study of the Fractional Order of an Alcoholism Model

**Yerra Shankar Rao,[1] Binayak Dihudi,[2] and Anasuya Nath[3]**
[1]*NIST University, Berhampur, Odisha, India*
[2]*Konark Institute Science and Technology (KIST), Bhubaneswar, Odisha, India*
[3]*Utkal University, Bhubaneswar, Odisha, India*

**Abstract**
Nowadays, alcoholism has become a global and challenging problem in society. Extreme use of alcohol increases the mortality rate as well as major health issues for millions of people around the globe. A mathematical model has been developed to analyze social behavior and to minimize alcohol addiction in society. This is the study of a non-integer model using discretion of the domain and the short memory principle. Here, the stability analysis of the proposed model is discussed. This analysis shows the efficiency and behavior of models. The fractional derivatives are used to derive the solution of the model. This model helps healthcare professionals and policymakers to develop prevention strategies for the affected.

## Introduction

Smoking and alcoholism have become major health problems in the world. Different factors can influence addiction to alcoholism. This may be happening internally or externally. The internal factors influence individual choices, genetics, drinking history, or psychological conditions. Similarly, external factors include job status, education, cultural or social programs, family parties, surroundings, and religious conviction. This addiction leads to various health and social problems. It can affect different organs of the body and cause death. People addicted to alcoholism face many health issues such as liver diseases, high blood pressure, liver cancer, liver cirrhosis, mouth, breast, and neck cancer, depression, dementia, sexual problems, etc. This has a major impact on socialism, financial problems, and health. According to the WHO [20], more than five million people die every year due to alcoholism.

In the modern age, women are increasingly becoming addicted to alcohol. During pregnancy, drinking alcohol causes many mental and physical problems. The child born may suffer from a mental disorder or other health issues. Hence, it is one of the most important topics in the world. These addictions also cause violence, traffic accidents, suicides, family disturbances, and an increase in divorce cases. Heavy drinkers or excessive alcohol consumption lead to death. The social and economic conditions of society and addicted individuals are adversely affected. For this reason, alcoholism is considered a global health problem that requires raising awareness about this serious issue. For a secure life and better economic conditions, a mathematical model is proposed. Mathematical models can play a vital role in studying alcoholic problems and examining the effects of addiction on different individuals. It is very important to understand these problems and their effects. Thus, mathematical models provide insight into the behavior and transmission of alcoholism among populations, helping to reduce or control it.

The epidemic model plays an important role in various fields of science and engineering, including the study of infectious diseases. A fractional-order smoking model was explained by Singh et al. [10]. In their paper, the authors focused on controlling the smoking problem using CPU computations. They derived the behavior of solutions over a long time. While different mathematical models have been developed to study epidemic models, few papers have addressed similar issues [6]. In one study, the authors discussed a realistic drinking model with two stages and a time-delay component. The stability of this model depends on the time delay, which influences the alcohol-free equilibrium [7]. Another study analyzed a mathematical model for liver complications in drinkers, focusing on minimizing complications at minimum cost using Pontryagin's principle [1].

In Ref. [3], the author derived the alcoholic reproduction number and studied equilibrium points as well as the existence of unique solutions for an SIR-type model. This model examined how diseases develop in the body due to alcoholism. A deterministic model was proposed to analyze the social and health problems caused by alcoholism, and the basic reproduction number was used to assess methods for controlling alcoholism [4]. Another study conducted at the Harvard School of Public Health investigated the influence of alcohol consumption among underage students and proposed policies to prevent underage drinking [17]. Swamali et al. developed a mathematical model for alcohol abusers, examining their behavior and providing a stability analysis for disease-free equilibrium and bifurcation points [10]. Similarly, Wang studied the drinking behavior of college students using a mathematical model, predicting whether the behavior would persist or spread on campus [21]. Mubayi proposed an alcohol-quitting model, explaining the impact of distributed time delays on the dynamic nature of alcoholism and the time delay in contact with infections [11]. Mulone analyzed the time spent in heavy and moderate drinker populations, investigating their distributions [12]. The authors developed a two-stage model to study stability and threshold values affecting the model's behavior [3].

A mathematical model was also proposed to study total recovery from alcoholism. However, the authors highlighted that complete recovery was often not possible after quitting alcohol. Fractional analysis has become highly relevant in many areas of science and technology, including physics, chemistry, mathematics, and signal processing. Fractional integrals and derivatives are critical concepts in fractional calculus. However, no existing work has employed fractional models specifically for drinking behavior, nor has any real work addressed the issue comprehensively.

To address these limitations, we propose a fractional epidemic model for alcoholism. In this model, the total population is divided into four subpopulations: potential drinkers, occasional drinkers, heavy drinkers, and quit drinkers (recovered from drinking). The model can represent a fractional dynamical system of ordinary differential equations as well as a classical dynamical system. Initially, basic definitions of fractional derivatives and integrals are introduced. Then, the boundedness of the models is discussed. The basic reproduction number is derived using the next-generation principle methods. This reproduction number helps determine the alcohol-free equilibrium and the presence of alcoholism in the population. Two equilibrium points are discussed in the model, providing insights into optimal control strategies to prevent alcoholism.

Lastly, our theoretical results are compared with simulated results obtained using MATLAB. These simulations provide insights into the effectiveness of proposed policies to control alcoholism and address the current problem.

## 12.1 Basic Terminologies

| Notation | Parametric description |
|---|---|
| $S(t)$ | Number of Potential drinkers (suspect to drink alcohol) |
| $I_1(t)$ | Number of Heavy drinkers |
| $I_2(t)$ | Number of occasional drinkers |
| $R(t)$ | Number of quit drinkers |
| $A$ | Influx the rate of drinkers into $S(t)$ |

| Notation | Parametric description |
|---|---|
| $\beta_1$ | The contact rate of the heavy drinker |
| $\beta_2$ | The contact rate of the occasional drinkers |
| $\gamma_1$ | The fraction of the recovered from the heavy drinkers |
| $\gamma_2$ | The fractions recovered from the occasional drinkers |
| $\delta_1$ | The Natural death of every population |
| $\delta_2$ | Death due to heavy drinkers |
| $\delta_3$ | The death rate due to occasional drinkers |

Basic assumptions:

i. Since all the people are normalized. So they have an equal probability of heavy drinkers or occasional drinkers.

ii. Heavy drinkers considered stopping drinking alcohol through proper treatment or counseling.

iii. The occasional drinkers may have to quit drinking by not going to friends who were drinking or stopping the parties and family functions

## 12.2 Preliminaries of the Basic Concept of Fractional Calculus

In this section, some basic definitions and properties of fractional calculus [1], [4], [5], [13] have been defined.

**Definition 12.2.1.** *A real function $f(x)$, $x \geq 0$ is said to be in the space $C_\mu$, $\mu \in R$ if there exists a real number $p \geq \mu$ such that $f(x) = x^p f_1(x)$, where $f_1(x) \in C[0, \infty)$. Clearly $C_\mu \leq C_\alpha$, if $\mu \leq \alpha$.*

**Definition 12.2.2.** *A function $f(x)$, $x \geq 0$ is said to be in space $C_\mu^m, m \in N \bigcup \{0\}$ if $f^{(m)} \in C_\mu$.*

The Riemann Liouville fractional integral operator of the order $\rho \geq 0$ of a function, $f \in C_\mu,\ \mu \geq -1$ is defined as

$$\left(J_a^\rho\ f\right)(x) = \frac{1}{\Gamma(\rho)} \int_a^x (x-\tau)^{\rho-1} f(\tau) d\tau, x \geq a, \tag{12.2.1}$$

$$\left(J^p \rho_a\ f\right)(x) = f(x). \tag{12.2.2}$$

All the properties of the operator $J^\rho$ can be found in Caputo's [2]. Here, we mention only the following:

For $f \in C_\mu, \mu \geq -1, a, \alpha \geq 0$ and $\gamma \geq -1$

$$\left(J_a^\rho J_a^\alpha f\right)(x) = \left(J_a^{\rho+\alpha} f\right)(x). \tag{12.2.3}$$

$$\left(J_a^\rho J_a^\alpha f\right)(x) = \left(J_a^\alpha J_a^\rho f\right)(x). \tag{12.2.4}$$

$$J_a^\rho X^\gamma = \frac{\Gamma(\gamma+1)}{\Gamma(\rho+\gamma+1)} X^{p+\gamma}. \tag{12.2.5}$$

**Definition 12.2.3.** *Grunwald Letnikov (GL) [14] derivative is given by*

$$_a{}^{D_t^p} g(t) = \lim_{h \longrightarrow 0} \frac{1}{h^p} \sum_{J=0}^{|n|} (-1)^j \binom{p}{j} g(t-jh). \tag{12.2.6}$$

*When $n = \frac{t-a}{h}$ $a$ is a real constant, and the integer part function is denoted by [.]. The general linear system of fractional order is given as*

$$_a{}^{D_t^p} x(t) = Ax(t) + Bu(t) y(t) = Cx(t). \tag{12.2.7}$$

*Using* (12.2.6), *the p-th order GL derivative at the point's kh (k=1,2,...) is defined as:*

$$\left(k - \frac{L}{h}\right) {}_a{}^{D_t^\rho} g(t) \approx \frac{1}{h^p} \sum_{J=0}^{k} (-1)^j \binom{p}{j} g(t_{k-j}). \tag{12.2.8}$$

*Where L is "memory length",$t_k = kh$ h denotes the step size of computation and the coefficients $C_j^{(p)}(j = 0, 1, 2, ...)$ are given by:*

$$C_j^{(p)} = \left(1 - \frac{1+p}{j}\right) C_{j-1}^{(p)} \tag{12.2.9}$$

*The numerical solution of the equation*

$$a^{D_t^p} y(t) = g(y(t), t), \tag{12.2.10}$$

*is given as*

$$y(t_k) = g(y(t_k), tk)h^p - \sum_{j=v}^{k} C^(p)_j y(t_{k-j}). \tag{12.2.11}$$

*Using the short memory principle, the lower index can be calculated as:*

$$v = \begin{cases} 1, & k < \frac{L}{h} \\ k - (\frac{L}{h}), & k > (\frac{L}{h}). \end{cases} \tag{12.2.12}$$

*For the L, we can use*

$$L \geq \left(\frac{M}{\varepsilon|1-p|}\right)^{\frac{1}{p}}. \tag{12.2.13}$$

*The equilibrium points of the fractional order systems*

$$a^{D_t^q} x_i(t) = 0, i = 1, 2, 3, 4, .., n. \tag{12.2.14}$$

*The Jacobian matrix for the proposed system of equations is:*

$$\begin{pmatrix} \frac{\partial f_1}{\partial x_1} & ..... & \frac{\partial f_1}{\partial x_n} \\ ..... & ...... & ...... \\ \frac{\partial f_n}{\partial x_1} & ..... & \frac{\partial f_n}{\partial x_n} \end{pmatrix}. \tag{12.2.15}$$

*The Jacobian matrix at the equilibrium point $(b_1, b_2, b_3, ..., b_n)$ is given by:*

$$\begin{pmatrix} \frac{\partial f_1}{\partial x_1} & \cdots\cdots & \frac{\partial f_1}{\partial x_n} \\ \cdots\cdots & \cdots\cdots & \cdots\cdots \\ \frac{\partial f_n}{\partial x_1} & \cdots\cdots & \frac{\partial f_n}{\partial x_n} \end{pmatrix}_{(b_1,b_2,b_3,\ldots,b_n)} . \qquad (12.2.16)$$

---

## 12.3 Mathematical Model Formations

In this chapter, the total number $N(t)$ population is closely divided into four subgroups. These are denoted as $S(t)$ as the potential drinker, $I_1(t)$ being heavy drinkers, $I_2(t)$ being the occasional drinker, and $R(t)$ being the drinker who quit drinking permanently. The assumption is based on natural death, death induced by drinks, and death due to heavy drinking in accidents. This assumption leads to a mathematical model in which the interaction between the subgroups.

A diagrammatical representation of the model is shown in Figure 12.1.

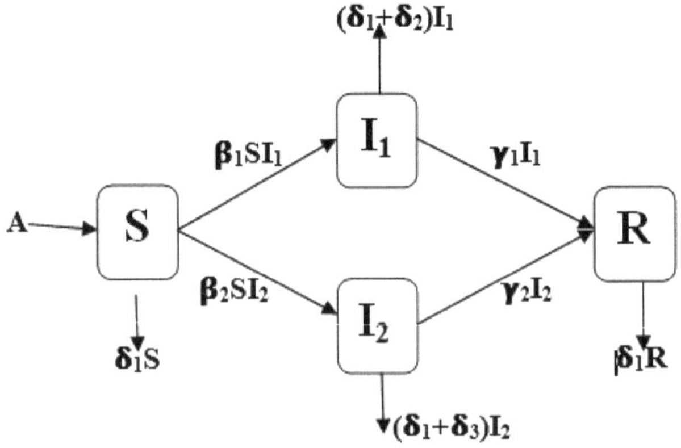


**FIGURE 12.1**. Transmission diagram of the alcoholism model.

### *12.3.1 Model Descriptions*

Let's begin the total populations $N(t) = S(t) + I_1(t) + I_1(t) + R(t)$. From the below system (12.3.1), in the absence of any type of drinkers, the results were obtained $\frac{dN(t)}{dt} = A - \delta_1 N$.

$$\begin{aligned}
\frac{dS(t)}{dt} &= A - \beta_1 S(t)I_1(t) - \beta_2 S(t)I_1(t) - \delta_1 S(t)\\
\frac{dI_1(t)}{dt} &= \beta_1 S(t)I_1(t) - (\delta_1 + \delta_2 + \gamma_1)I_1(t)\\
\frac{dI_2(t)}{dt} &= \beta_2 S(t)I_2(t) - (\delta_1 + \delta_2 + \gamma_2)I_2(t)\\
\frac{dR(t)}{dt} &= \gamma_1 I_1(t) + \gamma_2 I_2(t) - \delta_1 R(t)
\end{aligned} \tag{12.3.1}$$

This shows that the total populations remain constant due to $N(t) = \frac{A}{\delta_1}$. This above system (12.3.1) replaced the first derivative with the Caputo fractional derivatives. Moreover, the system of equations can be represented by the fractional differential equations as:

$$\begin{aligned}
aD_t^{p_1} S(t) &= A - \beta_1 S(t)I_1(t) - \beta_2 S(t)I_1(t) - \delta_1 S(t)\\
aD_t^{p_2} I_1(t) &= \beta_1 S(t)I_1(t) - (\delta_1 + \delta_2 + \gamma_1)\, I_1(t)\\
aD_t^{p_1} I_2(t) &= \beta_2 S(t)I_2(t) - (\delta_1 + \delta_3 + \gamma_2)\, I_2(t)\\
aD_t^{p_1} R(t) &= \gamma_1 I_1(t) + \gamma_2 I_2(t) - \delta_1 R(t).
\end{aligned} \tag{12.3.2}$$

With its initial conditions stratifying,

$$\begin{aligned}
&S(t), I_1(t), I_2(t), R(t) \in \Omega : S(0) \geq 0, I_1(0) \geq 0, I_2(0) \geq 0, R(0) \geq 0,\\
&S(t) + I_1(t) + I_2(t) + R(t) \leq \frac{A}{\delta_1},
\end{aligned} \tag{12.3.3}$$

where $0 \leq p_1, p_2, p_3, p_4 < 1$ and ${}_aD_t^p$ is the operator of the fractional order derivative of the order $p$, $(0 < p < 1)$ in the Caputo sense system of (12.3.2).

### 12.3.2 Proposed Method

Here we will use the proposed model to solve fractionally. The integer order alcoholic model is specified as:

$$\begin{aligned}
\frac{dS(t)}{dt} &= A - \beta_1 S(t)I_1(t) - \beta_2 S(t)I_2(t) - \delta_1 S(t)\\
\frac{dI_1(t)}{dt} &= \beta_1 S(t)I_1(t) - (\delta_1 + \delta_2 + \gamma_1)I_1(t)\\
\frac{dI_2(t)}{dt} &= \beta_2 S(t)I_2(t) - (\delta_1 + \delta_3 + \gamma_2)I_2(t)\\
\frac{dR(t)}{dt} &= \gamma_1 I_1(t) + \gamma_2 I_2(t) - \delta_1 R(t).
\end{aligned} \tag{12.3.4}$$

Integrating both sides of (12.3.4), we have

$$\begin{aligned}
S(t) &= \int_0^t [A - \beta_1 S(t)I_1(t) - \beta_2 S(t)I_2(t) - \delta_1 S(t)]dt\\
I_1(t) &= \int_0^t [\beta_1 S(t)I_1(t) - (\delta_1 + \delta_2 + \gamma_1)I_1(t)]dt\\
I_2(t) &= \int_0^t [\beta_2 S(t)I_2(t) - (\delta_1 + \delta_3 + \gamma_2)I_2(t)]dt\\
R(t) &= \int_0^t [\gamma_1 I_1(t) + \gamma_2 I_2(t) - \delta_1 R(t)]dt.
\end{aligned} \tag{12.3.5}$$

Integrate (12.3.5) with the help of (12.3.2)

$$\begin{aligned}
S(t) &= aD_t^{1-p_1}\left(\int_0^t (A - \beta_1 S(t)I_1(t) - \beta_2 S(t)I_2(t) - \delta_1 S(t)dt\right)\\
I_1(t) &= aD_t^{1-p_2}\left(\int_0^t (\beta_1 S(t)I_1(t) - (\delta_1 + \delta_2 + \gamma_1)I_1(t))dt\right)\\
I_2(t) &= aD_t^{1-p_3}\left(\int_0^t (\beta_2 S(t)I_2(t) - (\delta_1 + \delta_3 + \gamma_2)I_2(t))dt\right)\\
R(t) &= aD_t^{1-p_4}\left(\int_0^t (\gamma_1 I_1(t) + \gamma_2 I_2(t) - \delta_1 R(t)dt\right).
\end{aligned} \tag{12.3.6}$$

Using the GL definition in (12.3.6), we will get the following iterative relations

$$\begin{aligned}
S(t_k) &= (A - \beta_1 S(t_{k-j}) I_1(t_{k-j}) - \beta_2 S(t_{k-j}) I_2(t_{k-j}) \\
&\quad - \delta_1 S(t_{k-j}))h^{p_1} - \sum_{j=v}^{k} c_j^{(p_1)} S(t_{k-j}) \\
I_1(t_k) &= (\beta_1 S(t_k) I_1(t_{k-j}) - (\delta_1 + \delta_2 + \gamma_1) I_1(t_{k-j}))h^{p_2} - \\
&\quad \sum_{j=v}^{k} c_j^{(p_2)} I_1(t_{k-j}) \\
I_2(t_k) &= (\beta_2 S(t_k) I_2(t_{k-j}) - (\delta_1 + \delta_3 + \gamma_2) I_2(t_{k-j}))h^{p_3} - \\
&\quad \sum_{j=v}^{k} c_j^{(p_3)} I_2(t_{k-j}) \\
R(t_k) &= (\gamma_1 I_1(t_k) + \gamma_2 I_2(t_k) - \delta_1 R(t_{k-j}))h^{p_4} - \sum_{j=v}^{k} c_j^{(p_4)} R(t_{k-j}).
\end{aligned} \tag{12.3.7}$$

By solving the (12.3.7) iterative, we will get the accurate solution of the Alcoholic model. Minimizing the step size (h) will increase the accuracy results of the model. These results can be reflected in the simulations to get a better solution for the model.

---

## 12.4 Dynamics of the Fractional Alcoholic Model

### *12.4.1 Positivity and Boundedness of the Model*

**Theorem 12.4.1.** *Let's denote the region,* $R_+^4 = S(t), I_1(t), I_2(t), R(t) :$ $S(t) \geq 0, I_1(t) \geq 0, I_2(t) \geq 0, R(t) \geq 0\}$ *when* $R_+^4$ *in which all the solutions are uniformly nonnegative and bounded in the initial conditions, then the system* (12.3.2) *is positively invariant. Again this can be represented as:*

$$\begin{aligned}
&\lim_{t\to\infty}\ \sup\ S(t) \le S_\infty = \frac{A}{\delta_1}\\
&\lim_{t\to\infty}\ \sup\ I_1(t) \le I_{1\infty} = \frac{A}{\delta_1(\delta_1+\delta_2+\gamma_1)}\\
&\lim_{t\to\infty}\ \sup\ I_2(t) \le I_{2\infty} = \frac{A}{\delta_1(\delta_1+\delta_3+\gamma_2)}\\
&\lim_{t\to\infty}\ R(t) \le R_\infty = \frac{\gamma_1 I_{1\infty}+\gamma_1 I_{2\infty}}{\delta_1}.
\end{aligned} \tag{12.4.1}$$

*Proof.* The fractional system of differential (12.3.2) for the alcoholic model can be proven by the following results [9].

$$\begin{aligned}
&aD_1^{p_1}S(t)|_{S=0} = A > 0\\
&aD_t^{p_2}I_1(t)|_{I_1=0} = \frac{\beta_1 S(t)}{(\delta_1+\delta_2+\gamma_1)} > 0\\
&aD_t^{p_3}I_2(t)|_{I_2=0} = \frac{\beta_2 S(t)}{(\delta_1+\delta_1+\gamma_1)} > 0\\
&aD_t^{p_4}S(t)|_{I_1=0} = \frac{\gamma_1 I_1(t)+\gamma_2 I_2(t)}{\delta_1} > 0.
\end{aligned} \tag{12.4.2}$$

Here (12.4.2) and the generalized mean value theorem can be reduced so that all the populations are positive for $t > 0$. From (12.3.2) of the first, we have

$$aD_1^{p_1}S(t) \le A - \delta_1 S(t).$$

Using the fractional comparison principle, it is estimated first three equations of (12.3.4) that

$$aD_1^{p_1+p_2+p_3}(S(t)+I_1(t)+I_2(t))$$

$$\le (A - \delta_1 S(t) - (\delta_1+\delta_2+\gamma_1)I_1(t) - (\delta_1+\delta_2+\gamma_1)I_2(t)).$$

The above results establish that

$$\lim_{t\to\infty}\ \sup\ S(t) \le S_\infty,\ \lim_{t\to\infty}\ \sup\ I_1(t) \le I_{1\infty},\ \lim_{t\to\infty}\ \sup\ I_2(t) \le I_{2\infty}.$$

Again from the last equation of the system (12.3.2)

$$aD_t^{p_4}R(t) \leq \gamma_1 I_1(t) + \gamma_2 I_2(t) - \delta_1 R(t).$$

This result shows that $\lim_{t\to\infty} SupR(t) \leq R_\infty$ for t is very large.
Hence the system of equation (12.3.2) is positive and invariant.

---

## 12.5 Equilibrium Points and Calculation of Basic Reproduction Number

### *12.5.1 Stability Analysis of the Model*

Here there are two equilibrium points have been discussed
1. Alcoholic-free (drinking-free) Equilibrium Points $(S_0, 0, 0, 0)$
2. Endemic (Presence of Drinker) Equilibrium Points$(S^*, I_1^*, I_2^*, R^*)$.

For the steady state, the system of Equation (12.3.2)

$$\begin{aligned} A - \beta_1 S(t)I_1(t) - \beta_2 S(t)I_2(t) - \delta_1 S(t) &= 0 \\ \beta_1 S(t)I_1(t) - (\delta_1 + \delta_2 + \gamma_1)I_1(t) &= 0 \\ \beta_2 S(t)I_2(t) - (\delta_1 + \delta_3 + \gamma_2)I_2(t) &= 0 \\ \gamma_1 I_1(t) + \gamma_2 I_2(t) - \delta_1 R(t) &= 0. \end{aligned} \tag{12.5.1}$$

Solving algebraically in Equation (12.5.1). For alcoholic-free equilibrium points it becomes $S_0 = \frac{A}{\delta_1}, I_1 = 0, I_2 = 0, R = 0$. Similarly, for the endemic equilibrium points

$$\begin{aligned} S^* &= \frac{A}{(\delta_1 + \beta_1 I_1^* + \beta_2^*)} \\ I_1^* &= \frac{\delta_1(\delta_1 + \delta_2 + \gamma_1)I_1^*}{A\beta_1} \end{aligned}$$

$$I_2^* = \frac{\delta_1(\delta_1 + \delta_3 + \gamma_2)I_2^*}{A\beta_2}$$

$$R^* = \frac{\gamma_1 I_1^* + \gamma_2 I_2 *}{\delta_1}$$

### *12.5.2 Calculations of the Basic Reproduction Number*

The basic reproduction number is defined as the excepted number of heavy drinkers produced by the single drinker during the whole poetical drinker time. It can be calculated by the next-generation matrix method. This method was developed by Van den Driessche and Watmough [16]. The next-generation matrix can be expressed as the matrix form $a^{D_t^p(x)} = F(x) - V(x)$, where $F(x)$ is the rate of appearance of new Drinkers and $V(x)$ is the difference between rates of transfer from one group of drinkers to another group of drinkers.

$$F(x) = \begin{bmatrix} \beta_2 S & 0 & 0 \\ 0 & \beta_1 S & 0 \\ 0 & 0 & 0 \end{bmatrix}$$

$$V(x) = \begin{bmatrix} (\delta_1 + \delta_2 + \gamma_1) & 0 & 0 \\ 0 & (\delta_1 + \delta_3 + \gamma_2) & 0 \\ -\gamma_1 & -\gamma_2 & \delta_1 \end{bmatrix}$$

By calculation of the basic reproduction number of the (12.3.1) dominant Eigenvalues of $FV^{-1}$

$$\begin{aligned} R_0 = \rho(FV^{-1}) &= \frac{(\beta_1 + \beta_2)S_0}{(\delta_1 + \delta_2 + \gamma_1)((\delta_1 + \delta_3 + \gamma_2))} \\ &= \frac{(\beta_1 + \beta_2)A}{\delta_1(\delta_1 + \delta_2 + \gamma_1)(\delta_1 + \delta_3 + \gamma_2)}. \end{aligned} \tag{12.5.2}$$

## 12.6 Stability Analysis

### *12.6.1 Stability Analysis for Alcoholic-Free Equilibrium Point*

**Theorem 12.6.1.** *The alcoholic free equilibrium is locally asymptotical stable when $R_0 < 1$, otherwise equilibrium point is unstable.*

*Proof.* The Linearization of the Jacobian Matrix of (12.3.2) for the Alcoholic equilibrium point is

$$J_{AFE}(S_0, 0, 0, 0) = \begin{bmatrix} -\delta_1 & -\beta_1 S_0 & -\beta_2 S_0 & 0 \\ 0 & \beta_1 S_0 - (\delta_1 + \delta_2 + \gamma_1) & 0 & 0 \\ 0 & 0 & \beta_2 S_0 - (\delta_1 + \delta_3 + \gamma_2) & 0 \\ 0 & \gamma_1 & \gamma_2 & -\delta_1 \end{bmatrix}.$$

Therefore the eigenvalues of the characteristic polynomial equations of $J$ becomes

$$p(x) = (\lambda + \delta_1)^2 (\lambda^2 + T_1 \lambda + T_2) = 0.$$

The two eigenvalues are$\lambda = -\delta_1, \lambda = -\delta_1$ ,and other can be represented by the quadratic equations $(\lambda^2 + T_1\lambda + T_2) = 0$, where

$$T_1 = (-\beta_1 S_0 + (\delta_1 + \delta_2 + \gamma_1) - \beta_2 S_0 + (\delta_1 + \delta_3 + \gamma_2))$$
$$T_2 = (\beta_1 S_0 - (\delta_1 + \delta_2 + \gamma_1)(\beta_2 S_0 - (\delta_1 + \delta_3 + \gamma_2))).$$

Clearly it is verified that $T_1 > 0$ and $T_2 > 0$ when $R_0 < 1$. Hence the roots of the polynomial equations have the negative real parts. Therefore the real part of all four eigenvalues of $J_{AFE}$ are negative, consequently, $J_{AFE}$ is locally asymptotically stable. The proof is complete.

**Theorem 12.6.2.** *When $R_0 < 1$, the alcoholic free equilibrium point of* (12.3.2) *is not only locally stable but also it is uniformly asymptotically stable.*

*Proof.*

$$\begin{aligned} aD_t^{p_1} S(t) &= A - \beta_1 S(t) I_1(t) - \beta_2 S(t) I_2(t) - \delta_1 S(t) \\ aD_t^{p_2} I_1(t) &= \beta_1 S(t) I_1(t) - (\delta_1 + \delta_2 + \gamma_1) I_1(t) \\ aD_t^{p_3} I_2(t) &= \beta_2 S(t) I_2(t) - (\delta_1 + \delta_3 + \gamma_2) I_2(t). \end{aligned}$$

According to the invariant region, it is sufficient to take the model (12.3.2) defined by the region:

$$R_+^3 = \left\{ (S, I_1, I_2) : (S, I_1, I_2) > 0, S \leq \frac{A}{\delta_1} \right\}.$$

Let's take the Lyapunov Functions $V : R_4^+ \to R$ such that

$$\begin{aligned} V(S, I_1, I_2) = (S - S_0 \ln \frac{S}{S_0} - S_0) + &\frac{\beta_1 A I_1}{\delta_1(\delta_1 + \beta_1 I_1 + \beta_2 I_2)} \\ &+ \frac{\beta_2 A I_2}{\delta_1(\delta_1 + \beta_1 I_1 + \beta_2 I_2)}. \end{aligned}$$

Using the linearity property of the Caputo Derivatives [19] and algebraic manipulations, it is obtained that

$$aD_t^p V(t) \leq -(S - S_0)^2 + K_1 I_1 + K_2 I_2.$$

Where $K_1K_2 < 0$, simultaneously by the Lyapunov stability theorem of a fractional dynamic system, is uniformly stable. Hence the system of equation (12.3.2) is uniformly stable.

**Theorem 12.6.3.** *The Alcoholic free equilibrium is globally stable for the model* (12.3.2) *if* $R_0 > 0$.

*Proof.* It is obtained from the (12.3.2) that

$$aD_i^{p_t}S(t) \leq A - \delta_1 S(t).$$

For the compassion system, it is shown that $S_0$ is an equilibrium and global stable for the above equations. Then for $\epsilon > 0$, there exists a constant $B > 0$ such that $aD_i^{p_t}S(t) \leq S_0 + \epsilon$ for $t > B$. Thus, when $t > B$ from the (12.3.2), we have

$$aD_t^{p_2}I_1(t) < \beta_1 S(t)I_1(t) - (\delta_1 + \delta_2 + \gamma_1)I_1(t) \tag{12.6.1}$$
$$aD_t^{p_3}I_2(t) < \beta_2 S(t)I_2(t) - (\delta_1 + \delta_3 + \gamma_2)I_2(t). \tag{12.6.2}$$

When $R_0 < 1$, we can choose $\epsilon > 0$, small enough such that the positive solution of the characteristic equation of the system

$$aD_t^{p}I_1(t) = \beta_1(S_0 + \epsilon) - (\delta_1 + \delta_2 + \gamma_1)I_1(t)$$
$$aD_t^{p}I_2(t) = \beta_2(S_0 + \epsilon) - (\delta_1 + \delta_3 + \gamma_2)I_2(t).$$

As per the comparison, (12.6.1) and (12.6.2) tends to (0, 0) as $t \to \infty$. It follows that $I_1 \to 0$ as $t \to \infty$. Similarly $I_2 \to 0$ as $t \to \infty$. Therefore the Alcoholic free equilibrium is globally stable when $R_0 < 1$.

### 12.6.2 Locally Stable for the Endemic Equilibrium

**Theorem 12.6.4.** *The endemic equilibrium (Presents of alcoholism) point is locally asymptotically stable if* $R_0 > 1$ *otherwise it is unstable.*

*Proof.* At the alcoholism present equilibrium $(S^*, I_1^*, I_2^*, R^*)$ of the system (12.3.2), the variation matrix can be represented as:

$$J_{EE}(S^*, I_1^*, I_2^*, R^*) =$$

$$\begin{bmatrix} -\delta_1 - \beta_1 I_1^* - \beta_2 I_2^* & -\beta_1 S^* & -\beta_2 S^* & 0 \\ \beta_1 I_1^* & \beta_1 S^* - (\delta_1 + \delta_2 + \gamma_1) & 0 & 0 \\ \beta_2 I_2^* & 0 & \beta_2 S^* - (\delta_1 + \delta_3 + \gamma_2) & 0 \\ 0 & \gamma_1 & \gamma_2 & \delta_1. \end{bmatrix} \tag{12.6.3}$$

It is observed that the auxiliary equations of $J_{EE}$ have Eigenvalues $\lambda_1 = -\delta_1$ whose real part is negative. To determine the stability of the equilibrium of alcoholism present model (12.3.2) can be found by the following equations $\phi(\lambda)$, where

$$\phi(\lambda) = \lambda^3 + a_1\lambda^2 + a_2\lambda + a_3 = 0$$

$$a_1 = -(-\delta_1 - \beta_1 I_1^* - \beta_2 I_2^* + \beta_1 S^* - (\delta_1 + \delta_2 + \gamma_1) + \beta_2 S^* - (\delta_1 + \delta_3 + \gamma_2))$$

$$a_2 = \begin{bmatrix} (-\delta_1 - \beta_1 I_1^* - \beta_2 I_2^*)(\beta_1 S^* - (\delta_1 + \delta_2 + \gamma_1)) + \\ (-\delta_1 - \beta_1 I_1^* - \beta_2 I_2^*)(\beta_2 S^* - (\delta_1 + \delta_3 + \gamma_2)) \\ (\beta_1 S^* - (\delta_1 + \delta_2 + \gamma_1))(\beta_2 S^* - (\delta_1 + \delta_3 + \gamma_2)) + \\ \beta_1 S^* \beta_1 I_1^* + \beta_1 S^* \beta_2 I_2^* + \beta_2 S^* \beta_2 I_2^* \end{bmatrix}$$

$$a_3 = \begin{bmatrix} -(-\delta_1 - \beta_1 I_1^* - \beta_2 I_2^*)(\beta_1 S^* - (\delta_1 + \delta_2 + \gamma_1))(\beta_2 S^* \\ -(\delta_1 + \delta_3 + \gamma_2)) + (\beta_2 S^* - (\delta_1 + \delta_3 + \gamma_2))(\beta_1 S^* \beta_1 I_1^*) \\ +(\beta_1 S^* \beta_2 I_2^*)(\beta_2 S^* - (\delta_1 + \delta_3 + \gamma_2)) \\ +((\beta_2 S^* \beta_2 I_2^*)(\beta_1 S^* - (\delta_1 + \delta_2 + \gamma_1)) \end{bmatrix}$$

As per Routh-Hurwitz conditions, the system (12.3.2) is locally asymptotically stable when $a_1 > 0, a_2 > 0, a_3 > 0$ then $a_1 a_2 - a_3 > 0$. Then, by [8], Lassel's stability principle is stable. Therefore the Alcoholism present (endemic equilibrium) of (12.3.2) is locally asymptotically stable.

## 12.7 Numerical Simulation and Discussions of the Results

In this section, the numerical solution of the model (12.3.2) is illustrated. By considering different parametric values, the system of differential equations is shown to be stable within certain regions. Figure 12.2 demonstrates that the alcohol-free equilibrium is stable when the basic reproduction number is less than one. In Figure 12.3, the number of heavy drinkers decreases when appropriate guidance is provided to individuals.

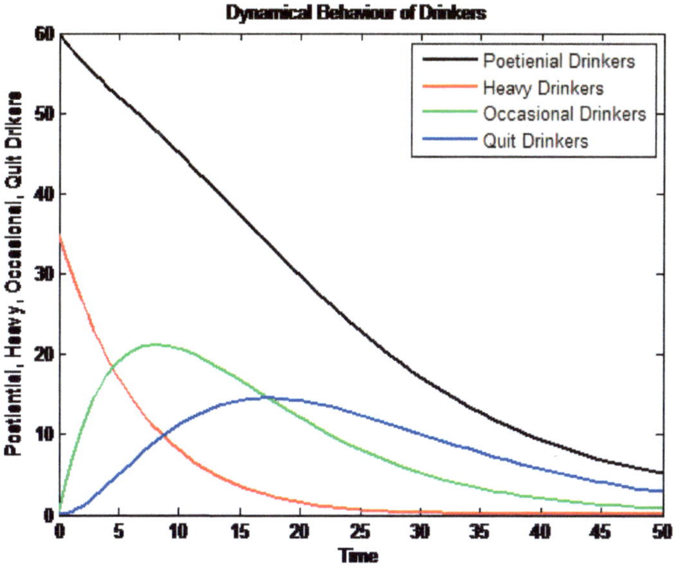


**FIGURE 12.2**. Dynamic behavior of the drinkers.

Figure 12.4 presents a 3D representation of heavy drinkers versus quit drinkers over time. It illustrates that the number of quit drinkers increases with proper treatment of heavy drinkers over time.

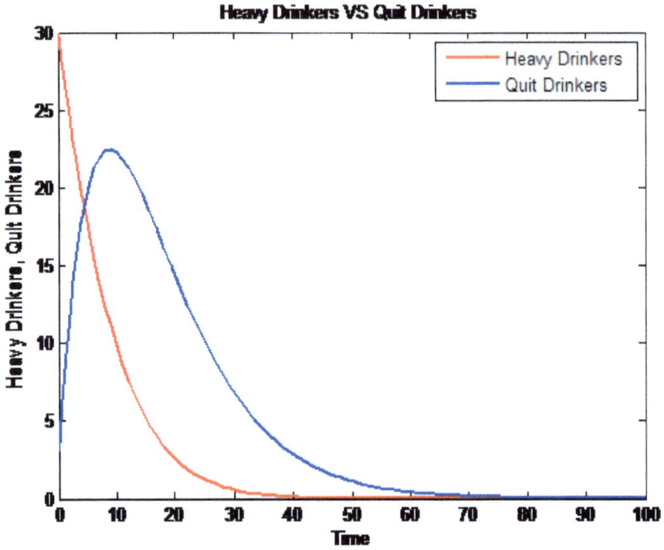


FIGURE 12.3. Heavy drinkers vs. quit drinkers.

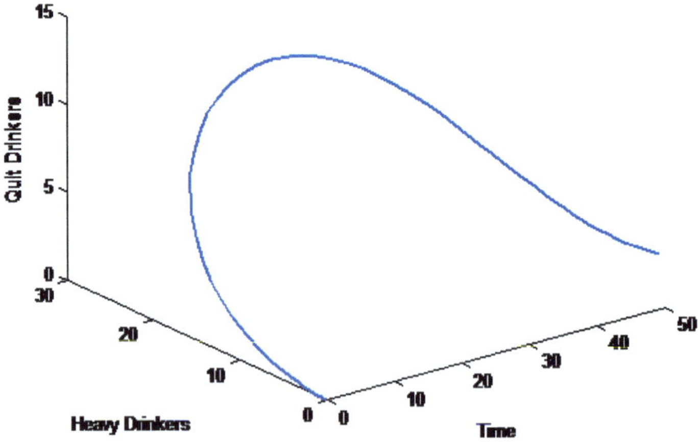


FIGURE 12.4. Three dimension pictures of heavy drinkers vs. quit drinkers.

As shown in Figure 12.5, the number of occasional drinkers decreases rapidly when avoiding contact with drinkers at social gatherings such as parties and family functions, while the number of quit drinkers increases over time. These simulated figures provide insights into the optimal control and minimization of alcohol consumption in society.

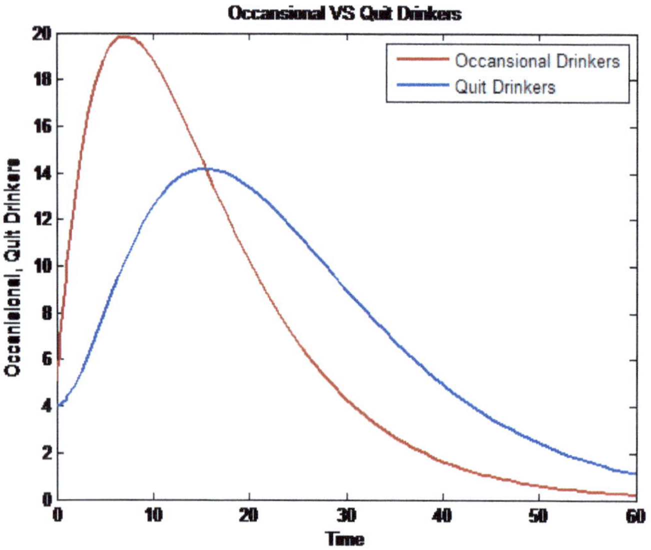


**FIGURE 12.5**. Occasional drinker vs. quit drinkers.

## 12.8 Conclusions

In this chapter, a fractional mathematical epidemic model for alcoholic drinking has been formulated. This model is explained by a system of fractional differential equations and solved successfully using Caputo's derivative. Firstly, the boundedness, positiveness, and calculation of the basic reproduction numbers have been derived. This number establishes

the stability analysis for alcohol-free equilibrium points and endemic equilibrium points. The numerical results are plotted through mathematical simulation using ODE45. A set of numerical results has been performed using MATLAB simulations supported by the theoretical results. The performance of the numerical results and the theoretical results is the best fit for the analysis of the problem. These results and stability analysis are determined by the parameters. The parameters show a decrease in alcoholism in society. Hence, it is recommended that policymakers and health organizations control or stop alcohol consumption.

# References


[1] Benedict, B., (2007). Modeling alcoholism as a contagious disease: how "infected" drinking buddies spread problem drinking, *SIAM News 40*, 3.

[2] Caputo, M., (1967). Linear models of dissipations whose Q is almost frequency Independent-II. *Geophys. J. Int. 13*, 529–539, doi: 10.1111/j.1365-246X.1967.tb02303.

[3] Claver, B., (2012). A mathematical analysis of alcoholism. *World J. Modell. Simul. 8*, 124–134.

[4] Daftardar-Geiji, V., (2019). *Fractional Calculus and Fractional Differential Equations.* Basel: Birkhauser.

[5] Driessche, Van den, P., & Watmough, J., (2002). Reproduction numbers and subthreshold endemic equilibria or compartmental models of disease transmission. *Math. Biosci. 180*, 29–48, doi: 10.1016/s0025-5564(02)00108-6 (Pubmed:12387915).

[6] Huo, Hai-Fang., Chen, Yong-Lan., & Xiang, Hong., (2017). Stability of a binge drinking model with delay. *J. Biol. Dyn. 11*, 210–225, doi: 10.1080/17513758.2017.1301579 (Pubmed:28300490).

[7] Khajji, B., Kouidere, A., Balatif, O., & Rachik, M., (2020), Mathematical modeling, analysis, and optimal control of an alcohol drinking model with liver complications. *Commun. Math. Biol. Neurosci.*, Article ID 32.

[8] Lasalle, J. P., (1976). The stability of dynamical systems. *SIAM, 21*, 3, doi: 10.1137/1021079.

[9] Lin, W., (2007). Global existence theory and chaos controls of fractions differential equations. *J. Math. Anal. Appl. 332*, 709–726, doi: 10.1016/j.j.maa. 2006.10.040.

[10] Manthey, J. L., Aidoo, A. Y., & Ward, K. Y., (2008). Campus drinking: an epidemiological model. *J. Biol. Dyn. 2*, 346–356, doi: 10.1080/17513750801911169 (Pubmed:22876873).

[11] Mubayi, A., Greenwood, P. E., Castillo-Chávez, C., Gruenewald, P. J., & Gorman, D. M., (2010). The impact of relative residence times on the distribution of heavy drinkers in highly distinct environments. *Socio Econ. Plann. Sci. 44*, 45–56, doi: 10.1016/j.seps.2009.02.002 (Pubmed:20161388).

[12] Mulone, G., & Straughan, B., (2012). Modeling binge drinking. *Int. J. Biomath., 05*(1), 1250005, doi: 10.1142/S1793524511001453.

[13] Oldham, K., & Spanier, J., (1974). *The fractional Calculus Theory and Applications of Differentiation and Integrations to Arbitrary Order.* Elsevier.

[14] Petráš, I., (2011). Modeling and numerical analysis of fractional-order Bloch equations. *Comput. Math. Appl. 61*(1), 341-–356, doi: 10.1016/j.camwa.2010.11.009.

[15] Rauta, A. K., Rao, Y. S., Behera, J., Dihudi, B., & Panda, T. C., (2021). *SIQR epidemic modeling and stability analysis of COVID-19* in (eds Khosla, P. K., Mittal, M., Sharma, D., Goyal, L. M.) (Singapore: Springer, 2021) (DOI 10.1007/978-981-33-4236-1-3). Predictive and Preventive Measures for Covid-19 Pandemic, 35–50.

[16] Samko, S. G., Kilbas, A. A., & Marichev, O., (1993). *Fractional Integrals and Derivatives.* Milton Park: Taylor Francis.

[17] Sharma, S., & Samanta, G. P., (2013). Drinking as an epidemic: A mathematical model with dynamic behavior. *J. Appl. Math. Inform. 31*, 1–25, doi:10.14317/jami.2013.001.

[18] Singh, H., Baleanu, D., Singh, J., & Dutta, H., (2021). Computational study of fractional order smoking model. *Chaos Solitons Fract. 142*, 110868, doi: 10.1016/j.chaos.2020.110440.

[19] Vargas-De-León, C., (2015). Volterra-type Lyapunov functions for fractional-order epidemic systems. *Commun. Nonlinear Sci. Numer. Simul. 24*, 27–85.

[20] WHO., (2022). Available from: https://www.who.int/news-room/fact-sheets/detais/alchols.

[21] Wang, X. Y., Hattaf, K., Huo, Hai-Feng, & Xiang, H., (2016). Stability analysis of a delayed social epidemics model with general contact rate and its optimal control. *J. Ind. Manag. Optim. 12*, 1267–1285, doi: 10.3934/jimo.2016.12.1267.

# Chapter 13

# A Statistical Application of the Erdélyi-Kober Fractional Integral in Epidemic Modeling Problems

**Hemant Kumar**
*Department of Mathematics, D. A.-V. Postgraduate College*
*Kanpur, Uttar Pradesh, India*

**Abstract**
This chapter reveals an analysis of the presence of an epidemic in a region performed by a deterministic model problem. Through previous research, we consider a deterministic epidemic model problem which has its solution in terms of the Kummer hypergeometric function. Here, to achieve our goal, we apply the Erdélyi-Kober fractional integrals to convert this Kummer hypergeometric function into a Wright-type density function. The Wright-type density function is a powerful tool to draw different graphs for various parameters occurring in this epidemic model problem. Therefore, with the application of our Wright-type density function, we draw some graphs to analyze the presence of an epidemic. Lastly, a specific model problem is also remarked upon.

## 13.1 Densities due to Erdélyi-Kober Fractional Integrals

In this section, we introduce some relations of Erdélyi-Kober fractional integrals with certain densities. In this regard, we claim that the theory and methods of Mathai and Haubold [[17], Chapter 2], in which all types of fractional derivatives and fractional integrals are represented through Erdélyi-Kober fractional integral operators and statistical distributions, are relevant. It is also remarked that the incomplete gamma function and the incomplete beta function [22, p.27], the improper integrals $\int_{-\infty}^{a} f(x)\mathrm{dx}$, $\int_{a}^{\infty} f(x)\mathrm{dx}$, and the proper integrals $\int_{0}^{a} f(x)\mathrm{dx}$, $\int_{a}^{b} f(x)\mathrm{dx}$ may be transformed into certain Erdélyi-Kober fractional integrals.

Therefore, to explore new ideas in the field of "A Statistical Oriented Application of the Erdélyi-Kober Fractional Integral in Epidemic Model Problems," we appeal to the above theory of fractional integral operators to transform the epidemic model problem into Erdélyi-Kober fractional integrals. In this connection, we again introduce the following formula for the first kind Riemann-Liouville fractional integral [2, 9, 20] of order $\alpha_1 \in \mathbb{C}$, $\Re(\alpha_1) > 0$, where the left limit is $a$, in the new notation [17, p.19]:

$$D_{1,(a,x)}^{-\alpha_1} f = \frac{1}{\Gamma(\alpha_1)} \int_a^x (x-y)^{\alpha_1 - 1} f(y) dy. \tag{13.1.1}$$

Again we put a generalization of the formula (13.1.1) in the first kind Erdélyi-Kober fractional integral of order $\alpha_1$ and with the parameter $\eta$, given in the notation [17, p.17]

$$K_{1,x,\eta}^{-\alpha_1} f = \frac{x^{-\alpha_1 - \eta}}{\Gamma(\alpha_1)} \int_a^x (x-y)^{\alpha_1 - 1} y^{\eta} f(y) dy, \Re(\alpha_1) > 0. \tag{13.1.2}$$

Further by (13.1.2), we present following useful Theorem due to Mathai and Haubold [17, p.19], in the form

**Theorem 13.1.1.** *[17, p.19] If the Erdélyi-Kober fractional integral of the first kind of order* $\alpha_1$ and with the parameter $\beta_1$ on $f$ is given as

$$K_{1,u,\beta_1}^{-\alpha_1} f = \frac{u^{-\alpha_1 - \beta_1}}{\Gamma(\alpha_1)} \int_{v=0}^u (u-v)^{\alpha_1 - 1} v^{\beta_1} f(v) dv, \tag{13.1.3}$$

$\Re(\alpha_1) > 0$, $\Re(\beta_1) > 0$. Then there exists a marginal density $g(u)$ of $u = \frac{x_2}{x_1}$ where $x_1$ has a type-I beta density with parameters $(\beta_1, \alpha_1)$ given by

$$h(x_1) = \frac{\Gamma(\alpha_1 + \beta_1)}{\Gamma(\alpha_1)\Gamma(\beta_1)} (x_1)^{\beta_1 - 1} (1 - x_1)^{\alpha_1 - 1}, \; 0 < x_1 < 1, \tag{13.1.4}$$

$\Re(\alpha_1) > 0$, $\Re(\beta_1) > 0$; $x_2$ has an arbitrary density $H(x_2) = f(x_2)$ for some density $f(x_2)$ such that $f(x_2) \geq 0$ for all $x_2$ and $\int_{x_2} f(x_2)\, dx_2 = 1$, and then the marginal density $g$ of $u$ is given by

$$g(u) = \int_v h\left(\frac{v}{u}\right) f(v) \left(-\frac{v}{u^2}\right) dv. \tag{13.1.5}$$

Also a relation of the Erdélyi-Kober fractional integral with the marginal density is found by

$$\frac{\Gamma(\beta_1)}{\Gamma(\alpha_1 + \beta_1)} g(u) = K_{1,u,\beta_1}^{-\alpha_1} f. \tag{13.1.6}$$

*By this Theorem 13.1.1, we obtain various results and graphs pertaining to the probability density function (PDF) of the Epidemic model problem to analyze the graphs due to the PDF of the model problem.*

---

## 13.2 An Epidemic Model Problem and Erdélyi-Kober Fractional Integral Representation of Its Solution

In this section, to derive the Erdélyi-Kober fractional integral representation of the solution of the epidemic model problem, we introduce a deterministic epidemic model problem [1, 6–8] due to Kumar [10] (see also in the literature of [5]), given as:

A population of $M(t)$ individuals, at time $t$, is partitioned into two classes $Y(t)$ and $I(t)$ that, in other words, let $M(t)$ be the population size of the continuous time variable $t$. It is changing on account of immigration, births, emigration, and deaths (due to disease in question or other causes). $Y(t)$ notify the susceptible, while $I(t)$ denotes infection at time $t$. Let $\beta$ be the rate at which the population is receiving new individuals due to immigration and birth and $\mu$ be the rate at which individuals are being removed on account of emigration and death. Hence all the new entrants are assumed to be susceptible. The population is assumed to be uniform or homogeneous. The daily contact rate $C(M)$, at which the number of susceptible become infected, is taken by (non-dimensional):

$$C(M) = \frac{[\alpha\ IY + \ (\beta - \mu)\ (\delta Y - nI)]\ M}{M}, \tag{13.2.1}$$

where, $n \in \{0,\ 1,\ 2,\ \ldots\}$ and $\alpha,\ \beta,\ \gamma,\ \delta$ are real such that $\beta > \mu > 0,\ \delta > 0$ .

Then by the above discussion, the deterministic model problem is presented by

$$\frac{\mathrm{dY}}{\mathrm{dt}} = \beta M - \mu Y - \alpha\ IY - (\beta - \mu)(\delta Y - nI), \tag{13.2.2}$$

$$\frac{dI}{dt} = \alpha IY - \mu I + (\beta - \mu)(\delta Y - nI), \tag{13.2.3}$$

$$M = I + Y; \tag{13.2.4}$$

Here, $n$ is a parameter such that $n \in \{0, 1, 2, \ldots\}$ and $\alpha,\ \beta,\ \gamma,\ \delta$ are real such that $\beta > \mu > 0,\ \delta > 0$ .

The initial values are taken as $\mathrm{M}(0) = \mathrm{M}_0 > 0$, $Y(0) = Y_0 > 0$, and $I(0) = I_0 > 0$.

By the formulae (13.2.1) – (13.2.4) we find that

$$M = \mathrm{M}_0 \exp[(\beta - \mu)t\ ], t > 0. \tag{13.2.5}$$

Again in above equations (13.2.1) – (13.2.5) on introducing new recovered susceptible or quarantine cases $U$ such that

$$\frac{dU}{dt} = -\alpha YU \text{ and } \frac{\alpha}{\beta - \mu} M = -H, H > 0,$$

we find a transformed equation in the form of confluent hypergeometric differential equation [3, 21, 22] as

$$H \frac{d^2U}{dH^2} + \left(\frac{\mu}{\beta - \mu} + \delta + n - H\right) \frac{\mathrm{dU}}{\mathrm{dH}} - \left(\frac{\beta}{\beta - \mu} + n\right) U = 0. \quad (13.2.6)$$

A general solution of (13.2.6) is found by (see also in [10, 12])

$$\begin{aligned} U &= U(\beta,\ \mu,\ \delta; M) \\ &= \sum_{n=0}^{\infty} C_n\ {}_1F_1 \left(\begin{matrix} \frac{\beta}{\beta-\mu} + n\ ; \\ \frac{\mu}{\beta-\mu} + \delta + n\ ; \end{matrix} \quad -\frac{\alpha}{\beta - \mu} M\right), \end{aligned} \quad (13.2.7)$$

where, $\alpha, \beta, \gamma, \delta$are real such that $\beta > \mu > 0, \delta > 1, \alpha > 0$ and all $C_n$, $n \in \{0, 1, 2, \ldots.\}$, are arbitrary constants.

**Theorem 13.2.1.** *If we introduce a general function*

$$V(\beta,\ \mu,\ \delta; y) = \sum_{n=0}^{\infty} C_n \frac{\Gamma\left(\frac{\mu}{\beta-\mu} + \delta + n\right) y^n}{\Gamma\left(\frac{\beta}{\beta-\mu} + n\right)}, \quad (13.2.8)$$

*then there exists an Erdélyi-Kober fractional integral representation of the solution of epidemical problem* (13.2.1) – (13.2.4) *as*

$$U(\beta,\ \mu,\ \delta; M, u) = K^{-\ (\delta-1)}_{1,u,\ \frac{\mu}{\beta-\mu}} f,$$

*where*

$$f(y/u) = e^{-\ \{(\frac{\alpha}{\beta-\mu})\frac{M}{u}\}y}\ V\left(\beta,\ \mu,\ \delta; \frac{y}{u}\right). \quad (13.2.9)$$

*Proof.* In the formula (13.2.7) use the integral formula [22, p.37]

$${}_1F_1(a; b; z) = \frac{\Gamma(b)}{\Gamma(a)\Gamma(b-a)} \int_0^1 e^{zy} y^{a-1} (1-y)^{b-a-1} dy, \quad (13.2.10)$$

where , $\Re(b-a)>0,\ \Re(b)>0$ to get a formula

$$U(\beta,\ \mu,\ \delta;M) = \frac{1}{\Gamma(\delta-1)}\int_0^1 y^{\frac{\beta}{\beta-\mu}-1}(1-y)^{\delta-2} \times e^{-\frac{\alpha}{\beta-\mu}My}\ \mathrm{V}\,(\beta,\ \mu,\ \delta;y)dy. \quad (13.2.11)$$

Here, the function V $(\beta,\ \mu,\ \delta;y)$ is given in (13.2.8).

Again if $u>0$, then by the (13.2.11) we find a transformation formula

$$U(\beta,\ \mu,\ \delta;M,u) = \frac{(u)^{\frac{\beta-2\mu}{\beta-\mu}-\delta}}{\Gamma(\delta-1)} \times \int_0^u (y)^{\frac{\mu}{\beta-\mu}}\ (u-y)^{\delta-2}e^{-\{(\frac{\alpha}{\beta-\mu})\frac{M}{u}\}y}\ \mathrm{V}\left(\beta,\ \mu,\ \delta;\frac{y}{u}\right)\mathrm{dy}. \quad (13.2.12)$$

Then on making an appeal to the *Theorem* 13.1.1, we find the required result (13.2.9).

---

## 13.3 A Wright Type Probability Density Function Due to Given Epidemic Model Problem

In this section, we make an appeal to the techniques of previous *Section* 13.2 in which the solution of the epidemic model problem is transformed into Erdélyi-Kober fractional integral representation.

Then, on making an appeal to the Section 13.1 and using this representation, we obtain a Wright-type probability density function to analyze that epidemic model problem.

**Theorem 13.3.1.** *If an epidemic model problem is stated in the Section 13.2, then* $\forall x\in(0,\infty)$, $\alpha,u>0,\delta>1,\beta>\mu>0,M>0$, *there exists a density function*

$$P\ (\beta,\ \mu,\ \delta, M; x, u) = \begin{cases} \frac{\{(\frac{\alpha}{\beta-\mu})\frac{M}{u^2}\}}{K(\beta,\ \mu,\ \delta; M)} \exp\left[-\left\{\left(\frac{\alpha}{\beta-\mu}\right)\frac{M}{u^2}\right\}x\right] V\ \left(\beta,\ \mu,\ \delta; \frac{x}{u^2}\right), \forall x \in (0,\infty), \\ 0, \quad elsewhere. \end{cases} \tag{13.3.1}$$

*Here in* (13.3.1), *the functions K and V are defined by the series*

$$K(\beta,\ \mu,\ \delta; M) = \sum_{n=0}^{\infty} C_n \frac{\Gamma\left(\frac{\mu}{\beta-\mu} + \delta + n\right)\Gamma(n+1)}{\Gamma\left(\frac{\beta}{\beta-\ \mu} +\ n\right)} \left(\frac{1}{\left(\frac{\alpha}{\beta-\mu}\right)M}\right)^n, \tag{13.3.2}$$

*and*

$$V\left(\beta,\ \mu,\ \delta; \frac{x}{u^2}\right) = \sum_{n=0}^{\infty} C_n \frac{\Gamma\left(\frac{\mu}{\beta-\mu} + \delta + n\right)\left(\frac{x}{u^2}\right)^n}{\Gamma\left(\frac{\beta}{\beta-\ \mu} +\ n\right)}. \tag{13.3.3}$$

*Proof.* Make an appeal to the theory of the *Theorem* 13.1.1 to find that

$$u = \frac{x}{y} \implies y = \frac{x}{u}, u > 0 \tag{13.3.4}$$

Then by the formula (13.2.12) we get a function that satisfies the equation given by

$$C\int_0^{\infty} e^{\{-(\frac{\alpha}{\beta-\mu})\frac{M}{u^2}\}x} V\left(\beta,\ \mu,\ \delta; \frac{x}{u^2}\right)dx + C\int_{-\infty}^{0} 0dx = 1 \tag{13.3.5}$$

Again, use of the formulae (13.3.2), (13.3.3) and (13.3.5), we get

$$C = \frac{\left(\frac{\alpha}{\beta-\mu}\right)\frac{M}{u^2}}{\sum_{n=0}^{\infty} C_n \frac{\Gamma(\frac{\mu}{\beta-\mu} + \delta+n)\Gamma(n+1)}{\Gamma(\frac{\beta}{\beta-\ \mu}+\ n)} \frac{1}{[(\frac{\alpha}{\beta-\mu})M]^n}} = \frac{\left(\frac{\alpha}{\beta-\mu}\right)\frac{M}{u^2}}{K(\beta,\ \mu,\ \delta; M)} \tag{13.3.6}$$

Finally, by making an appeal to the formulae (13.3.5) and (13.3.6), we obtain the probability density function (13.3.1).

**Corollary 13.3.1.** *In* (13.3.2) *and* (13.3.3) *of the* Theorem *13.1.1, if we set* $C_n = \frac{1}{n!}$, *then there exists Wright type series [14], (also see in [22, p. 50]), in the form*

$$K(\beta,\ \mu,\ \delta; M) = \sum_{n=0}^{\infty} \frac{\Gamma\left(\frac{\mu}{\beta-\mu} + \delta + n\right)\Gamma(n+1)}{\Gamma\left(\frac{\beta}{\beta-\ \mu} +\ n\right) n!} \left(\frac{1}{\left(\frac{\alpha}{\beta-\mu}\right) M}\right)^n$$

$$= {}_2\Psi_1\left(\left.\begin{matrix}\left(\frac{\mu}{\beta-\mu}\ +\ \delta, 1\right), (1,1)\\ \left(\frac{\beta}{\beta-\ \mu}\ ,\ 1\right)\end{matrix}\right| \frac{1}{\left(\frac{\alpha}{\beta-\mu}\right) M}\right), \tag{13.3.7}$$

*and*

$$V\left(\beta,\ \mu,\ \delta; \frac{x}{u^2}\right) = \sum_{n=0}^{\infty} \frac{\Gamma\left(\frac{\mu}{\beta-\mu}\ +\ \delta + n\right)}{\Gamma\left(\frac{\beta}{\beta-\ \mu} +\ n\right) n!} \left(\frac{x}{u^2}\right)^n$$

$$= {}_1\Psi_1\left(\left.\begin{matrix}\left(\frac{\mu}{\beta-\mu}\ +\ \delta, 1\right)\\ \left(\frac{\beta}{\beta-\ \mu}\ ,\ 1\right)\end{matrix}\right| \frac{x}{u^2}\right). \tag{13.3.8}$$

**Corollary 13.3.2.** *In the* Theorem *13.1.1, if we set* $C_n = \frac{1}{n!}$, *then there exists a Wright type probability density function*

$$P\ (\beta,\ \mu,\ \delta, M; x, u)$$

$$= \begin{cases} \dfrac{\left\{\left(\frac{\alpha}{\beta-\mu}\right)\frac{M}{u^2}\right\}}{{}_2\Psi_1\left(\left.\begin{matrix}\left(\frac{\mu}{\beta-\mu}\ +\ \delta, 1\right), (1,1)\\ \left(\frac{\beta}{\beta-\ \mu}\ ,\ 1\right)\end{matrix}\right| \frac{1}{\left(\frac{\alpha}{\beta-\mu}\right) M}\right)} \times \\ \times \exp\left[-\left\{\left(\frac{\alpha}{\beta-\mu}\right)\frac{M}{u^2}\right\} x\right] {}_1\Psi_1\left(\left.\begin{matrix}\left(\frac{\mu}{\beta-\mu}\ +\ \delta, 1\right)\\ \left(\frac{\beta}{\beta-\ \mu}\ ,\ 1\right)\end{matrix}\right| \frac{x}{u^2}\right), x \in (0,\infty), \\ 0, \quad \textit{elsewhere}. \end{cases} \tag{13.3.9}$$

On making and appeal to the probability, $P$ (13.3.9) we draw following figures with respect to $X$ and $M$:

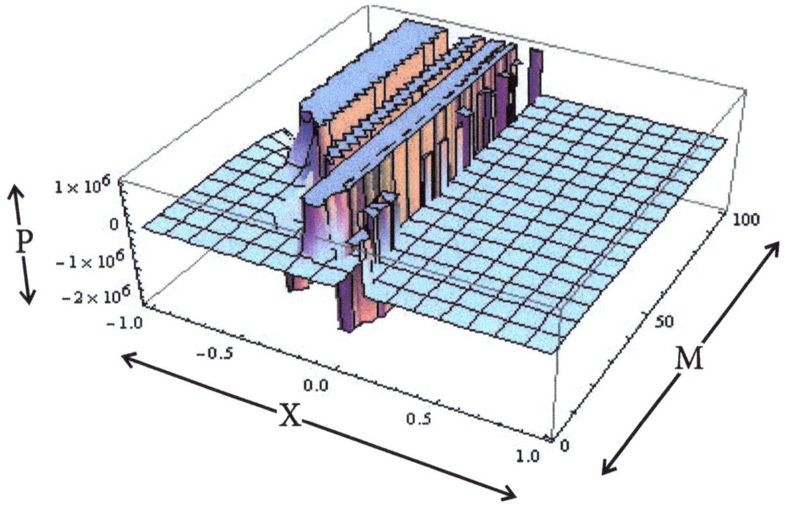


**FIGURE 13.1**. Distribution of $P$ with respect to $X$ and $M$ when $\alpha = 0.5, \mu = 5, \delta = 2, \mu = 0.1, \beta = 1.5, x = 10, \beta - \mu = X = -1$ to $+1$.

## 13.4 Analysis

In this chapter, the solution of the epidemic model problem is transformed into Kummer's hypergeometric function due to the techniques of Kumar [10]. Then we make an appeal to the theory of *Theorem* 13.1.1 due to Mathai and Haubold [17] for obtaining a probability density function (13.3.1). We again convert the formula (13.3.1) into a Wright-type probability density function (PDF) (13.3.9). By this PDF (13.3.9), various graphs are drawn. These graphs show the variation of population size by setting different values of the parameters under the conditions $\forall x \in (0, \infty), \alpha, u > 0, \delta > 1, \beta > \mu > 0, M > 0$. Figure 13.1 interprets that when the difference $\beta - \mu = X \in (-1, 0^-)$, in this case, the density

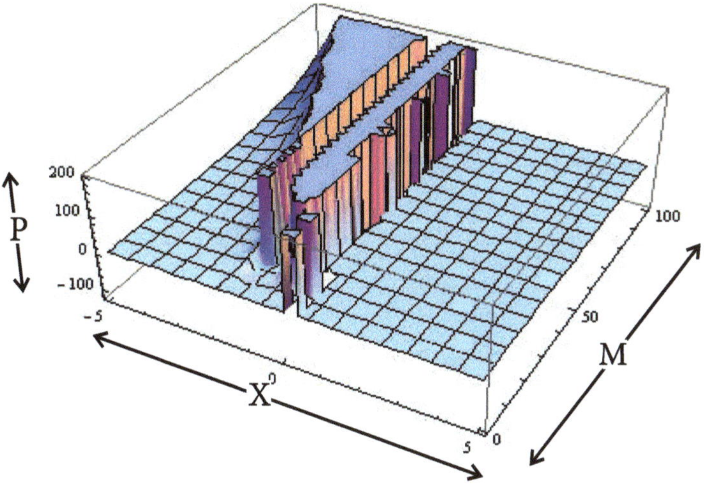


**FIGURE 13.2.** Distribution of $P$ with respect to $X$ and $M$ when $\alpha = 0.5, \mu = 5, \delta = 2, \mu = 0.1, \beta = 1.5, x = 10, \beta - \mu = X = -5$ to $+5$.

function $P(\beta, \mu, \delta, M; x, u)$ fluctuates under the range $(-2 \times 10^6, 1 \times 10^6)$. Hence, the epidemic is involved in the population.

Again, when $\beta - \mu = X \in (0^+, 1)$, the density function $P(\beta, \mu, \delta, M; x, u)$ seems to be zero and thus, due to the results (13.2.12), (13.3.1), and (13.3.9), the function $U$ (new individuals or quarantine cases) also tends to zero. Hence, according to the formulae given in (13.2.5) and (13.1.3), there exists a constant population in the case when $\beta - \mu = X \in (0, 1)$ so that the epidemic is about to be negligible.

Similarly, for the same values of the parameters, we set the difference $\beta - \mu = X \in (-5, +5)$. Then another Figure 13.2 is drawn, in which for $X \in (-5, 0^-)$, the density function $P(\beta, \mu, \delta, M; x, u)$ fluctuates in the range $(-200, 200)$, which is less than the previous case. Hence, in this case, the epidemic is spreading slowly. Again, for $X \in (0^+, +5)$, the epidemic is negligible.

On the other hand, for further directions in the field of the evaluation of connections between deterministic epidemic problems and hypergeometric functions, we suggest studying the research done in [11–13] and the literature cited in the books [2, 4, 9, 15–18, 20].

## 13.5 Specific Epidemic Model Problem

In the lecture notes [19], it is described that an infectious disease emerged from Wuhan city of China in December 2019 and spread worldwide. The World Health Organization (WHO) announced this infectious disease under the name COVID-19 (i.e., the coronavirus disease 2019) and declared it a pandemic in March 2020. It is also remarked that a few mathematical models were developed to interpret the dynamics of the evolution of COVID-19.

The spread of an epidemic disease depends on both the contact rate among individuals and the number of infected persons. That is, if each infected person is in contact with two other people before they recover, the disease will soon begin to spread rapidly. Therefore, there exists an exponential function like (13.2.5) to describe the spread of the disease, given by

$$M(t) = Ce^{at}, C > 0 \ \forall t \geq 0. \tag{13.5.1}$$

For $t > 0$, the exponential function (13.5.1) grows for $a > 0$, and decays for $a < 0$. When $a = 0$, the function becomes constant.

## 13.6 Algorithm to Plot 3D Graph (Figure 13.1) of Wright Type Probability Density Function, $P$, Due to MATHEMATICA

$$\text{Plot3D}\left[\frac{\left\{\frac{0.5}{\mathrm{X}}\frac{\mathrm{M}}{5^2}\right\}\exp\left(-\left\{\frac{0.5}{\mathrm{X}}\frac{\mathrm{M}}{5^2}\right\}10\right)\times\sum_{\mathrm{n}=0}^{5}\left(\frac{\Gamma\left(\frac{0.1}{\mathrm{X}}+2+\mathrm{n}\right)(10/5^2)^{\mathrm{n}}}{\Gamma\left(\frac{0.5}{\mathrm{X}}+\mathrm{n}\right)\Gamma(1+\mathrm{n})}\right)}{\sum_{\mathrm{n}=0}^{5}\left(\frac{\Gamma\left(\frac{0.1}{\mathrm{X}}+2+\mathrm{n}\right)(1/(0.5/\mathrm{X})\mathrm{M})^{\mathrm{n}}}{\Gamma(1.5/\mathrm{X}+\mathrm{n})}\right)}\right],$$

$$\{X,-1,1\},\{M,1,100\}$$

where $\alpha = 0.5$, $u = 5$, $\delta = 2$, $\mu = 0.1$, $\beta = 1.5$, $x = 10$, $\beta - \mu = X \in [-1, 1]$.

## 13.7 Algorithm to Plot 3D Graph (Figure 13.2) of Wright Type Probability Density Function, P, Due to MATHEMATICA

$$\text{Plot3D}\left[\frac{\left\{\frac{0.5}{\mathrm{X}}\frac{\mathrm{M}}{5^2}\right\}\exp\left(-\left\{\frac{0.5}{\mathrm{X}}\frac{\mathrm{M}}{5^2}\right\}10\right)\times\sum_{\mathrm{n}=0}^{5}\left(\frac{\Gamma\left(\frac{0.1}{\mathrm{X}}+2+\mathrm{n}\right)\left(\frac{10}{5^2}\right)^{\mathrm{n}}}{\Gamma\left(\frac{0.5}{\mathrm{X}}+\mathrm{n}\right)\Gamma(1+\mathrm{n})}\right)}{\sum_{\mathrm{n}=0}^{5}\left(\frac{\Gamma\left(\frac{0.1}{\mathrm{X}}+2+\mathrm{n}\right)\left(\frac{1}{(0.5/\mathrm{X})\mathrm{M}}\right)^{\mathrm{n}}}{\Gamma(1.5/\mathrm{X}+\mathrm{n})}\right)}\right],$$

$$\{X,-5,5\},\{M,1,100\}$$

when $\alpha = 0.5$, $u = 5$, $\delta = 2$, $\mu = 0.1$, $\beta = 1.5$, $x = 10$, $\beta - \mu = X \in [-5, 5]$.

# *References*


[1] Bailey, N. T. J., (1957). *The Mathematical Theory of Epidemics.* $1^{st}$ Edition, Griffin, London.

[2] Diethelm, K., (2010). *The Analysis of Fractional Differential Equations: An Application-Oriented Exposition Using Differential Operators of Caputo Type.* Springer-Verlag, Berlin, Heidelberg.

[3] Erdélyi, A., Magnus, W., Oberhettinger, F., & Tricomi, F. G. (1953). *Higher Transcendental Functions*, Vol. I. McGraw-Hill Book Co. Inc., New York.

[4] Hilfer, R., (2000). *Applications of Fractional Calculus in Physics.* World Scientific, Singapore.

[5] Joshi, D. C., (2004). Solution of the deterministic epidemiological model problem in terms of hypergeometric function ${}_0F_1(.)$. *Jñānābha 34*, 55–58.

[6] Kapur, J. N., (1998). *Mathematical Modeling.* New International (P) Limited Publishers, New Delhi.

[7] Kermack, W. O., & McKendrik, A. G., (1927a). Introduction to the mathematical theory of epidemics. Part I, *Proc. Roy. Soc. London Sect. A*, *115,* 700–721.

[8] Kermack, W. O., & McKendrik, A. G., (1927b). Introduction to the mathematical theory of epidemics. Part I, *Proc. Roy. Soc. London Sect. A*, *138*, 55–63.

[9] Kilbas, A. A., Srivastava, H. M., & Trujillo, J. J., (2006). *Theory and Applications of Fractional Differential Equations.* Elsevier, New York.

[10] Kumar, H., (2021). *Study on Some Inequalities of a Formula of Population Size due to Epidemics Model Problem.* In: Manuel Alberto M. Ferreira (Ed.) *Theory and Practice*

*of Mathematics and Computer Science*, Vol. 7; Book Publisher International, 51–56; https://www.bookpi.org; (http://doi.org/10.9734/bpi/tpmcs/v7/2800D).

[11] Kumar, H., & Kumari, Shilesh, (2008). On some measures, analogs to Carlson's Dirichlet measure, averages, generalized Erde′lyi-Kober integral operators and applications. *J. Tensor Society, 2*, 17–30.

[12] Kumar, H., & Kumari, Shilesh, (2010). An extension of an analytic formula of the deterministic epidemics model problem through the Lie group of operators. *Bull. Korean Math. Soc., 47*(6), 1131–1138.

[13] Luchko, Y., (2004). Operational rules for a mixed operator of the Erde′lyi-Kober type. *Fract. Calc. Appl. Analysis, 7*(3), 339–364.

[14] Luchko, Y., (2019). *The Wright function and its Applications.* In: Kochubei, A., Luchko, Y. (eds.) *Basic Theory*, De Gruyter, 241–268.

[15] Mathai, A. M., & Haubold H. J., (2008). *Special Functions for Applied Scientists.* Springer, New York.

[16] Mathai , A. M., & Haubold, H. J., (2018). *Erde′lyi-Kober Fractional Calculus: From a Statistical Perspective, Inspired by Solar Neutrino Physics.* Springer Briefs in Mathematical Physics *31*, Springer Nature Singapore Pte Ltd.

[17] Mathai, A. M., Saxena, R. K., & Haubold, H. J., (2010). *The H-function: Theory and Applications.* Springer, New York.

[18] Miller, K. S., & Ross, B., (1993). *An Introduction to Fractional Calculus and Fractional Differential Equations.* Wiley, New York.

[19] Okabe, Y., & Shudo A., (2020). A mathematical model of epidemics: A tutorial for students. *Mathematics, 8*, 1174, 1–16. DOI:10.3390/math8071174.

[20] Oldham, K. B., & Spanier, J., (2006). *The Fractional Calculus, Theory and Applications of Differentiation and Integration to Arbitrary Order.* Dover Publication Inc., New York.

[21] Rainville, E. D., (1971). *Special Functions.* Mac Milan, Chalsea Publication Co. Bronx, New York.

[22] Srivastava, H. M., & Manocha, H. L., (1984). *A Treatise on Generating Functions.* John Wiley and Sons, New York.

# Chapter 14

---

## A Comprehensive Qualitative Examination of a Fractional-Order Model for Divorce Dynamics

**Yasir Nadeem Anjam,**[1] **Mehmet Yavuz,**[2,3] **Amna Batool,**[1] **and Mati ur Rahman**[4]

[1]*Department of Applied Sciences, National Textile University Faisalabad, Pakistan*

[2]*Department of Mathematics and Computer Sciences, Faculty of Science, Necmettin Erbakan University, Meram Yeniyol, Meram, Konya, Türkiye*

[3]*Department of Applied Mathematics and Informatics, Kyrgyz-Turkish Manas University, Bishkek, Kyrgyzstan*

[4]*School of Mathematical Sciences, Jiangsu University, Zhenjiang, Jiangsu, P.R. China*

**Abstract**

This chapter proposes a non-linear deterministic mathematical model that encapsulates the dynamics of the prevailing degree of divorce in a population. The objectives are served by the launch of the divorce model through the employment of a fractional-order derivative scheme in the Caputo sense. The total population is divided into four compartments: Married class, Divorced class, Separated class, and the class facing Hardships due to divorce. The analytical solutions are achieved via the Laplace transform with the Adomian decomposition method (LADM) and the Homotopy perturbation method (HPM). The numerical environment is elaborated with respect to various parametric settings, including class size, time, and fractional orders. Additionally, numerical simulation results for the said schemes have been provided for various arbitrary fractional orders using MATLAB software, and the graphical representation is depicted. The validity of the issue at hand is also addressed through the simulated results, and thc compartmcnt dynamics are observed to be sensitive to various fractional orders, indicating the utility of the fractional approach.

---

## Introduction

Marriage, as a physical and moral connection, is regarded by the community as the foundation of a family. Marriage entails the distribution of rights and responsibilities among the contract's parties. Sometimes, it can be described as the union of matrimony or wedlock, where the relational sexual connection is biologically confessed [1]. Divorce (called the dissolution of marriage) is the optionally accessible manner of the breakdown of a wedding or marital union [2], which has a deformable indication on the family structure in our present epoch. Moreover, it is noticed that the current pace of rising weddings is proportionate to the rate of divorce caused by inter-cultural marriages. There are similarities between the spread of contagious disease and divorce. Alternatively, divorce can be viewed as a virus that spreads among compartments due to social influence and a variety of other factors [3, 7].

The most common reasons for divorce may not be limited to less education, early marriage, premarital cohabitation and pregnancy, physical assault, low income, lack of religious affiliation, and infidelity [5, 8]. On the other hand, the least common reasons for divorce are women's schooling, having distinctions or bringing up children, leaving the house for an extended period of time, and a refusal to cooperate for domestic chores [12]. Divorce is a prevalent issue that badly impacts the social and economic composition of modern society as much as any disease. It diminishes social support networks while undermining kids' future competency in the five key obligations or institutions of society: family, school, marketplace, religion, and government [6]. It also leads to more relationship problems, a higher risk of divorce, higher expectations of divorce early in adulthood, and a lower willingness to have kids. Approximately half of all initial marriages in our society end in divorce or separation, as documented in [9]. Second marriages have even higher divorce rates, as indicated in [10]. Notably, 75 percent of separations eventually culminate in divorce, as reported in [11]. Addressing this societal challenge is imperative to mitigate its impact on our community.

Mathematical frameworks are essential tools for predicting the influence of social issues and have been used in the past few decades to control epidemics. We can adopt them to control the spread of divorce in marital institutions and concentrate on the effect of the economy [13] and social contamination on divorce models. Since many global challenges are quasi-linear problems, they are often idealistic and unrealistic. As a result, a non-linear mathematical model explains real-world issues much better than linear models. The sustainability of the divorce model was assessed, and counseling was proposed as a feasible control strategy to sustain marital balance and tranquility. Consequently, through effective counseling, family harmony, peace, and satisfying life may be kept and maintained, which in turn minimizes the divorce possibilities among marriages [15]. Ranjitha and Akshatha [17] employed significant features to forecast the divorce rate by deleting duplicate aspects that do not assist in the prediction using an efficient machine learning algorithm for the standardized dataset provided. The Spatial Durbin Model (SDM) was used to estimate the spatial spillover effects of divorce in Senegal [18]. The cubic spline interpolation approach is used to estimate the functionality passing through provided points, and the extrapolation technique to evaluate marriage and divorce occurrences for the next year [19].

Several studies have modeled problems associated with the dynamics of divorce [4, 14, 16, 19]. In this study, we consider the proposed divorce mathematical model as having four compartments, namely married class (M), divorced class (D), separated class (S), and the class facing Hardships due to divorce (H). Examining this model enhances family role factors or family intervention and outside factors in an exertion to annihilate divorce. Therefore, the mathematical formulation of the considered problem is expressed as:

$$\begin{cases} \frac{dM}{dt} = \Lambda + \varepsilon_2 D + \varepsilon_1 S + \varepsilon_3 H - \rho\beta PM(S+D)(1-u_2) \\ \qquad -(1-\rho)\beta PM(S+D)(1-u_1) - \mu M, \\ \frac{dS}{dt} = (1-\rho)\beta PM(S+D)(1-u_1) - \varepsilon_1 S - \delta S - \mu S, \\ \frac{dD}{dt} = \rho\beta PM(S+D)(1-u_2) + \delta S - \varepsilon_2 D - \gamma D - \mu D, \\ \frac{dH}{dt} = \gamma D - \varepsilon_3 H - \mu H, \end{cases} \tag{14.0.1}$$

with the total population $N = M + D + S + H$.

To explore the intended model (14.0.1), we use the subsequent purely pertinent subsidiary conditions

$$M(0) = n_1, S(0) = n_2, D(0) = n_3, H(0) = n_4.$$

The used parameters along with a description are as follows: $\beta$ represents the average exchange between $M$ and people who are not in class $H$, $\Lambda$ represents the degree of input of entities who are predisposed to divorce, $\gamma$ is the proportion of people who suffer from depression or stress after divorce, $\delta$ shows the proportion of people getting a divorce after a separation, $u_1$ represents the percentage of people who are secure from becoming detached from their beds because of family intervention, $u_2$ is the percentage of people who vetoed divorce owing to mediators (Judges), $\varepsilon_1$ is the proportion of people who get married again after divorce, $\varepsilon_2$ shows the proportion of people who revert to marry again after separation, $\varepsilon_3$ is the proportion of people who return to the marital relationship after being depressed or stressed, $\rho$ is the percentage of $M$ who were divorced, (1-$\rho$) is the percentage of the number of $M$ entirely separated beds, $P$ is the possibility of effective contact among $M(t)$ and people other than $H(t)$, $\mu S$ is the percentage upon which the quantity of $S$ comes out naturally, $\mu M$ represents the percentage upon which the quantity of $M$ is extracted naturally, $\mu H$ is the percentage upon which the quantity of $H$ naturally comes out, $\mu D$ shows the percentage upon which the quantity of $D$ naturally comes out.

In the past few decades, fractional calculus has attained much attention from researchers on account of its many applications in numerous fields of scientific investigation such as engineering, physical sciences, biological sciences, and social sciences. Many mathematical models are constructed in the framework of differential and integral equations of integer order. It is also observed that the fractional differential equations (FDEs) of non-integer order have been used to construct natural phenomena with a better degree of exactitude and accuracy. Various methods have been used by many researchers for the qualitative analysis of fractional-order mathematical models; see, e.g., [24, 29, 30]. A new idea in fractional calculus for piecewise differential and integral operators has been introduced in [22]. Normally, it is not simple for non-linear fractional differential equations to get their general solution. To address this situation, the existing literature offers a range of approaches for approximating solutions to non-linear systems [25–27]. Runge-Kutta methods have found extensive application in mathematical analysis, as discussed in [32]. Furthermore, the Homotopy Perturbation Method (HPM), Laplace Adomian Decomposition Method (LADM), and the method of differences have been frequently utilized for addressing both linear and non-linear fractional-order equations, as documented in references [31, 33, 34].

Additionally, several types of fractional-order non-local derivatives have been established to achieve the advantages of integer-order derivatives (IOD). For example, Riemann-Liouville established the idea of fractional derivatives, which was afterward amended and developed by Caputo. The investigation of real problems in the sense of fractional derivatives normally generates singularities that are unrealistic for the dynamics of mathematical models. To avert such situations, new fractional derivatives, identified as Caputo and Caputo-Fabrizio operators, were proposed in [20, 35]. For more aspects about fractional-order derivatives and their illustrative applications, see [21, 36–45], and references therein.

In this research, we focus on achieving a more comprehensive qualitative analysis and numerical iterative analysis of the modified version of the developed model (14.0.1). Specifically, we consider the model in the context of the Caputo operator, which is as:

$$\begin{cases} {}^{c}D_t^{\theta} M(t) = \Lambda + \varepsilon_2 D + \varepsilon_1 S + \varepsilon_3 H - \rho\beta PM(S+D)(1-u_2) \\ \qquad\qquad -(1-\rho)\beta PM(S+D)(1-u_1) - \mu M, \\ {}^{c}D_t^{\theta} S(t) = (1-\rho)\beta PM(S+D)(1-u_1) - \varepsilon_1 S - \delta S - \mu S, \\ {}^{c}D_t^{\theta} D(t) = \rho\beta PM(S+D)(1-u_2) + \delta S - \varepsilon_2 D - \gamma D - \mu D, \\ {}^{c}D_t^{\theta} H(t) = \gamma D - \varepsilon_3 H - \mu H. \end{cases} \tag{14.0.2}$$

The rest of the chapter is organized as follows: Some basic results and definitions are given in Section 14.1. The series solutions of the model via the LADM are presented in Section 14.2. The numerical findings of the system are discussed in Section 14.3. In Section 14.4, the HPM is specified, and a comparison of both methods is provided. Furthermore, the obtained results are discussed accordingly. The conclusion of this work is presented in Section 14.5.

---

## 14.1 Analytical Preliminaries

This section provides fundamental background information and fractional calculus principles in alignment with [22, 23, 28].

### *14.1.1 Definition 1*

Let a function $f(x)$, with order $0 < \alpha < 1$, the Caputo derivative is written as:

$$ {}^{c}D_x^{\alpha} f(x) = \frac{d^{\alpha}}{dt^{\alpha}} f(t) = \frac{1}{\Gamma(\alpha)} \int_0^x (x-t)^{1-\alpha} \frac{d}{dt} f(t) dt, $$

where the function $\Gamma$ is expressed as

$$ \Gamma(n) = \int_0^{\infty} e^{-t} t^{n-1} dt. $$

### *14.1.2 Definition 2*

Let a function $g(t)$, then the Laplace transform is

$$ \mathscr{L}[{}^{c}D_t^{\alpha} g(t)] = s^{\alpha} G(s) - \sum_{j=0}^{m-1} s^{\alpha-j-1} g^{k}(0), \; m-1 < \alpha, m \in \mathbb{N}. $$

### *14.1.3 Definition 3*

The Homotopy perturbation method (HPM) is a quasi-approach for solving both linear and non-linear ODE and PDE equations. It has the potential to be employed within an equation that encompasses both linear and nonlinear components, allowing for the construction of a homotopy $v(l, p) : \Omega \times [0 \times 1] \to R$:

$$ H(v,p) = (1-p)[L(v) - L(u_0)] + p[L(v) + N(v) - f(l)] = 0, $$

here, we use $L$ to represent the linear component and $N$ to denote the nonlinear component. In this context, $l \in \Omega$, and $p \in [0,1]$ serves as the embedding parameter.

## 14.2 Series Solution of the Considered System (14.0.2) With (LADM)

This part is related to the general solution of the proposed model (14.0.2), which is constructed by employing the general technique of LADM with the initial conditions. Now, taking the Laplace transform of (14.0.2), we get

$$\begin{cases} \mathcal{L}[{}^cD_t^\alpha M(t)] = \mathcal{L}[\Lambda + \varepsilon_2 D + \varepsilon_1 S + \varepsilon_3 H - \rho\beta PM(S+D)(1-u_2) \\ \qquad -(1-\rho)\beta PM(S+D)(1-u_1) - \mu M], \\ \mathcal{L}[{}^cD_t^\alpha S(t)] = \mathcal{L}[(1-\rho)\beta PM(S+D)(1-u_1) - \varepsilon_1 S - \delta S - \mu S], \\ \mathcal{L}[{}^cD_t^\alpha D(t)] = \mathcal{L}[\rho\beta PM(S+D)(1-u_2) + \delta S - \varepsilon_2 D - \gamma D - \mu D], \\ \mathcal{L}[{}^cD_t^\alpha H(t)] = \mathcal{L}[\gamma D - \varepsilon_3 H - \mu H]. \end{cases} \tag{14.2.1}$$

Using the initial conditions yields

$$\begin{cases} \mathcal{L}[M(t)] = \frac{n_1}{s} + \frac{1}{s^\alpha}\mathcal{L}[\Lambda + \varepsilon_2 D + \varepsilon_1 S + \varepsilon_3 H - \rho\beta PM(S+D)(1-u_2) \\ \qquad -(1-\rho)\beta PM(S+D)(1-u_1) - \mu M], \\ \mathcal{L}[S(t)] = \frac{n_2}{s} + \frac{1}{s^\alpha}\mathcal{L}[(1-\rho)\beta PM(S+D)(1-u_1) - \varepsilon_1 S - \delta S - \mu S], \\ \mathcal{L}[D(t)] = \frac{n_3}{s} + \frac{1}{s^\alpha}\mathcal{L}[\rho\beta PM(S+D)(1-u_2) + \delta S - \varepsilon_2 D - \gamma D - \mu D], \\ \mathcal{L}[H(t)] = \frac{n_4}{s} + \frac{1}{s^\alpha}\mathcal{L}[\gamma D - \varepsilon_3 H - \mu H]. \end{cases} \tag{14.2.2}$$

The series solution of the linear terms $M$, $S$, $D$, and $H$ can be describe as:

$$M(t)=\sum_{n=0}^{\infty}M_n(t),\quad S(t)=\sum_{n=0}^{\infty}S_n(t),$$
$$D(t)=\sum_{n=0}^{\infty}D_n(t),\quad H(t)=\sum_{n=0}^{\infty}H_n(t),\tag{14.2.3}$$

and the nonlinear terms $M(t)S(t) = \sum_{n=0}^{\infty} X_n(t)$, and $M(t)D(t) = \sum_{n=0}^{\infty} Z_n(t)$

$$\begin{cases} X_n(t)=\frac{1}{n!}\frac{d^n}{d\lambda^n}\left[\sum_{k=0}^{n}\lambda^k M_k(t)\sum_{k=0}^{n}\lambda^k S_k(t)\right]\Bigg|_{\lambda=0} \\ Z_n(t)=\frac{1}{n!}\frac{d^n}{d\lambda^n}\left[\sum_{k=0}^{n}\lambda^k M_k(t)\sum_{k=0}^{n}\lambda^k D_k(t)\right]\Bigg|_{\lambda=0}. \end{cases}\tag{14.2.4}$$

Using equations (14.2.3) and (14.2.4) into (14.2.2) and comparing the terms of both sides, one can get

$$\begin{cases} \mathcal{L}[M_0(t)]=\frac{n_1}{s},\ \mathcal{L}[S_0(t)]=\frac{n_2}{s},\ \mathcal{L}[D_0(t)]=\frac{n_3}{s},\ \mathcal{L}[H_0(t)]=\frac{n_4}{s}, \\ \mathcal{L}[M_1(t)]=\frac{1}{s^\alpha}\mathcal{L}[\Lambda+\varepsilon_2 D_0+\varepsilon_1 S_0+\varepsilon_3 H_0-\rho\beta P(X_0+Z_0)(1-u_2) \\ \qquad -(1-\rho)\beta P(X_0+Z_0)(1-u_1)-\mu M_0], \\ \mathcal{L}[S_1(t)]=\frac{1}{s^\alpha}\mathcal{L}[(1-\rho)\beta P(X_0+Z_0)(1-u_1)-(\varepsilon_1+\delta+\mu)S_0], \\ \mathcal{L}[D_1(t)]=\frac{1}{s^\alpha}\mathcal{L}[\rho\beta P(X_0+Z_0)(1-u_2)+\delta S_0-(\varepsilon_2+\gamma+\mu)D_0], \\ \mathcal{L}[H_1(t)]=\frac{1}{s^\alpha}\mathcal{L}[\gamma D_0-(\varepsilon_3+\mu)H_0], \end{cases}\tag{14.2.5}$$

$$\vdots$$

$$\begin{cases} \mathcal{L}[M_{n+1}(t)] = \frac{1}{s^\alpha}\mathcal{L}[\Lambda + \varepsilon_2 D_n + \varepsilon_1 S_n + \varepsilon_3 H_n - \rho\beta P(X_n + Z_n)(1-u_2) \\ \qquad -(1-\rho)\beta P(X_n + Z_n)(1-u_1) - \mu M_n], \\ \mathcal{L}[S_{n+1}(t)] = \frac{1}{s^\alpha}\mathcal{L}[(1-\rho)\beta P(X_n + Z_n)(1-u_1) - (\varepsilon_1 + \delta + \mu)S_n], \\ \mathcal{L}[D_{n+1}(t)] = \frac{1}{s^\alpha}\mathcal{L}[\rho\beta P(X_n + Z_n)(1-u_2) + \delta S_n - (\varepsilon_2 + \gamma + \mu)D_n], \\ \mathcal{L}[H_{n+1}(t)] = \frac{1}{s^\alpha}\mathcal{L}[\gamma D_n - (\varepsilon_3 + \mu)H_n]. \end{cases}$$

Further, applying the inverse Laplace transform to (14.2.5), we obtain

$$M_0(t) = n_1,\ S_0(t) = n_2,\ D_0(t) = n_3,\ H_0(t) = n_4,$$

$$\begin{cases} M_1(T) = [\Lambda + \varepsilon_2 n_3 + \varepsilon_1 n_2 + \varepsilon_3 n_4 - \rho\beta P(n_1 n_2 + n_1 n_3)(1-u_2) \\ \qquad -(1-\rho)\beta P(n_1 n_2 + n_1 n_3)(1-u_1) - \mu n_1]\frac{t^\alpha}{\Gamma(\alpha+1)}, \\ S_1(T) = [(1-\rho)\beta P(n_1 n_2 + n_1 n_3)(1-u_1) - (\varepsilon_1 + \delta + \mu)n_2]\frac{t^\alpha}{\Gamma(\alpha+1)}, \\ D_1(T) = [\rho\beta P(n_1 n_2 + n_1 n_3)(1-u_2) + \delta n_2 - (\varepsilon_2 + \gamma + \mu)n_3]\frac{t^\alpha}{\Gamma(\alpha+1)}, \\ H_1(t) = [\gamma n_3 - (\varepsilon_3 + \mu)n_4]\frac{t^\alpha}{\Gamma(\alpha+1)}, \end{cases}$$

$$\begin{cases} M_2(T) = [\Lambda + \varepsilon_2 w_{11} + \varepsilon_1 u_{11} + \varepsilon_3 v_{11} \\ -\rho\beta P(n_1 u_{11} + n_2 x_{11} + n_1 w_{11} + n_3 x_{11})(1-u_2) - (1-\rho)\beta P \\ \times(n_1 u_{11} + n_2 x_{11} + n_1 w_{11} + n_3 x_{11})(1-u_1) - \mu x_{11}]\frac{t^{2\alpha}}{\Gamma(2\alpha+1)}, \\ S_2(T) = [(1-\rho)\beta P(n_1 u_{11} + n_2 x_{11} + n_1 w_{11} + n_3 x_{11})(1-u_1) \\ -(\varepsilon_1 + \delta + \mu)u_{11}]\frac{t^{2\alpha}}{\Gamma(2\alpha+1)}, \\ D_2(T) = [\rho\beta P(n_1 u_{11} + n_2 x_{11} + n_1 w_{11} + n_3 x_{11})(1-u_2) + \delta u_{11} \\ -(\varepsilon_2 + \gamma + \mu)w_{11}]\frac{t^{2\alpha}}{\Gamma(2\alpha+1)}, \\ H_2(t) = [\gamma w_{11} - (\varepsilon_3 + \mu)v_{11}]\frac{t^{2\alpha}}{\Gamma(\alpha+1)}. \end{cases}$$

Similarly, the renaming terms accordingly. Moreover, the unknown terms are mentioned as:

$$\begin{cases} w_{11} = \rho\beta P(n_1n_2 + n_1n_3)(1-u_2) + \delta n_2 - (\varepsilon_2 + \gamma + \mu)n_3, \\ u_{11} = (1-\rho)\beta P(n_1n_2 + n_1n_3)(1-u_1) - (\varepsilon_1 + \delta + \mu)n_2, \\ x_{11} = \Lambda + \varepsilon_2 n_3 + \varepsilon_1 n_2 + \varepsilon_3 n_4 - \rho\beta P(n_1n_2 + n_1n_3)(1-u_2) \\ -(1-\rho)\beta P(n_1n_2 + n_1n_3)(1-u_1) - \mu n_1, \\ v_{11} = \gamma n_3 - (\varepsilon_3 + \mu)n_4. \end{cases}$$

---

## 14.3 Numerical Scheme of the Model (14.0.2)

In this part, we conduct numerical simulations to validate the analytical results of our proposed model. To achieve this, we have assigned specific values to the parameters utilized in our model (14.0.2) as:

$$\begin{cases} n_1 = 4300, \ n_2 = 1, \ n_3 = 0, \ n_4 = 0.5, \\ \mu = 0.03, \ \beta = 0.08, \ \delta = 0.00001, \ \gamma = 0.00001, \\ \rho = 0.052, \ \varepsilon_1 = 0.99, \ \varepsilon_2 = 0.35, \ \varepsilon_3 = 0.001, \\ P = 0.021, \ u_1 = 0.9999, \ u_2 = 0, \ \Lambda = 100, \end{cases}$$

By employing these parameter values, we obtain the following terms within the model (14.0.2):

$$\begin{cases} M_0(t) = 4300, \ S_0(t) = 1, \ D_0(t) = 0, \ H_0(t) = 0.5, \\ M_1(t) = -28.3919\frac{t^\alpha}{\Gamma(\alpha+1)}, \ S_1(t) = -1.01932\frac{t^\alpha}{\Gamma(\alpha+1)}, \\ D_1(t) = 0.375658\frac{t^\alpha}{\Gamma(\alpha+1)}, \ H_1(t) = -0.0155\frac{t^\alpha}{\Gamma(\alpha+1)}, \\ M_2(t) = 99.9558\frac{t^{2\alpha}}{\Gamma(2\alpha+1)}, \ S_2(t) = 1.03927\frac{t^{2\alpha}}{\Gamma(2\alpha+1)}, \\ D_2(t) = -0.38705\frac{t^{2\alpha}}{\Gamma(2\alpha+1)}, \ H_2(t) = 0.00048425\frac{t^{2\alpha}}{\Gamma(2\alpha+1)}. \end{cases} \tag{14.3.1}$$

Moreover, solutions for the initial terms are presented as follows:

$$\begin{cases} M(t) = 4300 - 28.3858\frac{t^{\alpha}}{\Gamma(\alpha+1)} + 99.9558\frac{t^{2\alpha}}{\Gamma(2\alpha+1)}, \\ S(t) = 1 - 1.01932\frac{t^{\alpha}}{\Gamma(\alpha+1)} + 1.03927\frac{t^{2\alpha}}{\Gamma(2\alpha+1)}, \\ D(t) = 0 + 0.375658\frac{t^{\alpha}}{\Gamma(\alpha+1)} - 0.38705\frac{t^{2\alpha}}{\Gamma(2\alpha+1)}, \\ H(t) = 0.5 - 0.0155\frac{t^{\alpha}}{\Gamma(\alpha+1)} + 0.00048425\frac{t^{2\alpha}}{\Gamma(2\alpha+1)}. \end{cases} \tag{14.3.2}$$

Now, we evaluate equation (14.3.2) for various values of $\alpha$. Therefore, for $\alpha = 0.96$, we get

$$\begin{cases} M(t) = 4300 - 28.85490751t^{0.96} + 53.73832897t^{1.92}, \\ S(t) = 1 - 1.036165418t^{0.96} + 0.5587332916t^{1.92}, \\ D(t) = 0.3818661741t^{0.96} - 0.2080861764t^{1.92}, \\ H(t) = 0.5 - 0.01575615506t^{0.96} + 0.0002603429296t^{1.92}. \end{cases} \tag{14.3.3}$$

Similarly for the value $\alpha = 0.97$, the approximate solution for (14.3.2) is:

$$\begin{cases} M(t) = 4300 - 28.73973025t^{0.97} + 52.78523300t^{1.94}, \\ S(t) = 1 - 1.032029460t^{0.97} + 0.5488236711t^{1.94}, \\ D(t) = 0.3803419169t^{0.97} - 0.2043955872t^{1.94}, \\ H(t) = 0.5 - 0.01569326279t^{0.97} + 0.0002557255215t^{1.94}. \end{cases} \tag{14.3.4}$$

For $\alpha = 0.9998$, the approximate solution is:

$$\begin{cases} M(t) = 4300 - 28.38819996t^{0.9998} + 49.99634935t^{1.9996}, \\ S(t) = 1 - 1.019406181t^{0.9998} + 0.5198268234t^{1.9996}, \\ D(t) = 0.3756897611t^{0.9998} - 0.1935964398t^{1.9996}, \\ H(t) = 0.5 - 0.01550131049t^{0.9998} + 0.0002422143805t^{1.9996}, \end{cases} \tag{14.3.5}$$

and similarly, for $\alpha = 0.9999$ yields the following approximate solution for (14.3.2):

$$\begin{cases} M(t) = 4300 - 28.38700004t^{0.9999} + 49.98712422t^{1.9998}, \\ S(t) = 1 - 1.019363093t^{0.9999} + 0.5197309070t^{1.9998}, \\ D(t) = 0.3756738813t^{0.9999} - 0.1935607181t^{1.9998}, \\ H(t) = 0.5 - 0.01550065528t^{0.9999} + 0.0002421696880t^{1.9998}. \end{cases} \tag{14.3.6}$$

## 14.4 Series Solution for the Model (14.0.2) Using (HPM)

Next, we apply the Homotopy Perturbation Method (HPM) to derive the solution for the aforementioned model (14.0.2) as follows:

$$\begin{cases} (1-q)[{}^cD_t^\alpha(M(t)) - {}^cD_t^\alpha(M_0(t))] + q[{}^cD_t^\alpha(M(t)) - \Lambda - \varepsilon_2 D \\ -\varepsilon_1 S - \varepsilon_3 H + \rho\beta PM(S+D)(1-u_2) + (1-\rho)\beta PM(S+D) \\ \times(1-u_1) + \mu M] = 0, \\ (1-q)[{}^cD_t^\alpha(S(t)) - {}^cD_t^\alpha(S_0(t))] + q[{}^cD_t^\alpha(S(t)) - (1-\rho)\beta PM \\ \times(S+D)(1-u_1) + (\varepsilon_1 + \delta + \mu)S] = 0, \\ (1-q)[{}^cD_t^\alpha(D(t)) - {}^cD_t^\alpha(D_0(t))] + q[{}^cD_t^\alpha(D(t)) - \rho\beta PM(S+D) \\ \times(1-u_2) - \delta S + (\varepsilon_2 + \gamma + \mu)D] = 0, \\ (1-q)[{}^cD_t^\alpha(H(t)) - {}^cD_t^\alpha(H_0(t))] + q[{}^cD_t^\alpha(H(t)) - \gamma D \\ +(\varepsilon_3 + \mu)H] = 0. \end{cases} \tag{14.4.1}$$

When we set q equal to zero in (14.4.1), we obtain the subsequent set of fractional differential equations:

$$\begin{cases} {}^cD_t^\alpha(M(t)) - {}^cD_t^\alpha(M_0(t)) = 0, \\ {}^cD_t^\alpha(S(t)) - {}^cD_t^\alpha(S_0(t)) = 0, \\ {}^cD_t^\alpha(D(t)) - {}^cD_t^\alpha(D_0(t)) = 0, \\ {}^cD_t^\alpha(H(t)) - {}^cD_t^\alpha(H_0(t)) = 0, \end{cases} \tag{14.4.2}$$

the solution to the previous equation is straightforward. Now, when we substitute q = 1 into (14.4.1), we obtain a model similar to (14.0.2). We can assume the solution in the form of an infinite series as follows:

$$\begin{cases} M(t) = \sum_{n=0}^\infty q^n M_n(t), \;\; S(t) = \sum_{n=0}^\infty q^n S_n(t), \\ D(t) = \sum_{n=0}^\infty q^n D_n(t), \;\; H(t) = \sum_{n=0}^\infty q^n H_n(t). \end{cases} \tag{14.4.3}$$

Moreover, we can retrieve the initial system by setting q = 1 in (14.4.1). Next, by substituting (14.4.3) into (14.4.1) and examining the terms according to their powers of q, we derive the following:

$$q^0 : \begin{cases} M_0(t) = M(0) = n_1, \;\; S_0(t) = S(0) = n_2, \\ D_0(t) = D(0) = n_3, \;\; H_0(t) = H(0) = n_4. \end{cases} \tag{14.4.4}$$

Similarly,

$$q^1 : \begin{cases} M_1(t) = [\Lambda + \varepsilon_2 n_3 + \varepsilon_1 n_2 + \varepsilon_3 n_4 - \rho\beta P(n_1 n_2 + n_1 n_3)(1 - u_2) \\ \qquad -(1-\rho)\beta P(n_1 n_2 + n_1 n_3)(1 - u_1) - \mu n_1]\frac{t^\alpha}{\Gamma(\alpha+1)}, \\ S_1(t) = [(1-\rho)\beta P(n_1 n_2 + n_1 n_3)(1 - u_1) - (\varepsilon_1 + \delta + \mu)n_2]\frac{t^\alpha}{\Gamma(\alpha+1)}, \\ D_1(t) = [\rho\beta P(n_1 n_2 + n_1 n_3)(1 - u_2) + \delta n_2 - (\varepsilon_2 + \gamma + \mu)n_3]\frac{t^\alpha}{\Gamma(\alpha+1)}, \\ H_1(t) = [\gamma n_3 - (\varepsilon_3 + \mu)n_4]\frac{t^\alpha}{\Gamma(\alpha+1)}, \end{cases}$$

(14.4.5)

$$q^2:\begin{cases} M_2(t)=[\Lambda+\varepsilon_2 w_{11}+\varepsilon_1 u_{11}+\varepsilon_3 v_{11}-\rho\beta P(n_1u_{11}+n_2x_{11}\\ +n_1w_{11}+n_3x_{11})(1-u_2)-(1-\rho)\beta P(n_1u_{11}+n_2x_{11}+n_1w_{11}\\ +n_3x_{11})(1-u_1)-\mu x_{11}]\frac{t^{2\alpha}}{\Gamma(2\alpha+1)},\\ S_2(t)=[(1-\rho)\beta P(n_1u_{11}+n_2x_{11}+n_1w_{11}+n_3x_{11})(1-u_1)\\ -(\varepsilon_1+\delta+\mu)u_{11}]\frac{t^{2\alpha}}{\Gamma(2\alpha+1)},\\ D_2(t)=[\rho\beta P(n_1u_{11}+n_2x_{11}+n_1w_{11}+n_3x_{11})(1-u_2)+\delta u_{11}\\ -(\varepsilon_2+\gamma+\mu)w_{11}]\frac{t^{2\alpha}}{\Gamma(2\alpha+1)},\\ H_2(t)=[\gamma w_{11}-(\varepsilon_3+\mu)v_{11}]\frac{t^2\alpha}{\Gamma(2\alpha+1)}. \end{cases} \tag{14.4.6}$$

Similarly, the higher terms can be found, and the unknown terms are already stated in the preceding section. Consequently, we achieved the like terms as attained for the LADM approach. Therefore, both methods are effective in solving nonlinear fractional-order differential equations.

Figure 14.1 displays the dynamical behavior of the married class $M(t)$ at numerous fractional orders. The decline is greater at low fractional order and less at high fractional order, and the fractional simulation of low order becomes stable at around $M$=3333 and $t$ = 80 days, indicating convergence.

Figure 14.2 represents the behavior of the separated class *S(t)* versus time. It is noted that the separated class decreases fast at different arbitrary orders and eventually converges to zero. It also demonstrates that for small fractional orders, it becomes stable and converges to zero over time.

In Figure 14.3, it seems that the divorced class grows rapidly in several arbitrary orders and then begins to decline in small fractional order after one day. After 23 days, it reaches a stable value of zero. For high orders, it continues to rise over time until it reaches a plateau or begins to fall after 303 days.

Similarly, Figure 14.4 shows that over time, the proportion of persons who were depressed or stressed decreases at various arbitrary orders, and after 45 days, it converges to zero at small orders. However, at high orders, it steadily decreases over time.

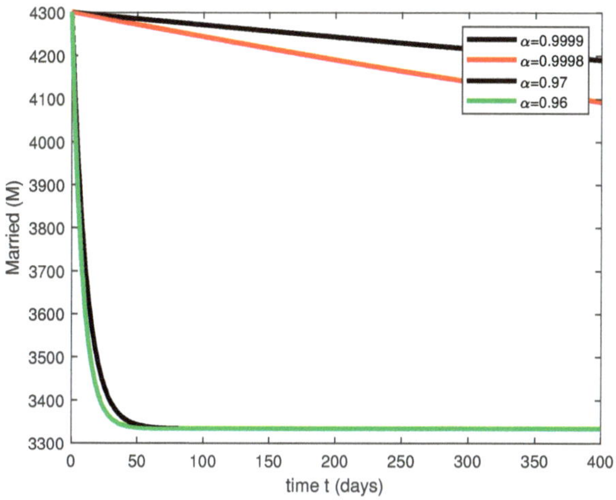


**FIGURE 14.1**. Dynamical behavior of the approximate solution of the model (14.0.2) for married *M(t)* class for different fractional order.

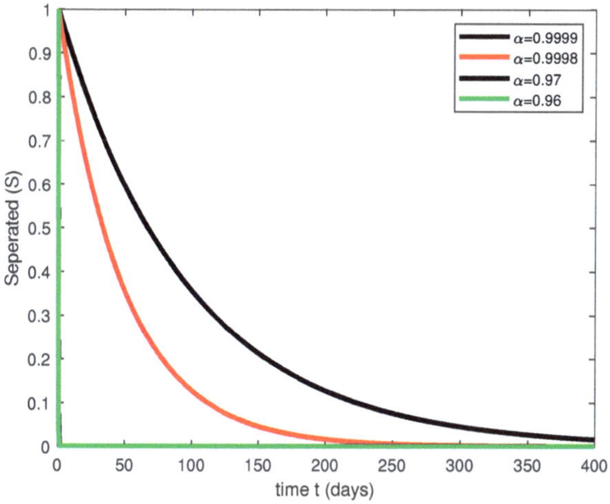


**FIGURE 14.2**. Dynamical behavior of the approximate solution of the model (14.0.2) for separated class *S(t)* at different fractional orders.

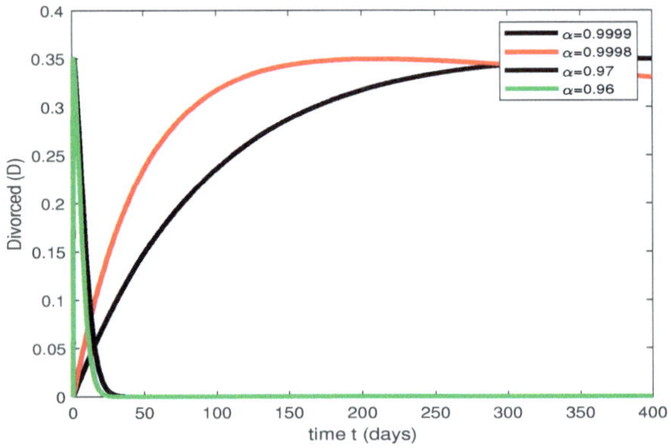


**FIGURE 14.3**. Dynamical behavior of the approximate solution of model (14.0.2) for divorced class *D(t)* at different arbitrary fractional orders.

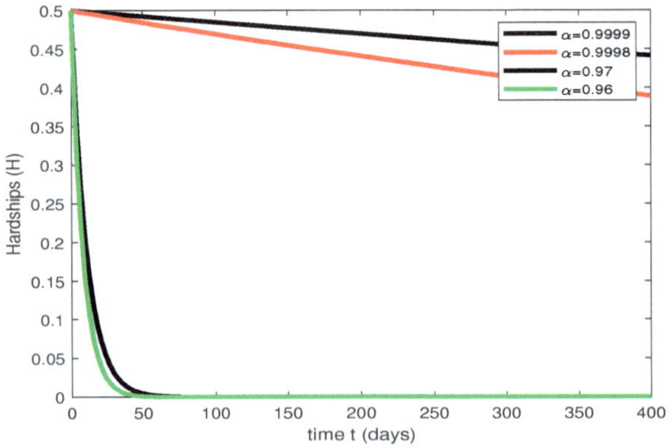


**FIGURE 14.4**. Dynamical behavior of the approximate solution of model (14.0.2) for hardships class *H(t)* at different arbitrary orders.

We plan to provide a comparison of numerical solutions for different compartments of the model under investigation using both the proposed HPM method and LADM. This comparison will focus on the initial terms that exhibit similar simulation results, as shown in Figures 14.5 to 14.8.

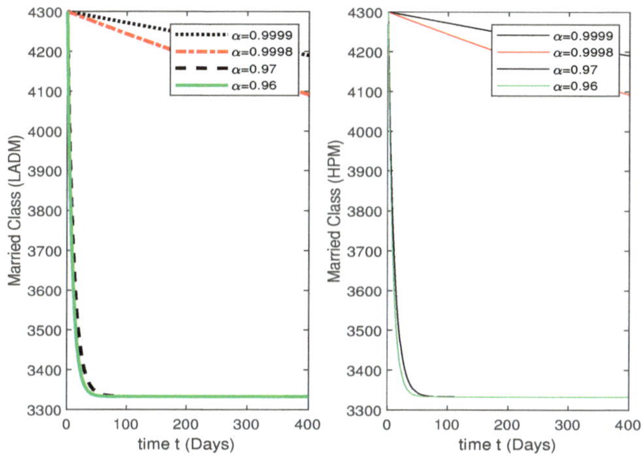


**FIGURE 14.5**. Comparing the approximate solution for $M$ of the system (14.0.2) at various orders by HPM and LADM.

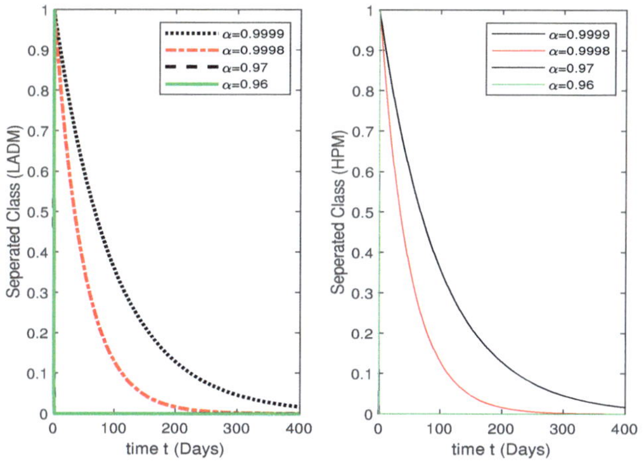


**FIGURE 14.6**. Comparing the approximate solution for $S$ of the system (14.0.2) at various orders by HPM and LADM.

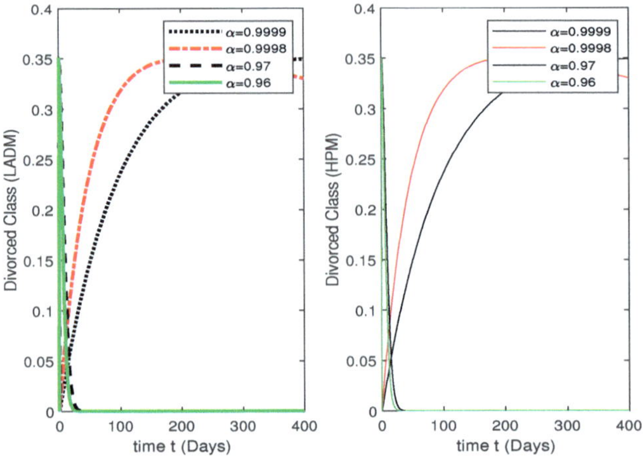


**FIGURE 14.7**. Comparing the approximate solution for $D$ of the system (14.0.2) at various orders by HPM and LADM.

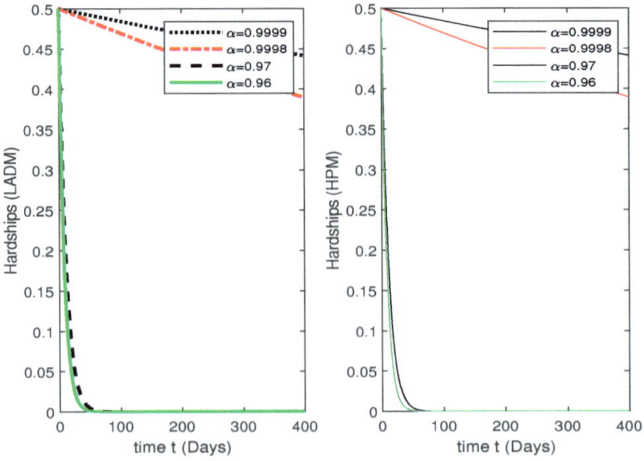


**FIGURE 14.8**. Comparing the approximate solution for $H$ of the system (14.0.2) at various orders by HPM and LADM.

## 14.5 Conclusion

This study delves into the examination of the dynamic characteristics of the divorce model utilizing the Caputo fractional operator. We have numerically investigated this model using both the Homotopy Perturbation Method (HPM) and the Laplace Adomian Decomposition Method (LADM), focusing on the initial three terms. Both approaches give comparable outcomes for the considered problem. The numerical results obtained from both techniques exhibit a high degree of similarity, offering robust validation for the studied model across a wide range of fractional-order derivatives. Moreover, numerous parameters influence the results, including the ratio of married persons who were divorced ($\rho$), the rate of average interaction between individuals with married and other persons ($\beta$), the possibility of successful interaction of a person with married and other persons ($P$), the ratio of persons who prevented divorce because of the presence of external parties (Judge) ($u_2$), the rate of death ($\mu$), the proportion of persons coming back to marital status after splitting ($\epsilon_2$), and the proportion of persons who suffer from depression or stress as a result of divorce ($\gamma$). The obtained results also help to reduce the cases of the spread of divorce in society. In general, we may conclude that the fractional-order differential equation outperforms the standard derivatives in terms of prediction.

# References


[1] Wimalasena, N. A. (2016). An analytical study of definitions of the term marriage. *International Journal of Humanities and Social Science*, *6*(1), 166–174.

[2] Gambrah, P. P., & Adzadu, Y. (2018). Mathematical model of divorce epidemic in Ghana. *International Journal of Statistics and Applied Mathematics*, *3*(2), 395–401.

[3] Gambrah, P. P., Rahaman, A. S., & Adzadu, Y. (2018). Divorce transmission model. *International Journal of Statistics and Applied Mathematics*, *3*(3), 102–106.

[4] Gambrah, P. P., Rahaman, A. S., & Adu, A. (2018). Mathematical model for minimizing divorce through counseling. *International Journal of Statistics and Applied Mathematics*, *3*(3), 218–225.

[5] Mekonnen, Y., Kassa, K., & Ayalew, M. (2019). Prevalence, causes and consequences of divorce in Bahir Dar city, Ethiopia. *African Journal of Social Work*, *9*(1), 3–78.

[6] Anderson, J. (2014). The impact of family structure on the health of children: Effects of divorce. *Linacre Quarterly*, *81*(4), 378–387.

[7] Bruze, G., Svarer, M., & Weiss, Y. (2015). The dynamics of marriage and divorce. *Journal of Labor Economics*, *33*(1), 123–170.

[8] Rezaei, V., Fallah, M. H., & Vaziri, S. (2020). Investigating the causes of divorce through narrative analysis in Yazd city and designing a prerequisite education based on the causes of divorce using a hidden learning approach on the basis of family, school, and student. *Journal of Toloo-e-Behdasht*.

[9] Bramlett, M. D., & Mosher, W. D. (2002). Cohabitation, marriage, divorce, and remarriage in the United States. *Division of Vital Statistics Series, 23*(22), 32.

[10] Cherlin, A. (2009). *Marriage, Divorce, Remarriage* (Revised and Enlarged Ed.). Harvard University Press.

[11] Hawkins, A. J., Willoughby, B. J., & Doherty, W. J. (2012). Reasons for divorce and openness to marital reconciliation. *Journal of Divorce & Remarriage, 53*(6), 453–463.

[12] Damota, M. D. (2019). *The Effect of Divorce on Families' Life.* Academia.edu, 46.

[13] Huang, J. T. (2003). Unemployment and family behavior in Taiwan. *Journal of Family and Economic Issues, 24*(1), 27–48.

[14] Duato, R., & Jódar, L. (2013). Mathematical modeling of the spread of divorce in Spain. *Mathematical and Computer Modelling, 57*(7–8), 1732–1737.

[15] Hugo, A., & Lusekelo, E. M. (2021). Mathematical control of divorce stress among marriages. *International Journal of Statistics and Applied Mathematics, 6*(5), 126–136.

[16] Lhous, M., Rachik, M., Laarabi, H., & Abdelhak, A. (2017). Discrete mathematical modeling and optimal control of the marital status: The monogamous marriage case. *Advances in Difference Equations, 2017*(1), 1–16.

[17] Ranjitha, P., & Prabhu, A. (2020). Improved divorce prediction using machine learning-particle swarm optimization (PSO). In *2020 International Conference for Emerging Technology (INCET)* (pp. 1–5). IEEE.

[18] Aw, A., & Cabral, E. N. (2019). Modeling spatial spillovers of divorce in Senegal using spatial Durbin model: A maximum likelihood estimation approach. *American Journal of Theoretical and Applied Statistics, 8*(1), 1–6.

[19] Ghazi, F. F., Khalaf, R. I., & Sabah, N. (2020). The estimating marriage and divorces and comparing them using numerical method. *Al-Qadisiyah Journal of Pure Science, 25*(2), 60–66.

[20] Goufo, E. F. D. (2016). Application of the Caputo-Fabrizio fractional derivative without singular kernel to Korteweg-de Vries-Burgers equation. *Mathematical Modelling and Analysis*, *21*(2), 188–198.

[21] Atangana, A. (2020). Extension of rate of change concept: From local to nonlocal operators with applications. *Results in Physics*, *19*, 103515.

[22] Atangana, A., & Araz, S. İ. (2021). New concept in calculus: Piecewise differential and integral operators. *Chaos, Solitons & Fractals*, *145*, 110638.

[23] Miller, K. S., & Ross, B. (1993). *An Introduction to the Fractional Calculus and Fractional Differential Equations.* John Wiley and Sons.

[24] Atangana, A. (2021). Mathematical model of survival of fractional calculus, critics and their impact: How singular is our world? *Advances in Difference Equations*, *2021*(1), 1–59.

[25] Bai, C. Z., & Fang, J. X. (2004). The existence of a positive solution for a singular coupled system of nonlinear fractional differential equations. *Applied Mathematics and Computation*, *150*(3), 611–621.

[26] Heydari, M. H., & Razzaghi, M. (2021). A numerical approach for a class of nonlinear optimal control problems with piecewise fractional derivative. *Chaos, Solitons & Fractals*, *152*, 11465.

[27] Heydari, M. H., & Atangana, A. (2023). A numerical method for nonlinear fractional reaction–advection–diffusion equation with piecewise fractional derivative. *Mathematical Sciences*, *17*(2), 169–181.

[28] Zeb, A., Atangana, A., Khan, Z. A., & Djillali, S. (2022). A robust study of a piecewise fractional order COVID-19 mathematical model. *Alexandria Engineering Journal*, *61*(7), 5649–5665.

[29] Khan, H., Li, Y., Khan, A., & Khan, A. (2019). Existence of solution for a fractional order Lotka-Volterra reaction-diffusion model with Mittag-Leffler kernel. *Mathematical Methods in the Applied Sciences*, *42*(9), 3377–3387.

[30] Khan, R. A., & Shah, K. (2015). Existence and uniqueness of solutions to fractional order multi-point boundary value problems. *Communications in Applied Analysis*, *19*(3–4), 515–526.

[31] Awawdeh, F., Adawi, A., & Mustafa, Z. (2009). Solutions of the SIR models of epidemics using HAM. *Chaos, Solitons & Fractals*, *42*(5), 3047–3052.

[32] Haq, F., Shah, K., Rahman, G. U., & Shahzad, M. (2017). Numerical analysis of fractional order model of HIV-1 infection of CD4+ T-cells. *Computational Methods for Differential Equations*, *5*(1), 1–11.

[33] Biazar, J. (2006). Solution of the epidemic model by Adomian decomposition method. *Applied Mathematics and Computation*, *173*(2), 1101–1106.

[34] Rafei, M., Ganji, D. D., & Daniali, H. (2007). Solution of the epidemic model by homotopy perturbation method. *Applied Mathematics and Computation*, *187*(2), 1056–1062.

[35] Caputo, M., & Fabrizio, M. (2015). A new definition of fractional derivative without singular kernel. *Progress in Fractional Differentiation and Applications*, *1*(2), 73–85.

[36] Chen, Y., & An, H. (2008). Numerical solutions of coupled Burgers equations with time and space fractional derivatives. *Applied Mathematics and Computation*, *200*(1), 87–95.

[37] Rossikhin, Y. A., & Shitikova, M. V. (1997). Applications of fractional calculus to dynamic problems of linear and nonlinear hereditary mechanics of solids. *Applied Mechanics Reviews*, *50*(1), 15–67.

[38] Eroğlu, B. B. İ. (2023). Two-dimensional Cattaneo-Hristov heat diffusion in the half-plane. *Mathematical Modeling and Numerical Simulation with Applications*, *3*(3), 281–296.

[39] Joshi, H., & Yavuz, M. (2023). Transition dynamics between a novel coinfection model of fractional-order for COVID-19 and tuberculosis via a treatment mechanism. *European Physical Journal Plus*, *138*(5), 468.

[40] Duran, S., Durur, H., Yavuz, M., & Yokus, A. (2023). Discussion of numerical and analytical techniques for the emerging fractional order Murnaghan model in materials science. *Optical and Quantum Electronics*, *55*(6), 571.

[41] Joshi, H., Yavuz, M., & Stamova, I. (2023). Analysis of the disturbance effect in intracellular calcium dynamics on fibroblast cells with an exponential kernel law. *Bulletin of Biomathematics, 1*(1), 24–39.

[42] ur Rahman, M., Arfan, M., & Baleanu, D. (2023). Piecewise fractional analysis of the migration effect in plant-pathogen-herbivore interactions. *Bulletin of Biomathematics, 1*(1), 1–23.

[43] Ahmed, I., Akgül, A., Jarad, F., Kumam, P., & Nonlaopon, K. (2023). A Caputo-Fabrizio fractional-order cholera model and its sensitivity analysis. *Mathematical Modeling and Numerical Simulation with Applications, 3*(2), 170–187.

[44] Yavuz, M., Akman, M., Usta, F., & Özdemir, N. (2022). Effect of the awareness parameter on a fractional-order tuberculosis model. In *AIP Conference Proceedings* (Vol. 2483, No. 1). AIP Publishing.

[45] Liu, X., Rahman, M. U., Ahmad, S., Baleanu, D., & Nadeem, A. (2022). A new fractional infectious disease model under the non-singular Mittag-Leffler derivative. *Waves in Random and Complex Media*, 1–27.

# Chapter 15

## A Numerical Study of the Jaulent-Miodek Model Using the Homotopy Perturbation Transform Technique with the Caputo-Fabrizio Operator

**Amit Prakash,[1] Ajay Kumar,[2] and Rahul[1]**

[1]*Department of Mathematics, National Institute of Technology, Kurukshetra, India*

[2]*Department of Basic Engineering and Applied Sciences, Mahamaya College of Agricultural Engineering and Technology, Ambedkar Nagar (A.N.D.U.A.T, Kumarganj), Ayodhya, India*

**Abstract**

This chapter discusses the Jaulent-Miodek equation using the homotopy perturbation transform method, which involves a combination of the fractional homotopy perturbation technique with the Laplace transform in the Caputo-Fabrizio derivative sense. The solution of the proposed equation is approximated using the fractional homotopy perturbation transform method. We applied the proposed approach to a test example and presented the solution graphically and in tabulated form.

## Introduction

Recently, research on fractional calculus has been found to be applicable in various fields of applied science and engineering. Fractional differential equations play a vital role in fluid dynamics, electromagnetism, biology, control theory, rheology, viscoelasticity, electrochemistry, diffusion, signal processing, and other areas [10, 17]. The present fractional calculus principle is furnished with excellent tools to describe dynamic behaviors, as well as related scientific systems and responses. Many researchers have used fractional differential equations for modeling and analyzing various scientific outcomes [8, 16, 23, 36]. The primary goal of exploring the idea of fractional calculus is to address asymmetric phenomena connected with problems [5, 9]. Several works on numerical solutions to fractional differential equations have been written in recent years, and researchers have employed various ideas to attain greater accuracy, convergence rate, and stability results. In a particularly linear sense, many different analytical methods have been presented so that the interested reader can see previous studies [1, 11, 25].

In the past few decades, fractional calculus has played a significant role in analyzing real-life problems. It is nearly impossible to solve fractional differential equations analytically. As a result, numerical techniques are employed to find the approximate solutions of fractional differential equations. Many different numerical methods have been presented by researchers to explore the field of fractional calculus, such as the homotopy analysis method [19], the Laplace Adomian decomposition technique [26], the homotopy perturbation method [37], the variational iteration method [14, 27], the homotopy perturbation Sumudu transform technique [28], the q-homotopy analysis transform technique [29], the homotopy perturbation transform method [22], the fractional Adams-Bashforth method [30, 34], and many more [2, 7, 12, 18, 31–33].

In fractional calculus, Riemann-Liouville [35] and Caputo derivatives [20, 24] with power-law kernels are widely used. But in the past few years, Caputo-Fabrizio [6] and Atangana-Baleanu [3] are two new fractional derivatives with nonsingular kernels that have been introduced by researchers and explored with various numerical techniques. These fractional derivatives encompass all appropriate themes and help us perceive natural phenomena in an orderly and dominant manner. The fractional Jaulent-Miodek equation is considered [4, 13, 21, 38, 39] as:

$$\begin{aligned} {}_0^{CF}D_t^{\alpha}\, v + v_{xxx} + aww_{xxx} + bw_x w_{xx} - cvv_x - dvww_x - ev_x w^2 &= 0, \\ {}_0^{CF}D_t^{\alpha}\, w + w_{xx} - fv_x w - gvw_x - hw_x w^2 &= 0, \end{aligned} \tag{15.0.1}$$

having initial conditions

$$\begin{aligned} v(x,0) &= f(x), \\ w(x,0) &= g(x). \end{aligned}$$

Where the variables $x, t$ are independent variables and v,w are dependent variables. In 2009 and 2012, Yildirim et al. and Zhang studied an exact solution and physical properties of (15.0.1) in [38, 39]; more recently, A.K. Gupta [13] solved Jaulent-Miodek equation by wavelet method and others [4, 21]. It is a veteran piece of equipment for solving many linear and nonlinear occurrences.

---

## 15.1 Basic Definitions of Fractional Calculus

**Definition 15.1.1.** *The Caputo derivative is defined as:*

$${}_0^{C}D_t^{\alpha}u(t) = \frac{1}{\Gamma(n-\alpha)} \int_0^t (t-\xi)^{(n-\alpha-1)} \frac{d^n}{dt^n} u(\xi) d\xi,$$

*where* $n - 1 < \alpha \leq n$ *and* $n \in \mathbb{N}$ *and* ${}_0^{C}D_t^{\alpha}u(t) = 0$, *if* $u$ *is a constant function.*

**Definition 15.1.2.** *The time-fractional Caputo-Fabrizio derivative is defined by:*

$${}_0^{CF}D_t^{\alpha}u(t) = \frac{M(\alpha)}{(1-\alpha)} \int_0^t \exp\left[-\frac{\alpha(1-\xi)}{1-\alpha}\right] u'(\xi) d\xi, t \geq 0, 0 < \alpha < 1,$$

*where, $M(\alpha)$ is a normalized function such that $M(0) = M(1) = 1$. and Caputo-Fabrizio integral is given by*

$$ {}_{0}^{CF}I_t^{\alpha}u(t) = \frac{2(1-\alpha)}{(2-\alpha)M(\alpha)}u(t) + \frac{2\alpha}{(2-\alpha)M(\alpha)}\int_0^t u(\xi)d\xi, t \geq 0. $$

**Definition 15.1.3.** *The Laplace transform for the Caputo-Fabrizio derivative of order $0 < \alpha < 1$. and $m \in N$ is given as:*

$$ \begin{aligned} \mathscr{L}\left[{}_{0}^{CF}D_t^{(m+\alpha)}\, u(t)\right](s) &= \frac{1}{1-\alpha}\mathscr{L}[u^{m+1}(t)]\mathscr{L}\left[exp\left(\frac{-\alpha}{(1-\alpha)}t\right)\right] \\ &= \frac{s^{m+1}\mathscr{L}[u(t)] - s^m u(0) - s^{m-1}u'(0)\ldots - u^m(0)}{s+\alpha(1-s)}, \end{aligned} $$

*for the several, we get*

$$ \mathscr{L}\left[{}_{0}^{CF}D_t^{(m+\alpha)}\, u(t)\right](s) = \frac{s\mathscr{L}(u(t)) - u(0)}{s+\alpha(1-s)}, \qquad m = 0, $$

$$ \mathscr{L}\left[{}_{0}^{CF}D_t^{(m+\alpha)}\, u(t)\right](s) = \frac{s^2\mathscr{L}(u(t)) - su(0) - u'(0)}{s+\alpha(1-s)}, \qquad m = 1. $$

---

## 15.2 Description of Homotopy Perturbation Transforms Method Using Caputo Fabrizio Derivative

The homotopy perturbation transforms method is a numerical technique used for finding numerical solutions of nonlinear partial differential equations and other mathematical problems. It is an iterative method that starts with an initial approximation and then perturbs it until a solution is found. We proposed the following Fractional differential equation with Caputo Fabrizio derivative:

$$ {}_{0}^{CF} D_t^{m+\alpha} u(x,t) + \beta u(x,t) + \varphi u(x,t) = \Theta(x,s), \qquad m < \alpha + m \leq m+1, \tag{15.2.1} $$

having initial conditions

$$ \frac{\partial^k u(x,0)}{\partial t^k} = f_k(x), \qquad k = 0,1,2,\ldots n-1. $$

Apply the Laplace transform on both side (15.2.1) we obtained:

$$ \mathscr{L}[u(x,t)] = \Theta(x,s) - \left(\frac{s+\alpha(1-s)}{s^{m+1}}\right) \mathscr{L}[\beta u(x,t) + \varphi u(x,t)]. \tag{15.2.2} $$

where

$$ \Theta(x,s) = \frac{1}{s^{m+1}}[s^m f_0(x) + s^{m-1} f_1(x) + \ldots + f_m(x)] + \frac{s+\alpha(1-s)}{s^{m+1}}\tilde{k}(x,s). \tag{15.2.3} $$

Applying inverse Laplace transform on (15.2.2), we have

$$ u(x,t) = \Theta(x,s) - \mathscr{L}^{-1}\left[\left(\frac{s+\alpha(1-s)}{s^{m+1}}\right) \mathscr{L}[\beta u(x,t) + \varphi u(x,t)]\right], \tag{15.2.4} $$

as a result of an infinite series

$$ u(x,t) = \sum_{n=0}^{\infty} p^n u_n(x,t). \tag{15.2.5} $$

and the nonlinear term is decomposable like

$$ \varphi u(x,t) = \sum_{n=0}^{\infty} p^n H_n(x,t). \tag{15.2.6} $$

We define the He's polynomial $H_m(x,t)$ as [15]

$$H_m(u_0, u_1, u_2, \ldots u_n) = \frac{1}{m!}\frac{\partial^m}{\partial p^m}\left[\left(\sum_{m=0}^{\infty} p^i u_i\right)\right]_{p=0}, \qquad m = 0, 1, 2, \ldots. \tag{15.2.7}$$

Substitute (15.2.5) & (15.2.6) into (15.2.4), we obtained

$$\begin{aligned}\sum_{m=0}^{\infty} u_m(x,t) &= \Theta(x,s) \\ &- p\mathscr{L}^{-1}\left[\left(\frac{s+\alpha(1-s)}{s^{m+1}}\right)\mathscr{L}\left[\beta\sum_{m=0}^{\infty} p^m u_m(x,t) + \sum_{n=0}^{\infty} p^m H_m\right]\right].\end{aligned} \tag{15.2.8}$$

We obtained the following approximations by equating the terms with similar powers in $p$ in (15.2.8)

$$\begin{aligned} p^0 : u_0(x,t) &= \Theta(x,s), \\ p^1 : u_1(x,t) &= -\mathscr{L}^{-1}\left[\left(\frac{s+\alpha(1-s)}{s^{m+1}}\right)\mathscr{L}[\beta u_0(x,t) + H_0(u)]\right] \\ p^2 : u_2(x,t) &= -\mathscr{L}^{-1}\left[\left(\frac{s+\alpha(1-s)}{s^{m+1}}\right)\mathscr{L}[\beta u_1(x,t) + H_1(u)]\right] \end{aligned}$$

$$\begin{aligned} p^3 : u_3(x,t) &= -\mathscr{L}^{-1}\left[\left(\frac{s+\alpha(1-s)}{s^{m+1}}\right)\mathscr{L}[\beta u_2(x,t) + H_2(u)]\right] \\ &\vdots \\ p^{m+1} : u_{m+1}(x,t) &= -\mathscr{L}^{-1}\left[\left(\frac{s+\alpha(1-s)}{s^{m+1}}\right)\mathscr{L}[\beta u_{m+1}(x,t) + H_{m+1}(u)]\right]. \end{aligned}$$

The solutions to (15.2.1) is defined as:

$$u(x,t) = \sum_{m=0}^{\infty} u_m(x,t). \tag{15.2.9}$$

## 15.3 Application

In the present section, we apply the homotopy perturbation transforms method for solving the fractional Jaulent-Miodek equation.
Taking $a = e = \frac{3}{2}, b = \frac{9}{2}, c = d = f = g = 6, h = \frac{15}{2}$ we consider the following nonlinear fractional Jaulent-Miodek equation

$$\begin{aligned} {}_0^{CF}D_t^{\alpha}\, v + v_{xxx} + \frac{3}{2}ww_{xxx} + \frac{9}{2}w_x w_{xx} - 6vv_x - 6vww_x - \frac{3}{2}v_x w^2 = 0, \\ {}_0^{CF}D_t^{\alpha}\, w + w_{xx} - 6v_x w - 6vw_x - \frac{15}{2}w_x w^2 = 0, \end{aligned} \tag{15.3.1}$$

having initial conditions

$$\begin{aligned} v(x,0) &= \frac{1}{8}\lambda^2(1 - 4\sec h^2(\frac{\lambda x}{2})), \\ w(x,0) &= \lambda \sec h(\frac{\lambda x}{2}). \end{aligned}$$

Applying Laplace transform on (15.3.1), we obtain

$$\begin{aligned} \mathscr{L}[v(x,t)] =& \frac{1}{s}v(x,0) - \left(\frac{s+\alpha(1-s)}{s}\right) \\ & \times \mathscr{L}\left[v_{xxx} + \frac{3}{2}ww_{xxx} + \frac{9}{2}w_x w_{xx} - 6vv_x - 6vww_x - \frac{3}{2}v_x w^2\right], \\ \mathscr{L}[w(x,t)] =& \frac{1}{s}w(x,0) - \left(\frac{s+\alpha(1-s)}{s}\right) \\ & \times \mathscr{L}\left[w_{xxx} - 6v_x w - 6vw_x - \frac{15}{2}w_x w^2\right]. \end{aligned} \tag{15.3.2}$$

Applying inverse Laplace transform to (15.3.2), we get

$$[v(x,t)] = v(x,0) - \mathscr{L}^{-1}\left[\left(\frac{s+\alpha(1-s)}{s}\right)\right.$$
$$\left.\times \mathscr{L}\left[v_{xxx} + \frac{3}{2}ww_{xxx} + \frac{9}{2}w_x w_{xx} - 6vv_x - 6vww_x - \frac{3}{2}v_x w^2\right]\right],$$
$$[w(x,t)] = w(x,0) - \mathscr{L}^{-1}\left[\left(\frac{s+\alpha(1-s)}{s}\right)\right.$$
$$\left.\times \mathscr{L}\left[w_{xxx} - 6v_x w - 6vw_x - \frac{15}{2}w_x w^2\right]\right]. \tag{15.3.3}$$

Now, using initial conditions, we apply the homotopy perturbation transforms method

$$\sum_{m=0}^{\infty} v_m = \frac{1}{8}\lambda^2(1 - 4\sec h^2\left(\frac{\lambda x}{2})\right) - p\mathscr{L}^{-1}\left[\left(\frac{s+\alpha(1-s)}{s}\right)\right.$$
$$\left.\times \mathscr{L}\left[\left(\sum_{m=0}^{\infty} p^m v_m\right)_{xxx} - \left(\sum_{i=0}^{\infty} p^n H_m(v)\right)\right]\right],$$

$$\sum_{m=0}^{\infty} w_m = \lambda \sec h\left(\frac{\lambda x}{2}\right) - p\mathscr{L}^{-1}\left[\left(\frac{s+\alpha(1-s)}{s}\right)\right.$$
$$\left.\times \mathscr{L}\left[\left(\sum_{m=0}^{\infty} p^m w_m\right)_{xxx} - \left(\sum_{m=0}^{\infty} p^n H_m(w)\right)\right]\right]. \tag{15.3.4}$$

where, $H_m(v)$ & $H_m(w)$ defined in (15.2.7) represents the nonlinear term. The $H_m(v)$ & $H_m(w)$ are calculated in this form:

$$H_0(v) = \frac{3}{2}w_0(w_0)_{xxx} + \frac{9}{2}(w_0)_x(w_0)_{xx} - 6v_0(v_0)_x - 6v_0w_0(w_0)_x$$
$$- \frac{3}{2}(v_0)_x w_0^2,$$
$$H_0(w) = -6(v_0)_x w_0 - 6v_0(w_0)_x - \frac{15}{2}(w_0)_x w_0^2,$$

$$H_1(v) = \frac{\partial}{\partial p}\left[\begin{array}{c} \frac{3}{2}(w_0+pw_1)(w_0+pw_1)_{xxx} \\ +\frac{9}{2}(w_0+pw_1)_x(w_0+pw_1)_{xx} \\ -6(v_0+pv_1)(v_0+pv_1)_x \\ -6(v_0+pv_1)(w_0+pw_1)(w_0+pw_1)_x \\ -\frac{3}{2}(v_0+pv_1)_x(w_0+pw_1)^2 \end{array}\right]_{p=0},$$

$$\begin{aligned} H_1(v) =& \frac{3}{2}w_0(w_1)_{xxx} + \frac{3}{2}w_1(w_0)_{xxx} + \frac{9}{2}(w_0)_x(w_1)_{xx} + \frac{9}{2}(w_1)_x(w_0)_{xx} \\ &- 6v_0(v_1)_x - 6v_1(v_0)_x - 6(v_0w_1)(w_0)_x - 6v_0w_0(w_0)_x \\ &- 6v_0w_0(w_1)_x - 3(v_0)_x(w_0w_1) - 3(v_1)_xw_0^2, \end{aligned}$$

$$H_1(w) =) = \frac{\partial}{\partial p}\left[\begin{array}{c} -6(v_0+pv_1)_x(w_0+pw_1) \\ -6(v_0+pv_1)(w_0+pw_1)_x \\ -\frac{15}{2}(w_0+pw_1)_x(w_0+pw_1)^2 \end{array}\right]_{p=0},$$

$$\begin{aligned} H_1(w) =& - 6(v_1)_xw_0 - -6(v_0)_xw_1 - 6v_0(w_1)_x - 6v_1(w_0)_x \\ &- 6v_1(w_0)_x - \frac{15}{2}(w_1)_xw_0^2 - 15(w_0)_x(w_0w_1), \end{aligned}$$

$$H_2(v) = \frac{1}{2!}\frac{\partial^2}{\partial p^2}\left[\begin{array}{c} \frac{3}{2}(w_0+pw_1)(w_0+pw_1)_{xxx} \\ +\frac{9}{2}(w_0+pw_1)_x(w_0+pw_1)_{xx} \\ -6(v_0+pv_1)(v_0+pv_1)_x \\ -6(v_0+pv_1)(w_0+pw_1)(w_0+pw_1)_x \\ -\frac{3}{2}(v_0+pv_1)_x(w_0+pw_1)^2 \end{array}\right]_{p=0},$$

$$\begin{aligned} H_2(v) =& \frac{3}{2}w_0(w_2)_{xxx} + \frac{3}{2}w_1(w_1)_{xxx} + \frac{3}{2}w_2(w_0)_{xxx} + \frac{9}{2}(w_0)_x(w_2)_{xx} \\ &+ \frac{9}{2}(w_1)_x(w_1)_{xx} + \frac{9}{2}(w_2)_x(w_0)_{xx} - 6v_0(v_2)_x - 6v_1(v_1)_x \\ &- 6v_2(v_0)_x - 12v_2w_0(w_0)_x - 12v_1w_1(w_0)_x - 12v_0w_2(w_0)_x \\ &- 12v_1w_0(w_1)_x - 12v_0w_1(w_1)_x - 12v_0w_0(w_2)_x \\ &- 3(v_0)_x(w_0)^2 - 6(v_0)_xw_0w_1 - 3(v_0)_x(w_1)^2 - 6(v_0)_xw_0w_2, \end{aligned}$$

$$H_2(w) = \frac{1}{2!}\frac{\partial^2}{\partial p^2}\left[\begin{array}{c} 3\left(v_0+pv_1+p^2v_2\right)_x\left(w_0+pw_1+p^2w_2\right) \\ +3\left(v_0+pv_1+p^2v_2\right)\left(w_0+pw_1+p^2w_2\right)_x \\ +15\left(w_0+pw_1+p^2w_2\right)_x\left(w_0+pw_1+p^2w_2\right)^2 \end{array}\right]_{p=0},$$

$$\begin{aligned} H_2(w) =& 6\left(v_2w_0\right)_x + 6\left(v_1\right)_x w_1 + 6\left(v_0w_2\right)_x + 3v_0\frac{\partial}{\partial x}\left(w_2\right)_x + 3v_1\left(w_1\right)_x \\ &+ 3v_2\left(w_0\right)_x + 45\left(w_0\right)_x\left(w_1\right)^2 + 45\left(w_0\right)_x^2 w_2. \end{aligned} \tag{15.3.5}$$

Thus, $H_0(v) = \frac{1}{8}\lambda^2\left(1-4\sec h^2\left(\frac{\lambda x}{2}\right)\right), \quad H_0(w) = \lambda\sec h\left(\frac{\lambda x}{2}\right).$
Substituting all these values in (15.3.4) and then comparing the coefficient of like power of $p$, we obtained

$$\begin{aligned} p^0 : v_0(x,t) =& \frac{1}{8}\lambda^2\left(1-4\sec h^2\left(\frac{\lambda x}{2}\right)\right), \\ p^0 : w_0(x,t) =& \lambda\sec h\left(\frac{\lambda x}{2}\right), \end{aligned}$$

$$\begin{aligned} p^1 : v_1(x,t) =& \left(\frac{1}{4}\lambda^5 - \frac{1}{4}\lambda^5\alpha + \frac{1}{4}\lambda^5\alpha t\right)\sec h^4\left(\frac{\lambda x}{2}\right)\tanh\left(\frac{\lambda x}{2}\right) \\ &- \left(\frac{1}{4}\lambda^3 - \frac{1}{4}\lambda^3\alpha + \frac{1}{4}\lambda^3\alpha t\right)\sec h\left(\frac{\lambda x}{2}\right)\tan h^2\left(\frac{\lambda x}{2}\right) \\ &- \left(\frac{1}{2}\lambda^5 - \frac{1}{2}\lambda^5\alpha + \frac{1}{2}\lambda^5\alpha t\right)\sec h^2\left(\frac{\lambda x}{2}\right)\tan h^3\left(\frac{\lambda x}{2}\right), \\ p^1 : w_1(x,t) =& \left(\frac{3}{8}\lambda^4 - \frac{3}{8}\lambda^4\alpha + \frac{3}{8}\lambda^4\alpha t\right)\sec h\left(\frac{\lambda x}{2}\right)\tan h\left(\frac{\lambda x}{2}\right) \\ &- \left(\frac{47}{8}\lambda^4 - \frac{47}{8}\lambda^4\alpha + \frac{47}{8}\lambda^4\alpha t\right)\sec h^3\left(\frac{\lambda x}{2}\right)\tan h\left(\frac{\lambda x}{2}\right) \\ &+ \left(\frac{1}{8}\lambda^4 - \frac{1}{8}\lambda^4\alpha + \frac{1}{8}\lambda^4\alpha t\right)\sec h\left(\frac{\lambda x}{2}\right)\tan h^3\left(\frac{\lambda x}{2}\right), \end{aligned}$$

$$
\begin{aligned}
p^2 : v_2(x,t) =& \frac{1}{256}\Bigg[\Big( - 216\lambda^5 + 312\lambda^5\alpha - 312\lambda^5\alpha t \\
& - 96\lambda^5\alpha^2 + 192\lambda^5\alpha^2 t - 48\lambda^5\alpha^2 t^2\Big)\sec h^3\left(\frac{\lambda x}{2}\right) \\
& + \Big(144\lambda^6 - 144\lambda^6\alpha + 144\lambda^6\alpha t + 6\lambda^8 - 6\lambda^8\alpha \\
& + 6\lambda^8\alpha t\Big)\sec h^4\left(\frac{\lambda x}{2}\right) \\
& + \Big(150\lambda^5 - 308\lambda^5\alpha + 3008\lambda^5\alpha t + 1504\lambda^5\alpha^2 \\
& - 3008\lambda^5\alpha^2 t + 752\lambda^5\alpha^2 t^2 - \ldots\Big)\sec h^5\left(\frac{\lambda x}{2}\right) \\
& + \Big(48\lambda^6 - 128\lambda^6\alpha + 128\lambda^6\alpha t + 80\lambda^6\alpha^2 - 160\lambda^6\alpha^2 t \\
& + 40\lambda^6\alpha^2 t^2 + 90\lambda^8 - \ldots\Big)\sec h^6\left(\frac{\lambda x}{2}\right) \\
& - \Big(3162\lambda^7 - 6324\lambda^7\alpha + 6324\lambda^7\alpha t + 3162\lambda^7\alpha^2 \\
& - 6324\lambda^7\alpha^2 t + 1581\lambda^7\alpha^2 t^2\Big)\sec h^7\left(\frac{\lambda x}{2}\right) \\
& + \Big(112\lambda^8 - 320\lambda^8\alpha + 320\lambda^8\alpha t + 208\lambda^8\alpha^2 \\
& - 416\lambda^8\alpha^2 t + 104\lambda^8\alpha^2 t^2\Big)\sec h^8\left(\frac{\lambda x}{2}\right) - \ldots\Bigg], \\
p^2 : w_2(x,t) =& \frac{1}{64}[(-6159\lambda^5\alpha + 6159\lambda^5\alpha^2 + 18\lambda^7 - 18\alpha\lambda^7 - 96491\lambda^5) \\
& \sec h^3\left(\frac{\lambda x}{2}\right) \\
& - (96491\lambda^5 - 192982\lambda^5\alpha + 96491\lambda^5\alpha^2 - 6003\lambda^7 \\
& - 18\lambda^5\alpha)\sec h^5\left(\frac{\lambda x}{2}\right) \\
& + (48\lambda^6 - 96\lambda^5\alpha + 48\lambda^6\alpha^2)\sec h^6\left(\frac{\lambda x}{2}\right) + \ldots].
\end{aligned}
\tag{15.3.6}
$$

Following the same procedure, the remaining iterates $v_n(x,t)$ and $w_n(x,t)$ can be calculated easily. Hence, the numerical solution solution of (15.3.1) is given by

$$\left.\begin{aligned} v(x,t) &= v_0(x,t) + v_1(x,t) + v_2(x,t) + \ldots, \\ w(x,t) &= w_0(x,t) + w_1(x,t) + w_2(x,t) + \ldots \end{aligned}\right\}. \qquad (15.3.7)$$

The approximate solution (15.3.7) is used to show the graphical behavior of the fractional Jaulent-Miodek equation for different values of fractional order and its absolute error. Here, upto second order iterations are used for approximate solution. The absolute error shows the efficiency of the proposed numerical technique, which can be improved by using the higher order iterates.

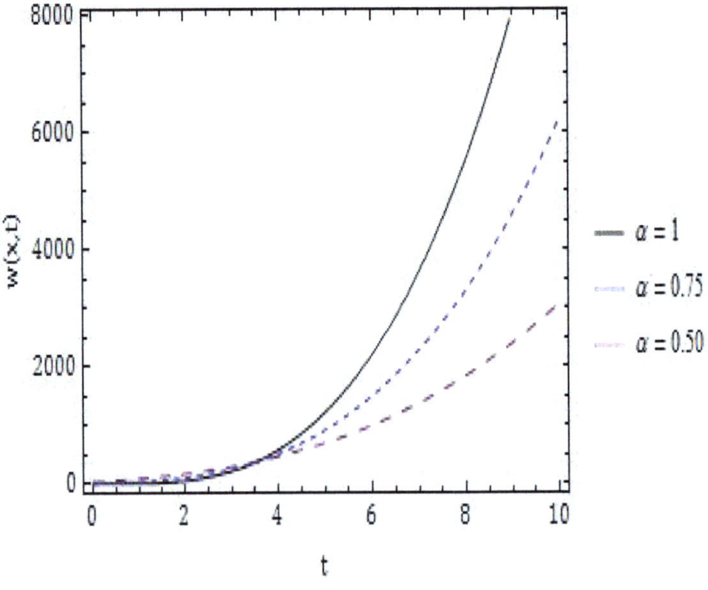


**FIGURE 15.1**. Plot of HPTM solution $v(x,t)$ vs. t when $\lambda = 0.5, x = 1$.

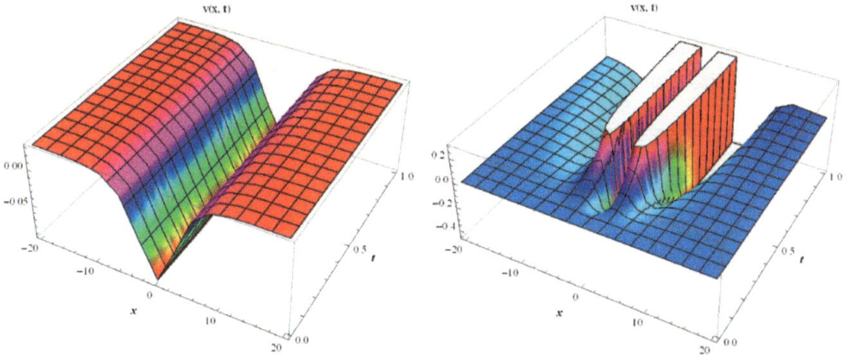


**FIGURE 15.2**. (a) The exact solution (b) Absolute error $= |v_{(exa.)} - v_{HPTM}|$, when $\lambda = 0.5, \alpha = 1$.

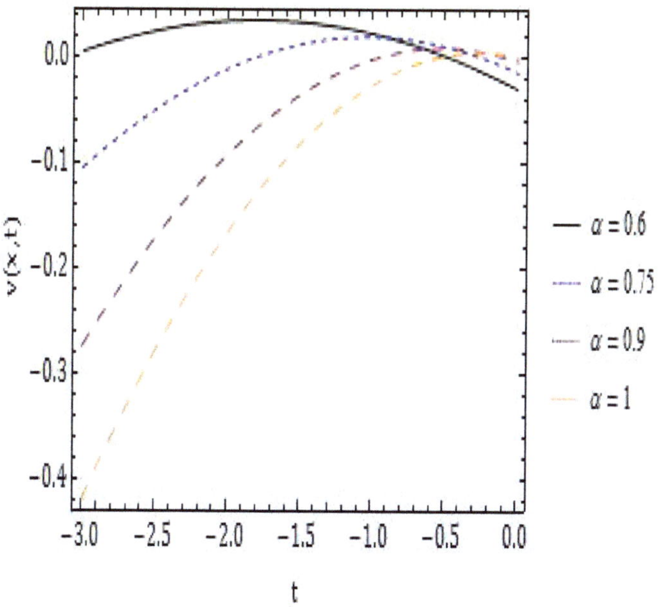


**FIGURE 15.3**. Plot of HPTM solution $w(x,t)$ vs. t when $\lambda = 0.5, x = 1$.

**TABLE 15.1.** Absolute Error for $v(x,t)$ of the System of (15.3.1), When $\lambda = 0.5$ for Different Value of $\alpha$ with Diverse $x$ and $t$

| $x$ | $t$ | $\alpha = 0.6$ | $\alpha = 0.75$ | $\alpha = 0.9$ | $\alpha = 1$ |
|---|---|---|---|---|---|
| 0.2 | 0.2 | 0.063031 | –0.202246 | –0.176785 | –0.109324 |
| | 0.4 | –0.599689 | –0.699361 | –0.442811 | –0.201462 |
| | 0.6 | –1.49661 | –1.45568 | –0.928384 | –0.438141 |
| | 0.8 | –2.64248 | –2.50001 | –1.68329 | –0.887652 |
| | 1 | –4.05205 | –3.86116 | –2.75731 | –1.61828 |
| 0.4 | 0.2 | –0.141947 | –0.247586 | –0.171383 | –0.104305 |
| | 0.4 | –0.776481 | –0.700573 | –0.393488 | –0.166501 |
| | 0.6 | –1.60359 | –1.36101 | –0.779736 | –0.320574 |
| | 0.8 | –2.63225 | –2.24639 | –1.36036 | –0.608001 |
| | 1 | –3.8714 | –3.37421 | –2.16561 | –1.07026 |
| 0.6 | 0.2 | –0.321622 | –0.27784 | –0.160872 | –0.0983904 |
| | 0.4 | –0.892966 | –0.663227 | –0.328912 | –0.128823 |
| | 0.6 | –1.60675 | –1.19539 | –0.599173 | –0.19563 |
| | 0.8 | –2.46585 | –1.87996 | –0.981389 | –0.312166 |
| | 1 | –3.47317 | –2.72257 | –1.4853 | –0.491783 |
| 0.8 | 0.2 | –0.469049 | –0.293865 | –0.146849 | –0.0920116 |
| | 0.4 | –0.953087 | –0.597465 | –0.256701 | –0.0917974 |
| | 0.6 | –1.52624 | –0.984544 | –0.406458 | –0.0747517 |
| | 0.8 | –2.1856 | –1.44941 | –0.586287 | –0.0273882 |
| | 1 | –2.92825 | –1.98638 | –0.786359 | 0.0637792 |
| 1 | 0.2 | –0.469049 | –0.293865 | –0.146849 | –0.0920116 |
| | 0.4 | –0.953087 | –0.597465 | –0.256701 | –0.0917974 |
| | 0.6 | –1.52624 | –0.984544 | –0.406458 | –0.0747517 |
| | 0.8 | –2.1856 | –1.44941 | –0.586287 | –0.0273882 |
| | 1 | –2.92825 | –1.98638 | –0.786359 | 0.0637792 |

**TABLE 15.2.** Absolute Error for $w(x,t)$ of the System of (15.3.1), When $\lambda = 0.5$ for Distinct Value of $\alpha$ with Diverse $x$ and $t$

| $x$ | $t$ | $\alpha = 0.6$ | $\alpha = 0.75$ | $\alpha = 0.9$ | $\alpha = 1$ |
|---|---|---|---|---|---|
| 0.2 | 0.2 | 5.1829 | 4.84664 | 2.92154 | 1.30077 |
| | 0.4 | 7.83954 | 8.27969 | 6.39215 | 4.19243 |
| | 0.6 | 10.7437 | 12.2593 | 10.8798 | 8.52896 |
| | 0.8 | 13.864 | 16.7238 | 16.278 | 14.1644 |
| | 1 | 17.1687 | 21.6117 | 22.4803 | 20.9526 |
| 0.4 | 0.2 | -2.40433 | 2.12894 | 2.38473 | 1.23258 |
| | 0.4 | -2.61685 | 3.75343 | 5.07462 | 3.82687 |
| | 0.6 | -2.98595 | 5.45099 | 8.32724 | 7.49244 |
| | 0.8 | -3.57436 | 7.09909 | 11.9309 | 11.9389 |
| | 1 | -4.44482 | 8.57521 | 15.6738 | 16.8757 |
| 0.6 | 0.2 | -9.29833 | -0.434334 | 1.82309 | 1.14073 |
| | 0.4 | -12.1756 | -0.546623 | 3.70296 | 3.38442 |
| | 0.6 | -15.5891 | -1.0383 | 5.6967 | 6.31205 |
| | 0.8 | -19.6294 | -2.08613 | 7.49886 | 9.50462 |
| | 1 | -24.3868 | -3.86688 | 8.80398 | 12.5431 |
| 0.8 | 0.2 | -15.0937 | -2.69805 | 1.26579 | 1.02947 |
| | 0.4 | -20.2846 | -4.38251 | 2.34737 | 2.88641 |
| | 0.6 | -26.3481 | -6.85536 | 3.12165 | 5.04544 |
| | 0.8 | -33.3977 | -10.3382 | 3.20564 | 6.98119 |
| | 1 | -41.5468 | -15.0528 | 2.21632 | 8.16826 |
| 1 | 0.2 | -19.5176 | -4.55698 | 0.738177 | 0.903662 |
| | 0.4 | -26.5654 | -7.5775 | 1.06932 | 2.35539 |
| | 0.6 | -34.7641 | -11.734 | 0.718414 | 3.75012 |
| | 0.8 | -44.2444 | -17.2818 | -0.755629 | 4.48279 |
| | 1 | -55.1372 | -24.4762 | -3.7939 | 3.94833 |

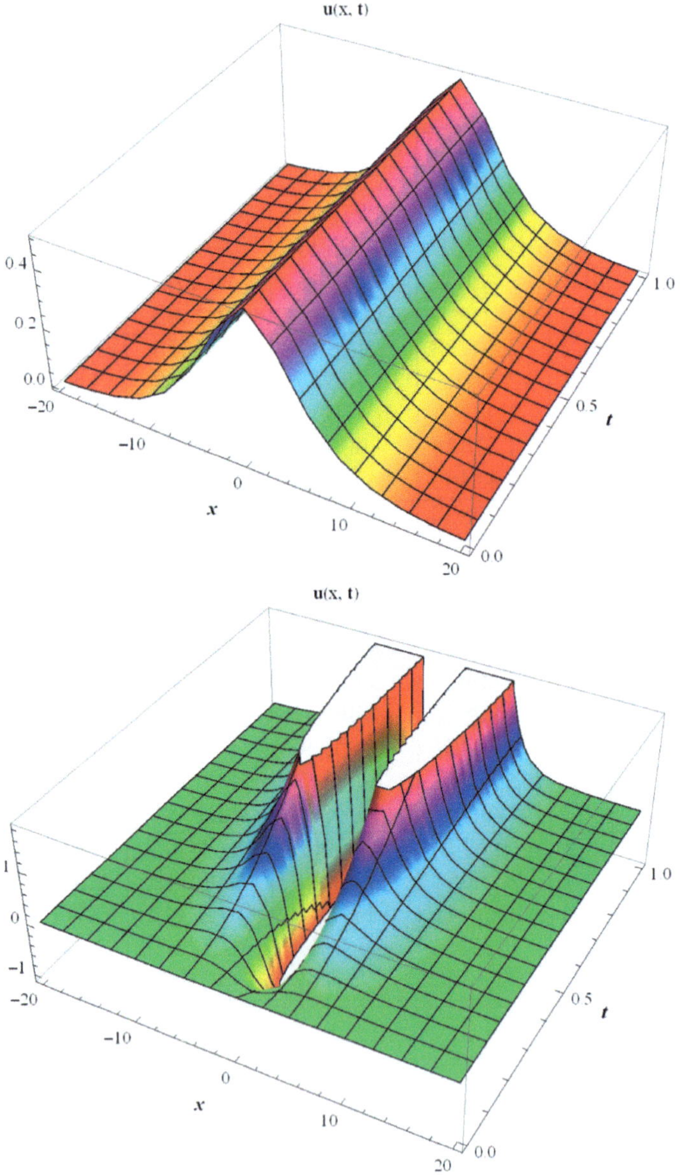


**FIGURE 15.4**. (a) The exact solution (b) Absolute error $= |w_{(exa.)} - w_{HPTM}|$, when $\lambda = 0.5, \alpha = 1$.

## 15.4 Conclusion

In this chapter, the homotopy perturbation transform method is successfully applied to find a numerical solution of the coupled fractional Jaulent-Miodek equation. Using the homotopy perturbation transform method, we acquired a solution in terms of an infinite series for the fractional Jaulent-Miodek equation via the Caputo-Fabrizio operator. The result obtained demonstrates that the recommended technique is trustworthy and powerful in obtaining approximate solutions for a broad range of linear and nonlinear fractional differential equations without the need for any limiting assumptions.

# References

[1] Abbasbandy, S., (2006). Application of He's homotopy perturbation method for Laplace transform. *Chaos,Solitons and Fractals 30*(5), 1206–1212.

[2] Algahtani, O. J. J., (2016). Comparing the Atangana-Baleanu and Caputo-Fabrizio derivative with fractional order: Allen Cahn model. *Chaos, Solitons Fractals 89*, 552–559.

[3] Atangana. A., & Baleanu, D., (2016). New fractional derivatives with nonlocal and non-singular kernel: theory and application to heat transfer model. *Therm. Sci. 20*(2), 763–769.

[4] Atangana, A., & Baleanu, D., (2013). Nonlinear fractional Jaulent-Miodek and Whitham-Broer-Kaup equations within Sumudu transform. *Abstr Appl Anal.*, pp. 1–8, Article ID 160681.

[5] Baskonus, H. M., Guirao, J. L. G., Kumar, A., Causanilles, F. S. V., & Bermudez G. R., (2020). Complex mixed dark bright wave patterns to the modified $\alpha$ and modified Vakhnenko-Parkes equations. *Alexandria Engineering Journal 59*(4), 2149–2160.

[6] Caputo, M., & Fabrizio, M., (2015). A new definition of fractional derivative without singular kernel. Prog *Fract. Differ Appl. 1*(2), 73–85.

[7] Goyal, M., Prakash, A., & Baleanu D., (2022). An efficient hybrid computational technique for the time-dependent Lane-Emden equation of arbitrary order. *Journal of Ocean Engineering and Science 7*(2), 131–142.

[8] Guirao, J. L. G., Baskonus, H. M., Kumar, A., Rawat, M. S., & Yel, G., (2020). Complex Patterns to the (3+1)-Dimensional B-type

Kadomtsev-Petviashvili-Boussinesq Equation. *Symmetry 12*(1), 17, 1–10.

[9] Guirao, J. L. G., Baskonus, H. M., & Kumar, A., (2020). Regarding New Wave Patterns of the Newly Extended Nonlinear (2+1)-Dimensional Boussinesq Equation with Fourth Order. *Mathematics 8*(3), 341, 1–9.

[10] Gupta, S., Kumar, D., & Singh, J., (2015). Analytical solutions of convection-diffusion problems by combining Laplace transform method and homotopy perturbation method. *Alexandria Engineering Journal 54*(3), 645–651.

[11] Gupta, P. K., (2011). Approximate analytical solutions of fractional Benney-Lin equation by reduced differential transform method and the homotopy perturbation method. *Computers Mathematics with Applications 61*(9), 2829–2842.

[12] Gupta, S., Goyal, M., & Prakash, A., (2021). A hybrid computational scheme with convergence analysis for the dependent Rosenau-Hyman equation of arbitrary order via Caputo-Fabrizio operator. *Int. J. Appl. Comput. Math. 7*, 259.

[13] Gupta, A. K., & Ray, S. S., (2015). An investigation with Hermite Wavelets for accurate solution of Fractional Jaulent–Miodek equation associated with energy-dependent Schrödinger potential. *Applied Mathematics and Computation 270*(1), 458–471.

[14] He, J. H., (1997). Variational iteration method for delay differential equations. *Communications in nonlinear Science and Numerical Simulation 2*, 235–236.

[15] He, J. H., (1999). Homotopy perturbation technique. *Comput. Methods. Appl.Mech.Eng. 178*(3–4), 257–262.

[16] Jleli, M., Kumar, S., Kumar, R., & Samet B., (2019). Analytical approach for time fractional wave equations in the sense of Yang-Abdel-Aty-Cattani via the homotopy Perturbation transform method. *Alexandria Engineering Journal 59*(5), 1–5.

[17] Khan, Y., & Wu, Q., (2011). Homotopy perturbation transform method for nonlinear equations using He's polynomials. *Computers Mathematics with Applications 61*(8), 1963–1967.

[18] Kumar, A., Prakash, A., & Baskonus, H. M., (2022). The epidemic COVID-19 model via Caputo-Fabrizio fractional operator. *Waves in Random and Complex Media*, pp. 1–15 https://doi.org/10.1080/17455030.2022.2075954.

[19] Liao, S. J., (2004). On the homotopy analysis method for nonlinear problems. *Applied Mathematics and Computation 147*, 499–513.

[20] Mainardi, F., (1996). Fundamental solutions for the fractional diffusion-wave equations. *Appl. Math. Lett. 9*(6), 23–28.

[21] Majlesi, A., Ghehsareha, H. R., & Zaghian A., (2017). On the fractional Jaulent-Miodek equation associated with energy-dependent Schrodinger potential: Lie symmetry reductions, explicit exact solutions and conservation laws. *Eur. Phys J Plus 132*, 1–13.

[22] Mohan, L., & Prakash, A., (2023). Stability and numerical analysis of the generalized time-fractional Cattaneo model for heat conduction in porous media. *Eur. Phys. J. Plus 138*, 294.

[23] N-Golshan, A., Nourazar, S. S., G-Fard, H., Yildirim, A., & Campo, A., (2013). A modified homotopy perturbation method coupled with the Fourier transform for nonlinear and singular Lane-Emden equations. *Applied Mathematics Letters 26*(10), 1018–1025.

[24] Podlubny I., (1999). *Fractional differential equations, mathematics in science and engineering*, San Diego: Academic Press, 198.

[25] Prakash, A., Goyal, M., Baskonus, H. M., & Gupta S., (2020). A reliable hybrid numerical method for a time dependent vibration model of arbitrary order. *AIMS Mathematics 5*(2), 979–1000.

[26] Prakash, A., & Verma V., (2019). Numerical method for Fractional model of Newell Whitehead-Segel equation. *Frontiers in Physics 7*(15), 1–10. doi: 10.3389/fphy.2019.00015.

[27] Prakash, A., and Kumar, M., (2016). He's Variational iteration method for the solution of nonlinear Newell-Whitehead-Segel equation. *Journal of Applied Analysis and Computation 6*(3), 738–748.

[28] Prakash, A., Kumar, M., & Baleanu D., (2018). A new iterative technique for a fractional model of nonlinear Zakharov-Kuznetsov equations via Sumudu transform. *Applied Mathematics and Computation 334*, 30–40.

[29] Prakash, A., & Kaur, H., (2017). Numerical solution for a fractional model of Fokker-Planck equation by using q-HATM. *Chaos, Solitons and Fractals 105*, 99–110.

[30] Prakash, A., & Rahul, (2022). Analysis and numerical simulation of fractional biological population model with singular and non-singular kernels. *Proc. Inst. Math Mech. 48*, 178–193.

[31] Prakash, A., & Kaur, H., (2018). An efficient hybrid computational technique for solving nonlinear local fractional partial differential equations arising in fractal media. *Nonlinear Engineering-Modelling and Application 7*(3), 229-235, doi.org/10.1515/nleng-2017-0100.

[32] Prakasha, D. G., Prakash, A., Nagaraja, M., & Veeresha P., (2022). On Lorentzian para-Sasakian manifolds. *Iranian Journal of Mathematical Sciences and Informatics 17*(2), 243-252.

[33] Prakash, A., & Kaur, H., (2023). Numerical simulation of coupled fractional-order Whitham-Broer-Kaup equations arising in shallow water with Atangana-Baleanu derivative. *Mathematical Methods in the Applied Sciences, 46*, 11583–11602.

[34] Rahul, & Prakash, A., (2023). Numerical simulation of SIR childhood diseases model with fractional Adams–Bashforth method. *Math Methods Appl Sci. 46*(12), 12340–12360.

[35] Samko, G., Kilbas, A. A., & Marichev O. I., (1993). *Fractional integrals and derivatives: theory and applications.* Switzerland, Gordon and Breach Science Publishers.

[36] Shirkhani, M. R., Hoshyara, H. A., Rahimipetroudi, I., Akhavan, H., & Ganji, D. D., (2018). Unsteady time-dependent incompressible Newtonian fluid flow between two parallel plates by homotopy analysis method, homotopy perturbation method and collocation method. *Propulsionand Power Research 7*(3), 247–256.

[37] Yildinm, A., (2009). An Algorithm for Solving the Fractional Nonlinear Schrödinger equation by means of the Homotopy perturbation method. *International Journal of nonlinear Sciences and Numerical Simulation 10*(4), 445–450.

[38] Yildirim, A., & Kelleci, A., (2009). Numerical simulation of the Jaulent–Miodek equation by He's homotopy perturbation method. *World Appl Sci J. 7*, 84–89.

[39] Zhang, Y., Liu, X., & Wang, G., (2012). Symmetry reductions and exact solutions of the (2 + 1) dimensional Jaulent–Miodek equation. *Applied Mathematics and Computation 219*(3), 911–916.

# Chapter 16

# *Fractional Kinetic Equation with Incomplete Wright Function*

**Dharmendra Kumar Singh**
*Department of Mathematics, School of Basic Sciences,*
*CSJM University, Kanpur, India*

**Abstract**
The object of this chapter is to represent the utility of the incomplete generalized Wright function with fractional kinetic equations. The obtained results are the product of the incomplete Wright function and the Mittag-Leffler function.

## 16.1 Introduction

### *16.1.1 Laplace Transform of the Fractional Integral*

The Laplace transform of a function $f(t)$, denoted by $F(s)$, is defined by the equation [11]

$$F(s) = L\left(f(t); s\right) = \int_0^{\infty} e^{-st} f(t)dt, \tag{16.1.1}$$

where $\mathrm{Re}(s) > 0$ and $f(t)$ is the inverse Laplace transform of $F(s)$. The inverse Laplace transform of $s^{-p}$ is

$$\mathrm{L}^{-1}\left\{s^{-p};t\right\} = \frac{t^{p-1}}{\Gamma(p)}. \tag{16.1.2}$$

Riemman-Liouville fractional integral [12] is defined by

$${}_0I_x^{\nu} f(x) = \frac{1}{\Gamma(\nu)} \int_0^x (x-t)^{\nu-1} f(t)dt, \tag{16.1.3}$$

where $\mathrm{Re}(\nu) > 0$.
Application of the convolution theorem of the Laplace transform gives

$$L\left({}_0I_x f(x);s\right) = L\left(\frac{t^{\nu-1}}{\Gamma(\nu)}\right) L\left(f(t);s\right) = s^{-\nu}F(s). \tag{16.1.4}$$

### *16.1.2 Fractional Kinetic Equation*

The usefulness of the fractional kinetic equation has attracted the attention of many scientists and mathematicians. The implication of its utility can be seen in various fields of science such as kinematics in viscoelastic media, relaxation processes in viscoelastic materials, glassy materials, synthetic polymers, diffusion in porous media with fractal geometry, etc. Seeing its usefulness and results in science many researchers started working with different types of special functions and achieved useful results [1, 5, 7, 13]. The concept of the fractional kinetic equation has been provided by Haubold and Matahi [2]. To get this concept integrating the standard kinetic equation

$$\frac{d}{dt}N_i(t) = -c_iN_i(t), (c_i > 0) \tag{16.1.5}$$

and obtained

$$N_i(t) - N_0 = -c_{i\,0}D_t^{-1}N_i(t), \tag{16.1.6}$$

where $i$ is the number of density of species, ${}_0D_1^{-1}$ is R-L Integral operator, $N_i = N_i(t)$ is a function of time and $N_i(t=0) = N_0$ is the number density of species $i$ at time $t = 0$. By dropping the index i (16.1.6) Haubold and Mathai [2] present an extended form which as follows

$$N(t) - N_0 = -c_0^\nu D_t^{-\nu} N(t). \tag{16.1.7}$$

The solution of (16.1.3) is obtained as

$$N(t) = N_0 E_\nu(-c^\nu t^\nu) \tag{16.1.8}$$

For more details see [2, 8, 9].

### *16.1.3 Incomplete Generalized Hypergeometric Wright Function*

Wright [15] Introduced the following series, known as the generalized hypergeometric Wright function

$$ {}_p\Psi_q(z) = \sum^{\infty} \frac{\prod_{i=1}^{p} \Gamma(\alpha_i + \beta_i k)}{\prod_{j=1}^{q} \Gamma(\alpha_j + \beta_j k} \frac{z^k}{k!}, \tag{16.1.9}$$

where $\alpha_i, \beta_j \in R = (-\infty, \infty), (\alpha_i, \beta_j \neq 0,\ i = 1, 2, ...p\ \ \&\ j = 1, 2, ...q)$. Equation (16.1.9) an entire function, if

$$\sum_{j=1}^{q} \beta_j - \sum_{i=1}^{p} \alpha_i > -1. \tag{16.1.10}$$

Here $z \in C$ is the set of complex numbers condition for the existence of equation (16.1.9) together with its representation in terms of Mellin Barnes integral and of $H$-function where established in [3, 4, 16]. This function has been an interesting topic for many mathematicians for many years and it has been presented in various ways and its applications. Singh and

Porwal [10] generalized the Wright function with the help of an incomplete Pochhammer symbol, say $(\lambda; x)_n$ and $[\lambda; x]_n$, which is defined as

$$(\lambda; x)_v = \frac{\gamma(\lambda + v; x)}{\Gamma(\lambda)}, \quad (\lambda, v \in C;\ x \geq 0) \tag{16.1.11}$$

$$[\lambda; x]_v = \frac{\Gamma(\lambda + v; x)}{\Gamma(\lambda)}, \quad (\lambda, v \in C;\ x \geq 0) \tag{16.1.12}$$

and these incomplete Pochhammer symbols satisfy the following decomposition relation

$$(\lambda; x)_v + [\lambda; x]_v = (\lambda)_v; \quad (\lambda, v \in C;\ x \geq 0) \tag{16.1.13}$$

and where incomplete gamma function $\gamma(s, x)$ and $\Gamma(s, x)$ defined by:

$$\gamma(s, x) = \int_0^x t^{s-1} e^{-t} dt; \quad (\operatorname{Re}(s) > 0;\ x \geq 0) \tag{16.1.14}$$

and

$$\Gamma(s, x) = \int_x^\infty t^{s-1} e^{-t} dt; \quad (\operatorname{Re}(s) > 0;\ x \geq 0). \tag{16.1.15}$$

Applying the above concept of the incomplete pochhammer symbol, Singh and Porwal [10] introduced the incomplete generalized hypergeometric Wright function by defining the function,

$$ {}_p\overline{\Psi}_q z = {}_p\overline{\Psi}_q \left[ \begin{matrix} [a_1, \alpha_1; x], (a_i, \alpha_i)_{(2,p)} \\ [b_1, \beta_1; x], (b_j, \beta_j)_{(2,q)} \end{matrix} \, |z \right]$$

$$= \sum_{k=0}^{\infty} \frac{\Gamma(a_1 + \alpha_1 k; x) \prod_{i=2}^{p} \Gamma(a_i + \alpha_i k)}{\Gamma(b_1 + \beta_1 k; x) \prod_{j=2}^{q} \Gamma(b_j + \beta_j k)} \frac{z^k}{k!}. \tag{16.1.16}$$

$$ {}_p\underline{\Psi}_q z = {}_p\underline{\Psi}_q \left[ \begin{matrix} (a_1, \alpha_1; x), (a_i, \alpha_i)_{(2,p)} \\ (b_1, \beta_1; x), (b_j, \beta_j)_{(2,q)} \end{matrix} \, |z \right]$$

$$= \sum_{k=0}^{\infty} \frac{\gamma(a_1 + \alpha_1 k; x) \prod_{i=2}^{p} \Gamma(a_i + \alpha_i k)}{\gamma(b_1 + \beta_1 k; x) \prod_{j=2}^{q} \Gamma(b_j + \beta_j k)} \frac{z^k}{k!}. \tag{16.1.17}$$

Provided that defining the infinite series in each case is absolutely convergent if satisfy the condition

$$|(\lambda;x)_n| \leq |(\lambda)_n| \text{and} |[\lambda;x]_n| \leq |(\lambda)_n|. \tag{16.1.18}$$

By decomposition formula (16.1.13), (16.1.16) and (16.1.17) can be written in terms of generalized hypergeometric Wright function (16.1.9).

### *16.1.4 Mittag-Leffler Function*

In 1903, the Swedish mathematician Mittag-Leffler [6] defined the generalization of the exponential function as

$$E_\alpha(z) = \sum_{n=0}^{\infty} \frac{z^n}{\Gamma(\alpha n+1)}, (\alpha \in C, \operatorname{Re}(\alpha) > 0). \tag{16.1.19}$$

where $z$ is the complex variable called the Mittag-Leffler function of order $\alpha$. Wiman [14] established new form of Mittag-Leffler function embedded parameter $\beta$

$$E_{\alpha,\beta}(x) = \sum_{n=0}^{\infty} \frac{x^n}{\Gamma(\alpha n+\beta)}, \tag{16.1.20}$$

where $(\alpha, \beta \in C, \operatorname{Re}(\alpha) > 0, \operatorname{Re}(\beta) > 0)$.

---

## 16.2 Main Results

**Theorem 16.2.1.** *Let* $\operatorname{Re}(\nu) > 0, -d > 0$, *and condition (16.1.18) satisfied, then the fractional kinetic equation*

$$N(t) - N_{0p}\overline{\Psi}_q(t) = -d_0^\nu D_t^{-\nu} N(t) \tag{16.2.1}$$

*is solvable, and there exists a solution given by the formula*

$$N(t) = N_0 \; {}_{p+1}\overline{\Psi}_q \left[ \begin{matrix} [a_1, \alpha_1; x], (a_i, \alpha_i)_{(2,p)}, (1,1) \\ [b_1, \beta_1; x], (b_j, \beta_j)_{(2,q)} \end{matrix} \Big| t \right] E_{\nu,k+1}(-d^\nu t^\nu). \tag{16.2.2}$$

*Proof.* Taking Laplace on both sides of (16.2.1) and using (16.1.4), we get

$$\mathrm{L}\{N(t); s\} - N_0 \mathrm{L}\left\{{}_p\overline{\Psi}_q(t); s\right\} = -d^\nu s^{-\nu} \mathrm{L}\{N(t); s\}$$

$$(1 + d^\nu s^{-\nu}) \mathrm{L}\{N(t); s\} = N_0 \mathrm{L}\left\{{}_p\overline{\Psi}_q(t); s\right\}$$

Using (16.1.16) in the above expression, we obtain

$$= N_0 \sum_{k=0}^{\infty} \frac{\Gamma(a_1 + \alpha_1 k; x) \prod_{i=2}^{p} \Gamma(a_i + \alpha_i k)}{\Gamma(b_1 + \beta_1 k; x) \prod_{j=2}^{q} \Gamma(b_i + \beta_j k)} \frac{1}{k!} \int_0^\infty e^{-st} t^{k+1-1} dt$$

$$\mathrm{L}\{N(t); s\} = N_0 \sum_{k=0}^{\infty} \frac{\Gamma(a_1 + \alpha_1 k; x) \prod_{i=2}^{p} \Gamma(a_i + \alpha_i k)}{\Gamma(b_1 + \beta_1 k; x) \prod_{j=2}^{q} \Gamma(b_i + \beta_j k)} \frac{\Gamma(k+1)}{k!} \frac{1}{s^{k+1}} \frac{1}{(1 + d^\nu s^{-\nu})}$$

$$= N_0 \sum_{k=0}^{\infty} \frac{\Gamma(a_1 + \alpha_1 k; x) \prod_{i=2}^{p} \Gamma(a_i + \alpha_i k)}{\Gamma(b_1 + \beta_1 k; x) \prod_{j=2}^{q} \Gamma(b_i + \beta_j k)} \frac{\Gamma(k+1)}{k!} \frac{1}{s^{k+1}} \sum_{r=0}^{\infty} (-1)^r d^{\nu r} s^{-\nu r}.$$

Taking inverse Laplace transformation on both sides, we have

$$= N_0 \sum_{k=0}^{\infty} \frac{\Gamma(a_1 + \alpha_1 k; x) \prod_{i=2}^{p} \Gamma(a_i + \alpha_i k)}{\Gamma(b_1 + \beta_1 k; x) \prod_{j=2}^{q} \Gamma(b_i + \beta_j k)} \frac{\Gamma(k+1)}{k!} \mathrm{L}^{-1} \left\{ \sum_{r=0}^{\infty} (-1)^r d^{\nu r} s^{-\nu r - k - 1} \right\}$$

$$= N_0 \sum_{k=0}^{\infty} \frac{\Gamma(a_1 + \alpha_1 k; x) \prod_{i=2}^{p} \Gamma(a_i + \alpha_i k)}{\Gamma(b_1 + \beta_1 k; x) \prod_{j=2}^{q} \Gamma(b_i + \beta_j k)} \frac{\Gamma(k+1)}{k!} \sum_{r=0}^{\infty} (-1)^r d^{\nu r} \mathrm{L}^{-1}\left\{s^{-\nu r - k - 1}\right\}.$$

Now, using the formula (16.1.2) in the above expression, we achieved

$$= N_0 \sum_{k=0}^{\infty} \frac{\Gamma(a_1 + \alpha_1 k; x) \prod_{i=2}^{p} \Gamma(a_i + \alpha_i k)}{\Gamma(b_1 + \beta_1 k; x) \prod_{j=2}^{q} \Gamma(b_i + \beta_j k)} \Gamma(k+1) \frac{t^k}{k!} \sum_{r=0}^{\infty} \frac{(-1)^r d^{\nu r} t^{\nu r}}{\Gamma(\nu r + k + 1)}$$

$$N(t) = N_0 \ {}_{p+1}\overline{\Psi}_q \left[ \begin{matrix} [a_1, \alpha_1; x], (a_i, \alpha_i)_{(2,p)}, (1,1) \\ [b_1, \beta_1; x], (b_j, \beta_j)_{(2,q)} \end{matrix} \middle| t \right] E_{\nu,k+1}(-d^\nu t^\nu). \tag{16.2.3}$$

This is the required result

**Theorem 16.2.2.** *Let* $\mathrm{Re}(\nu) > 0, d > 0$ *and condition (16.1.18) satisfied, then the fractional kinetic equation*

$$N(t) - N_0 \ {}_p\overline{\Psi}_q(d^\nu t^\nu) = -d_0^\nu D_t^{-\nu} N(t) \tag{16.2.4}$$

*is solvable, and there exists a solution given by the formula*

$$N(t) = N_0 \ {}_{p+1}\overline{\Psi}_q \left[ \begin{matrix} [a_1, \alpha_1; x], (a_i, \alpha_i)_{(2,p)}, (1,\nu) \\ [b_1, \beta_1; x], (b_j, \beta_j)_{(2,q)} \end{matrix} \middle| d^\nu t^\nu \right] E_{\nu,\nu k+1}(-d^\nu t^\nu). \tag{16.2.5}$$

*Proof.* Taking Laplace on both side (16.2.4), we get

$$\mathrm{L}\{N(t); s\} - N_0 \mathrm{L}\left\{ {}_p\overline{\Psi}_q(d^\nu t^\nu); s \right\} = \mathrm{L}\left\{ -d_0^\nu D_t^{-\nu} N(t).; s \right\}.$$

Using (16.1.4) in above the expression, we have

$$\mathrm{L}\{N(t); s\} - N_0 \mathrm{L}\left\{ {}_p\overline{\Psi}_q(d^\nu t^\nu); s \right\} = -d^\nu s^{-\nu} \mathrm{L}\{N(t); s\}$$

$$(1 + d^\nu s^{-\nu}) \mathrm{L}\{N(t); s\} = N_0 \mathrm{L}\left\{ {}_p\overline{\Psi}_q(d^\nu t^\nu); s \right\}.$$

With (16.1.16), we obtained

$$(1 + d^\nu s^{-\nu}) \mathrm{L}\{N(t); s\} =$$

$$N_0 \sum_{k=0}^{\infty} \frac{\Gamma(a_1 + \alpha_1 k; x) \prod_{i=2}^{p} \Gamma(a_i + \alpha_i k)}{\Gamma(b_1 + \beta_1 k; x) \prod_{j=2}^{q} \Gamma(b_i + \beta_j k)} \frac{d^{\nu k}}{k!} \int_0^\infty e^{-st} t^{\nu k + 1 - 1} dt$$

$$\mathrm{L}\{N(t); s\} =$$

$$N_0 \sum_{k=0}^{\infty} \frac{\Gamma(a_1 + \alpha_1 k; x) \prod_{i=2}^{p} \Gamma(a_i + \alpha_i k)}{\Gamma(b_1 + \beta_1 k; x) \prod_{j=2}^{q} \Gamma(b_i + \beta_j k)} \frac{d^{\nu k} \Gamma(\nu k + 1)}{k!} \frac{1}{s^{\nu k + 1}} \frac{1}{(1 + d^\nu s^{-\nu})}$$

$$= N_0 \sum_{k=0}^{\infty} \frac{\Gamma(a_1 + \alpha_1 k; x) \prod_{i=2}^{p} \Gamma(a_i + \alpha_i k)}{\Gamma(b_1 + \beta_1 k; x) \prod_{j=2}^{q} \Gamma(b_i + \beta_j k)} \frac{d^{\nu k} \Gamma(\nu k + 1)}{k!} \frac{1}{s^{\nu k + 1}}$$

$$\sum_{r=0}^{\infty} (-1)^r d^{\nu r} s^{-\nu r}.$$

Taking inverse Laplace transformation, we obtain

$$= N_0 \sum_{k=0}^{\infty} \frac{\Gamma(a_1 + \alpha_1 k; x) \prod_{i=2}^{p} \Gamma(a_i + \alpha_i k)}{\Gamma(b_1 + \beta_1 k; x) \prod_{j=2}^{q} \Gamma(b_i + \beta_j k)} \frac{d^{\nu k} \Gamma(\nu k + 1)}{k!} \mathrm{L}^{-1}$$

$$\left\{ \sum_{r=0}^{\infty} (-1)^r d^{\nu r} s^{-\nu r - \nu k - 1} \right\}$$

$$= N_0 \sum_{k=0}^{\infty} \frac{\Gamma(a_1 + \alpha_1 k; x) \prod_{i=2}^{p} \Gamma(a_i + \alpha_i k)}{\Gamma(b_1 + \beta_1 k; x) \prod_{j=2}^{q} \Gamma(b_i + \beta_j k)} \frac{d^{\nu k} \Gamma(\nu k + 1)}{k!}$$

$$\sum_{r=0}^{\infty} (-1)^r d^{\nu r} \mathrm{L}^{-1} \left\{ s^{-\nu r - \nu k - 1} \right\}$$

with (16.1.2), we arrived at

$$= N_0 \sum_{k=0}^{\infty} \frac{\Gamma(a_1 + \alpha_1 k; x) \prod_{i=2}^{p} \Gamma(a_i + \alpha_i k)}{\Gamma(b_1 + \beta_1 k; x) \prod_{j=2}^{q} \Gamma(b_i + \beta_j k)} \Gamma(\nu k + 1) \frac{d^{\nu k} t^{\nu k}}{k!} \sum_{r=0}^{\infty} \frac{(-1)^r d^{\nu r} t^{\nu r}}{\Gamma(\nu r + \nu k + 1)}$$

$$N(t) = N_0 \ {}_{p+1}\overline{\Psi}_q \left[ \begin{matrix} [a_1, \alpha_1; x], (a_i, \alpha_i)_{(2,p)}, (1, \nu) \\ [b_1, \beta_1; x], (b_j, \beta_j)_{(2,q)} \end{matrix} \middle| d^\nu t^\nu \right] E_{\nu, \nu k + 1}(-d^\nu t^\nu). \tag{16.2.6}$$

This is the required result.

**Theorem 16.2.3.** *Let* $\mathrm{Re}(\nu) > 0, d > 0, a > 0$ *and condition (16.1.18) satisfied, then the fractional kinetic equation*

$$N(t) - N_0 \ {}_p\overline{\Psi}_q(d^\nu t^\nu) = -a_0^\nu D_t^{-\nu} N(t) \tag{16.2.7}$$

*is solvable, and there exists a solution given by the formula*

$$N(t) = N_0 \ {}_{p+1}\overline{\Psi}_q \left[ \begin{matrix} [a_1, \alpha_1; x], (a_i, \alpha_i)_{(2,p)}, (1, \nu) \\ [b_1, \beta_1; x], (b_j, \beta_j)_{(2,q)} \end{matrix} \middle| d^\nu t^\nu \right] E_{\nu,\nu k+1}(-a^\nu t^\nu). \tag{16.2.8}$$

*Proof.* Taking Laplace on both sides (16.2.7), we get

$$\mathrm{L}\{N(t); s\} - N_0 \mathrm{L}\left\{{}_p\overline{\Psi}_q(d^\nu t^\nu); s\right\} = \mathrm{L}\left\{-a_0^\nu D_t^{-\nu} N(t); s\right\}.$$

By using (16.1.4), we achieved

$$\mathrm{L}\{N(t); s\} - N_0 \mathrm{L}\left\{{}_p\overline{\Psi}_q(d^\nu t^\nu); s\right\} = -a^\nu s^{-\nu} \mathrm{L}\{N(t); s\}$$

$$(1 + a^\nu s^{-\nu}) \mathrm{L}\{N(t); s\} = N_0 \mathrm{L}\left\{{}_p\overline{\Psi}_q(d^\nu t^\nu); s\right\}.$$

With (16.1.2), we achieved

$$(1 + a^\nu s^{-\nu}) \mathrm{L}\{N(t); s\} =$$

$$N_0 \sum_{k=0}^{\infty} \frac{\Gamma(a_1 + \alpha_1 k; x) \prod_{i=2}^{p} \Gamma(a_i + \alpha_i k)}{\Gamma(b_1 + \beta_1 k; x) \prod_{j=2}^{q} \Gamma(b_i + \beta_j k)} \frac{d^{\nu k}}{k!} \int_0^\infty e^{-st} t^{\nu k+1-1} dt$$

$$\mathrm{L}\{N(t); s\} =$$

$$N_0 \sum_{k=0}^{\infty} \frac{\Gamma(a_1 + \alpha_1 k; x) \prod_{i=2}^{p} \Gamma(a_i + \alpha_i k)}{\Gamma(b_1 + \beta_1 k; x) \prod_{j=2}^{q} \Gamma(b_i + \beta_j k)} \frac{d^{\nu k} \Gamma(\nu k + 1)}{k!} \frac{1}{s^{\nu k+1}} \frac{1}{(1 + a^\nu s^{-\nu})}$$

$$= N_0 \sum_{k=0}^{\infty} \frac{\Gamma(a_1 + \alpha_1 k; x) \prod_{i=2}^{p} \Gamma(a_i + \alpha_i k)}{\Gamma(b_1 + \beta_1 k; x) \prod_{j=2}^{q} \Gamma(b_i + \beta_j k)} \frac{d^{\nu k} \Gamma(\nu k + 1)}{k!} \frac{1}{s^{\nu k+1}} \sum_{r=0}^{\infty} (-1)^r a^{\nu r} s^{-\nu r}.$$

Taking inverse Laplace transformation, we obtain

$$N(t) = N_0 \sum_{k=0}^{\infty} \frac{\Gamma(a_1 + \alpha_1 k; x) \prod_{i=2}^{p} \Gamma(a_i + \alpha_i k)}{\Gamma(b_1 + \beta_1 k; x) \prod_{j=2}^{q} \Gamma(b_i + \beta_j k)} \frac{d^{\nu k} \Gamma(\nu k + 1)}{k!}$$

$$\times \mathrm{L}^{-1} \left\{ \sum_{r=0}^{\infty} (-1)^r a^{\nu r} s^{-\nu r - \nu k - 1} \right\}$$

$$= N_0 \sum_{k=0}^{\infty} \frac{\Gamma(a_1+\alpha_1 k;x)\prod_{i=2}^{p}\Gamma(a_i+\alpha_i k)}{\Gamma(b_1+\beta_1 k;x)\prod_{j=2}^{q}\Gamma(b_i+\beta_j k)} \frac{d^{\nu k}\Gamma(\nu k+1)}{k!}$$

$$\sum_{r=0}^{\infty}(-1)^r a^{\nu r} \mathrm{L}^{-1}\left\{s^{-\nu r-\nu k-1}\right\}.$$

With (16.1.2), we arrived at

$$= N_0 \sum_{k=0}^{\infty} \frac{\Gamma(a_1+\alpha_1 k;x)\prod_{i=2}^{p}\Gamma(a_i+\alpha_i k)}{\Gamma(b_1+\beta_1 k;x)\prod_{j=2}^{q}\Gamma(b_i+\beta_j k)} \Gamma(\nu k+1)\frac{d^{\nu k}t^{\nu k}}{k!} \sum_{r=0}^{\infty} \frac{(-1)^r a^{\nu r} t^{\nu r}}{\Gamma(\nu r+\nu k+1)}$$

$$N(t) = N_0 \;{}_{p+1}\overline{\Psi}_q \left[ \begin{matrix} [a_1,\alpha_1;x],(a_i,\alpha_i)_{(2,p)},(1,\nu) \\ [b_1,\beta_1;x],(b_j,\beta_j)_{(2,q)} \end{matrix} \middle| d^{\nu}t^{\nu} \right] E_{\nu,\nu k+1}(-a^{\nu}t^{\nu}). \tag{16.2.9}$$

This is the required result.

---

## 16.3 Conclusion

All obtained results can be represented as incomplete Gauss hypergeometric function, incomplete Bessel Maitland function, incomplete H function, and incomplete G function, by using (16.1.16) and (16.1.17), respectively. Obtained results and their special cases can make a significant contribution to the field of physics and engineering.

# References


[1] Gupta, R. K., Atangana, A., Shaktawat, B. S., & Purohit, S. D., (2019). On the solution of generalized fractional kinetic equations involving generalized M-series. *Caspian Journal of Applied Mathematics, Ecology and Economics 7*(1), 88–98.

[2] Haubold, H. J., & Mathai, A. M., (2000). The fractional kinetic equation and thermonuclear functions. *Astrophysics and Space Science, 273*, 53–63.

[3] Kilbas, A. A., Saigo, M., & Trujillo, J. J., (2002). On the generalized Wright function. *Fract. Calc. Appl. Anal., 4*, 437–460.

[4] Kilbas, A. A., (2005). Fractional calculus of generalized Wright function. *Fract. Calc. Appl. Anal., 8*(2), 113–126.

[5] Komorowski, T., Stefano, O., & Ryzhik, L., (2020). Fractional diffusion limit for a kinetic equation with an interface. *Ann. Probab., 48*(5), 2290–2320.

[6] Mittag-Leffler, G. M., (1905). Sur la representation analytiqie d'une fonction monogene cinquieme note. *Acta Math., 29*, 101–181.

[7] Nisar, K. S., (2020). Generalized Mittag-Leffler Type Function: Fractional Integrations and Application to Fractional Kinetic Equations. *Frontiers in Physic*, 1–7.

[8] Saxena, R. K., Mathai, A. M., & Haubold, H. J., (2002). On fractional Kinetic Equations. *Astrophysics and Space Science, 282*, 281–287.

[9] Saxena, R. K., Mathai, A. M., & Haubold, H. J., (2004). Unified Fractional Kinetic Equation and a Fractional Diffusion Equation. *Astrophysics and Space Science, 290*, 299–310.

[10] Singh, D., & Porwal, S., (2013). Incomplete Mittag-Leffler Function. *Acta Universitatis Apulensis, 34*, 151–162.

[11] Sneddon, I. (1979). *The Use of Integral Transfrom*, Tata McGraw Hill.
[12] Srivastava, H. M., Saxena, R. K. (2001). Operators of fractional integration and applications, *Appl. Math. Comput.*, *118*, 1–52.
[13] Suthar, D. L., Purohit, S. D., & Araci, Serkan (2020). Solution of Fractional Kinetic Equations Associated with the(p, q)-Mathieu-Type Series, *Discrete Dynamics in Nature and Society*, 1–7.
[14] Wiman, A. (1905). Uber den fundamental satz in der theorie der funktionen $E_\alpha(z)$, *Acta Math.*, *29*, 191–201.
[15] Wright, E. M. (1935). The asymptotic expansion of the generalized hypergeometric function, *J. London Math. Soc.*, *10*, 287–293.
[16] Wright, E. M. (1940). The generalized Bessel function of order greater than one, *Quart. J. Math.* *11*(1), 36–48.

# Index

**H**

**Q**



**R**

9 781779 643353

Printed by Libri Plureos GmbH in Hamburg,
Germany